AF581732

Gas Transport in Glassy Polymers

Gas Transport in Glassy Polymers

Editors

Maria Grazia De Angelis
Giulio C. Sarti

MDPI • Basel • Beijing • Wuhan • Barcelona • Belgrade • Manchester • Tokyo • Cluj • Tianjin

Editors
Maria Grazia De Angelis
University of Bologna
University of Edinburgh
UK

Giulio C. Sarti
Alma Mater Studiorum
Universita di Bologna
Italy

Editorial Office
MDPI
St. Alban-Anlage 66
4052 Basel, Switzerland

This is a reprint of articles from the Special Issue published online in the open access journal *Membranes* (ISSN 2077-0375) (available at: https://www.mdpi.com/journal/membranes/special_issues/gas_transport_glassy_polymers).

For citation purposes, cite each article independently as indicated on the article page online and as indicated below:

LastName, A.A.; LastName, B.B.; LastName, C.C. Article Title. *Journal Name* **Year**, *Volume Number*, Page Range.

ISBN 978-3-0365-0212-0 (Hbk)
ISBN 978-3-0365-0213-7 (PDF)

Cover image courtesy of Francesco M. Benedetti.

Contents

About the Editors

Maria Grazia De Angelis is a Professor of Thermodynamics of Materials and Processes at the University of Edinburgh (UK) since 2020. Since 2014 she holds an Associate Professorship in Chemical Engineering at the University of Bologna where she obtained her Ph.D. in 2002. She was Visiting Researcher or Lecturer at the North Carolina State University in Raleigh, USA, the National Technical University of Athens, Greece, the Universidad Nacional del Sur in Bahia Blanca, Argentina and the University of Melbourne, Australia. Her research activity deals with the experimental and theoretical analysis of fluid transport in solid materials, with applications in the separation of gaseous and liquid mixtures with membranes, CO_2 capture, packaging. The modeling efforts include the macroscopic equation of state approaches as well as molecular and multiple scale methods. She is the co-author of around 80 publications with an h-index of 26, and a member of the Editorial Board of *Membranes*. She is an elected member of the European Membrane Society Council (2019–2022) and a member of the Working Party on Thermodynamics and Transport Properties of the European Federation of Chemical Engineers (EFCE).

Giulio C. Sarti is currently Emeritus Professor at the University of Bologna, where he was a Full Professor in Chemical Engineering between 1986 and 2018. He also worked at the University of Naples, North Carolina State University, University of California at Davis and University of Texas at Austin. His research activity spans fundamental studies of mass transport in polymeric materials, thermodynamics of polymeric solutions and membrane-based separations of fluid mixtures and biomolecules. One of his most popular contributions was the development of the non-equilibrium thermodynamics for glassy polymers (NET-GP) approach for the solubility of fluids in glassy polymers. He co-authored around 200 documents with an h-index of 41. He serves as an editor for *Membranes*.

Editorial

Gas Transport in Glassy Polymers

Maria Grazia De Angelis * and Giulio C. Sarti

Department of Civil, Chemical, Environmental and Materials Engineering, University of Bologna, I-40131 Bologna, Italy; giulio.sarti@unibo.it

* Correspondence: grazia.deangelis@unibo.it

Received: 26 November 2020; Accepted: 2 December 2020; Published: 7 December 2020

This Special Issue of Membranes provides an updated and comprehensive overview of the state of fundamental knowledge on the fluid sorption and transport in glassy polymers, combining original experimental and modeling works, as well as reviews, prepared by renowned experts. Sophisticated experimental insights are given at the transport in glassy polymeric materials in challenging operative ranges such as in mixed gas conditions, in the presence of humidity and of strongly interacting species, as well as at high pressure and temperatures. A thorough investigation of the transport in complex novel structures, based on high performance polymers incorporating fillers of different types (Mixed Matrix Membranes, MMMs) is also presented. Macroscopic models suitable for the interpretation of competition effects during the mixed gas sorption in glassy polymers are analyzed, while a comprehensive review of the modeling of facilitated transport in membranes is presented. Finally, a complete review is provided of the most recent and effective molecular and multiscale models for the simulation of complex glassy polymeric structures and for the fluid sorption and transport therein. The present issue is mainly devoted to the applications of membranes for fluid separations, although the results obtained are valid in general.

The experimental analysis of transport in glassy polymers often requires dedicated techniques because experiments are time-consuming, due to the long relaxation times typical of glassy polymers, and they might show history-dependent properties that require appropriate pretreatment protocols. In many cases, connection with theoretical models is required to support the experimentation and the analysis of the results.

The experimental analysis is complicated when strong interacting species such as polar compounds are considered, for which specific models are required. In the paper by Mensitieri et al. [1], the sorption thermodynamics of CO_2, H_2O and CH_3OH interacting with a glassy Poly (ether imide) (PEI) were studied; gravimetric analysis and vibrational spectroscopy were used for the collection of the data, and non-equilibrium thermodynamics modeling was applied for their interpretation. The molecular information gathered from the Fourier-transform infrared (FTIR) spectroscopic analysis is used to tailor thermodynamics modeling. The investigated penetrants display different interactions with the polymer, which reflects in different sorption thermodynamic properties. For the specific case of water, the outcomes from molecular dynamics simulations are compared with the results of the analysis.

The sorption of CO_2 in PEI could be described with the NELF model in view of the weak interactions established with groups present on the polymer backbone. For the case of H_2O sorption, data were interpreted using the Non Equilibrium Thermodynamics for Glassy Polymers-Non Random Hydrogen Bonding (NET-GP-NRHB) model in order to cope with specific H-bonding interactions occurring in the PEI–H_2O system. FTIR spectroscopy revealed the occurrence of two water species, one self-interacting via H-bonding and the other cross-interacting with carbonyl groups of PEI, making it possible to adjust the H-bonding terms in the NET-GP-NRHB model.

In glassy polymeric membranes for gas separation, membranes' multicomponent effects have been proven to be particularly important, especially because of competition between penetrants, which affects sorption and sorption-selectivity. Mixed gas effects can completely reverse the performance of membranes characterized under pure gas conditions only.

In the paper by Genduso et al. [2], the nonideal behavior of CO_2-CH_4 mixtures was elucidated in 6FDA-mPDA, a polyimide of remarkable performance. To quantify the deviation from ideality, sorption and diffusion contributions to permeation were decoupled. Experimental data of mixed-gas solubility revealed a decrease in both CO_2 and, more markedly, CH_4 solubility due to mixture effects. CO_2 versus CH_4 mixed-gas solubility coefficients of 6FDA-mPDA and other glassy polymers previously studied follow a linear trend regardless of equilibrium concentration. Since the CO_2/CH_4 solubility selectivity of 6FDA-mPDA improved under mixed-gas conditions, the observed decline in mixed-gas permeability selectivity from the corresponding ideal values cannot be attributed to competitive sorption, as frequently assumed in the literature, but rather to a depression of the size-sieving capability of 6FDA-mPDA induced by the presence of CO_2 in the polymeric film matrix. The authors performed kinetic measurements which reinforced the idea that CO_2 addition lowers the effective diffusion coefficient of CH_4 in the polymer.

In the scientific literature on gas separation membranes, most data are reported in pure gas conditions, at room temperature and relatively low pressure. However, many separations would preferably occur at high temperatures and pressures, such as in the case of pre-combustion CO_2 capture where H_2/CO_2 separation is relevant, and natural gas treatment where the CO_2/CH_4 mixture is involved. In addition, temperature and pressure have a dramatic impact on the membranes' performance, because the selectivity can decrease or even reverse with temperature, and the high pressure of mixtures containing CO_2 causes plasticization.

In the paper by Lasseuguette et al. [3], the transport properties of CO_2, N_2, CH_4 and H_2 in a Polymer of Intrinsic Microporosity (PIM-EA(H_2)-TB), with high CO_2 and H_2 permeability and good ideal selectivity over N_2, were determined for upstream pressures up to 20 bar and temperatures up to 200 °C. No increase in CO_2 permeability due to plasticization was noted over the range of pressure tested. The permeability coefficient of N_2, CH_4 and H_2 increase with temperature, while for CO_2 the permeability decreases with temperature, due to its high solubility, which is an exothermic process. Therefore, the separation performance of PIM-EA(H_2)-TB for H_2/CO_2 is reversed at high temperature and maintained also at high pressure. This suggests that, after further development to enhance absolute selectivity of H_2 over CO_2, PIMs could become good candidates for membrane materials for use in pre-combustion CO_2 capture.

A significant interest was created in the last 20 years around Mixed Matrix Membranes (MMM), namely polymers to which particles of different structure and morphology were added, with the aim of improving their performance in fluid separations. The glassy polymers, due to their rigidity, are particularly suitable for incorporation of such fillers, which normally reduce the flexibility of softer materials like the rubbery ones. The interactions between the polymer and the filler play an important role in determining the final performance.

In the paper by Dai et al. [4], Poly(1-trimethylsilyl-1-propyne) (PTMSP), a polymer with an exceptionally high gas permeation rate but a serious aging problem and low selectivity, was combined with different selective fillers with the aim of increasing its selectivity. The gas permeation of the hybrid membranes was evaluated using a mixed gas permeation test with the presence of water vapour to simulate the flue gas conditions. The addition of ZIF-L improves the CO_2/N_2 selectivity at the expenses of CO_2 permeability, while the addition of TiO_2, ZIF-7 and ZIF-8 increases the CO_2 permeability but the CO_2/N_2 selectivity decreases. The hybrid membranes are characterized by high thermal stability, while the poor affinity with the polymer phase causes the formation of interfacial voids. The solvent used for the membrane preparation has a strong effect on the final performance, as is usually the case with glassy polymers. For all the hybrid membranes, increasing the water vapour content in the gaseous streams lowers the CO_2 permeability.

Escorihuela and coworkers [5] studied the effect of adding different fillers on various polyimides suitable for gas and liquid separations, particularly for biogas upgrading and organic solvent nanofiltration (OSN). The most promising result was obtained for Matrimid®—10 wt.% $BaCe_{0.2}Zr_{0.7}Y_{0.1}O_3$ (BCZY) MMM, which showed improvement in CO_2 and H_2O permeabilities accompanied by increased CO_2/CH_4 selectivity. It was also observed that, in general, the inorganic fillers could produce small rigidification in the polymer matrix, although they do not exhibit higher T_g. Regarding the temperature effects, some changes were observed in the activation energy of the process, while the incorporation of inorganic fillers did not significantly affect the permeation mechanism determined by the polymer transport properties. Water permeability was first reported for several polyimides and MMMs of inorganic particles with polyimides, reaching relatively high values.

In the review by Golemme and Santaniello [6], the focus is on MMMs formed by molecular sieves dispersed in perfluorinated glassy polymers. First, the issue of compatibilization of ceramic molecular sieves with the polymers is considered, examining the effect of the surface treatment on the gas transport properties of the filler. Then, the preparation of the defect-free hybrid membranes and their gas separation capabilities are described. Finally, modelling of the gas transport properties of the perfluoropolymer MMMs is reviewed, with a four-phase approach implemented in the frame of the Maxwell model and finite element simulation. The four-phase approach is a convenient representation of the transport in MMMs when more than one single interfacial effect is present. Grafting perfluorinated tails on the outer surface of zeolitic molecular sieves improved the surface permeability and allowed the preparation of defect-free MMMs with perfluoropolymers.

Macroscopic models for describing fluid transport in glassy polymers are computationally inexpensive but their accuracy can be limited by the absence of the experimental information required to parametrize them. Two important cases of study are reported: the simulation of mixed gas sorption, and the facilitated transport process.

As pointed out above, the mixed gas sorption in glassy polymers is complicated by the presence of competition effects and it is important to rely on predictive models which could estimate the mixed gas performance in the absence of specific experimental data. The paper by Ricci and De Angelis [7] aims at assessing the performance of a model used to predict the mixed-gas solubility and selectivity of glassy membranes, the Dual Mode Sorption (DMS) model in its multicomponent extension.

The data of CO_2/CH_4 mixtures in three high free volume glassy polymers, PTMSP, PIM-1 and tetrazole-modified PIM-1 (TZ-PIM), were taken as test cases. These systems exhibit deviations from the ideal pure-gas behavior, especially due to competition. The DSM model parameters retrieved from the best fit of pure-gas sorption isotherms provided a good qualitative picture of sorption in mixed-gas conditions, displaying a reduction in solubility that was experimentally observed and due to competition. A sensitivity analysis carried out on the DMS model parameters revealed that a small uncertainty in the pure-gas data propagates greatly in the mixed-gas calculation, limiting the quantitative accuracy of the mixed-gas prediction in the absence of experimental mixed-gas data to use for validation. The authors concluded that the DSM model is a useful tool for a first estimate of the mixed-gas effects, due to its simplicity, but that for quantitative accuracy one might resort to other models such as the Non Equilibrium Lattice Fluid (NELF) model.

The Facilitated Transport (FT) process is one of the most promising processes when it comes to CO_2 capture with membranes, because the reaction taking place between diffusing CO_2 and certain aminic species present in the membrane boosts the CO_2 selectivity and permeability. The process is composed of pure diffusion and chemical reaction between solute and carrier and requires a specific modeling approach. The review by Rea et al. [8] elucidates the various models devoted to explaining the FT mechanism in the two main families of carrier-mediated membranes: the mobile carrier (MC) systems where the carrier is free to diffuse across the membrane, and the fixed site carrier (FSC) where the carrier is fixed to the polymeric backbone.

For the MC systems, the models by Teramoto and by Morales-Cabrera et al. seem to be the more flexible ones as they can be applied in a wide range of operative conditions, without strong assumptions but with the numerical solution of a system of equations. The simpler models, provided by Smith and Quinn, Noble et al. and Jeema and Noble, have a good descriptive ability coupled with a very simple mathematical form, but they are only valid under certain assumptions. For the FSC systems, interesting approaches are those by Noble (1992), Kang et al. (1996) and Zarca et al. (2017). Despite the number of modeling tools available, the main issue is the knowledge of the physico-chemical parameters of the models, which are difficult to determine experimentally.

The modeling of fluid transport in glassy polymers can be made less reliant on experimentally-derived parameters by using more predictive modeling tools like atomistic simulations. However, glassy materials are challenging for such methods, due to the long relaxation times that make the calculations computationally expensive. The review by Vergadou and Theodorou [9] aims at giving an overview of the most recent efforts dedicated to the development of molecular and multiscale methods for the simulation of sorption and diffusion in glassy polymers. The basic concepts of equilibrium and non-equilibrium Molecular Dynamics (MD) and Monte Carlo (MC) techniques for simulating polymers and sorption and diffusion of small molecules are introduced. For diffusion, the focus is placed on multiscale methods that adopt infrequent event analysis of elementary diffusive jumps, such as the Transition State Theory (TST) and the Transition Path Sampling (TPS) Method. Indeed, penetrant jumps between sorption sites in glassy polymers cannot be simulated with MD, due to the excessively long computational times required. Multiscale methods such as those based on coarse-graining/reverse mapping for the generation of realistic polymeric configurations and for the sorption and diffusion phenomena in polymers per se are also discussed. Recent approaches designed to cut down on the computational cost, such as mesoscopic kinetic MC techniques that utilize atomistic information and coarse-grained MD simulations with appropriate time mapping, are discussed. Finally, open challenges and future perspectives of molecular simulations of transport in dense membranes are illustrated.

In conclusion, it can be said that glassy polymers, and their use in applications where fluid transport is important, still have open issues that need to be investigated. We hope that this Special Issue provides inspiration to new and established scientists in the field, and a source of information about the different aspects of the problem that often need to be considered altogether.

Author Contributions: Conceptualization, writing, M.G.D.A.; writing, editing, G.C.S. Both authors have read and agreed to the published version of the manuscript.

Funding: This research received no external funding

Acknowledgments: We thank all the authors for the excellent quality of their work which contributed tremendously to the great success of this Special Issue.

Conflicts of Interest: The authors declare no conflict of interest.

References

1. Mensitieri, G.; Scherillo, G.; La Manna, P.; Musto, P. Sorption Thermodynamics of CO_2, H_2O, and CH_3OH in a Glassy Polyetherimide: A Molecular Perspective. *Membranes* **2019**, *9*, 23. [CrossRef] [PubMed]
2. Genduso, G.; Ghanem, B.S.; Pinnau, I. Experimental Mixed-Gas Permeability, Sorption and Diffusion of CO_2-CH_4 Mixtures in 6FDA-mPDA Polyimide Membrane: Unveiling the Effect of Competitive Sorption on Permeability Selectivity. *Membranes* **2019**, *9*, 10. [CrossRef] [PubMed]
3. Lasseuguette, E.; Malpass-Evans, R.; Carta, M.; McKeown, N.; Ferrari, M.C. Temperature and Pressure Dependence of Gas Permeation in a Microporous Tröger's Base Polymer. *Membranes* **2018**, *8*, 132. [CrossRef] [PubMed]
4. Dai, Z.; Løining, V.; Deng, J.; Ansaloni, L.; Deng, L. Poly(1-trimethylsilyl-1-propyne)-Based Hybrid Membranes: Effects of Various Nanofillers and Feed Gas Humidity on CO_2 Permeation. *Membranes* **2018**, *8*, 76. [CrossRef] [PubMed]
5. Escorihuela, S.; Valero, L.; Tena, A.; Shishatskiy, S.; Escolástico, S.; Brinkmann, T.; Serra, J.M. Study of the Effect of Inorganic Particles on the Gas Transport Properties of Glassy Polyimides for Selective CO_2 and H_2O Separation. *Membranes* **2018**, *8*, 128. [CrossRef] [PubMed]
6. Golemme, G.; Santaniello, A. Perfluoropolymer/Molecular Sieve Mixed-Matrix Membranes. *Membranes* **2019**, *9*, 19. [CrossRef] [PubMed]
7. Ricci, E.; De Angelis, M.G. Modelling Mixed-Gas Sorption in Glassy Polymers for CO_2 Removal: A Sensitivity Analysis of the Dual Mode Sorption Model. *Membranes* **2019**, *9*, 8. [CrossRef] [PubMed]

8. Rea, R.; De Angelis, M.G.; Giacinti Baschetti, M. Models for Facilitated Transport Membranes: A Review. *Membranes* **2019**, *9*, 26. [CrossRef] [PubMed]
9. Vergadou, N.; Theodorou, D.M. Molecular Modeling Investigations of Sorption and Diffusion of Small Molecules in Glassy Polymers. *Membranes* **2019**, *9*, 98. [CrossRef] [PubMed]

Article

Sorption Thermodynamics of CO_2, H_2O, and CH_3OH in a Glassy Polyetherimide: A Molecular Perspective

Giuseppe Mensitieri [1,2,*], Giuseppe Scherillo [1], Pietro La Manna [2] and Pellegrino Musto [2]

1 Department of Chemical, Materials and PrNoduction Engineering, University of Naples Federico II, Piazzale Tecchio 80, 80125 Naples, Italy; gscheril@unina.it

2 Institute for Polymers, Composites and Biomaterials, National Research Council of Italy, Viale Campi Flegrei 34, 80078 Pozzuoli (Na), Italy; pietro.lamanna@unina.it (P.L.M.); pellegrino.musto@cnr.it (P.M.)

* Correspondence: mensitie@unina.it; Tel.: +39-081-7682512

Received: 15 November 2018; Accepted: 18 January 2019; Published: 1 February 2019

Abstract: In this paper, the sorption thermodynamics of low-molecular-weight penetrants in a glassy polyetherimide, endowed with specific interactions, is addressed by combining an experimental approach based on vibrational spectroscopy with thermodynamics modeling. This modeling approach is based on the extension of equilibrium theories to the out-of-equilibrium glassy state. Specific interactions are accounted for in the framework of a compressible lattice fluid theory. In particular, the sorption of carbon dioxide, water, and methanol is illustrated, exploiting the wealth of information gathered at a molecular level from Fourier-transform infrared (FTIR) spectroscopy to tailor thermodynamics modeling. The investigated penetrants display a different interacting characteristic with respect to the polymer substrate, which reflects itself in the sorption thermodynamics. For the specific case of water, the outcomes from molecular dynamics simulations are compared with the results of the present analysis.

Keywords: glassy polymers; sorption thermodynamics; lattice fluid theory; polyetherimide; carbon dioxide; water; methanol

1. Introduction

Sorption thermodynamics of low-molecular-weight compounds within synthetic polymers, as well as the associated interactional issues, is a subject of considerable technological and fundamental interest, as proven by the recent literature reporting on experimental characterization techniques, theoretical approaches, and molecular simulations [1–13]. Sorption of gases and vapors in polymers has important technological implications as, for example, durability of polymer matrices for composites [14], membranes for separation of liquid and gaseous mixtures [15–18], and barrier polymers for packaging applications [19].

Depending on the chemical structure of the polymer and the penetrant, the system can be endowed with specific interactions, such as hydrogen bonding or other acid–base type interactions. We previously investigated and theoretically modeled sorption thermodynamics of gases and vapors in glassy polymers characterized by different levels of interactions [20–23] by complementing the theoretical modeling with the wealth of information gathered from vibrational spectroscopy and gravimetric measurements. These techniques provided a comprehensive molecular characterization of the systems under scrutiny, which was used to tailor the structure of the model. In detail, the experimental results were successfully interpreted using the so-called non-equilibrium theory for glassy polymers with non-random hydrogen bonding (NETGP-NRHB) [22], a model based on a compressible lattice fluid framework which is able to account both for the specific interactions and the non-equilibrium nature of glassy polymers. This approach is based on an extension of the equilibrium NRHB equation of state (EoS) theory [24,25], originally developed by Panayiotou et al. to deal with rubbery polymers. This extension was performed following the procedure introduced by Sarti and Doghieri [26,27].

Polyimides find relevant applications as membrane materials in view of their permeability and selectivity properties, as well as of their strong resistance to organic solvents [28]. Polyetherimide (PEI) belongs to this class of polymers. It is an amorphous engineering thermoplastic with a relatively high glass transition temperature (≅216 °C), a chain flexibility provided by ether groups, and high heat resistance imparted by aromatic imide groups. Thanks to its excellent chemical, mechanical, thermal, and transport properties, and to its durability [29–31], it is the material of choice for several separation processes [19,32–35]. Understanding at a molecular level the interactions between polymer backbone and penetrants is crucial for the optimal design of separation processes. In particular, hydrogen-bonding (HB) interactions significantly affect membrane mass transport and separation performances in those processes involving water and alcohols [36–45].

We present here a theoretical analysis of the sorption of low-molecular-weight penetrants in PEI combined with experimental investigations. The present contribution puts together unpublished results on the CO_2–PEI system with results on the H_2O–PEI and CH_3OH–PEI systems obtained by our group and already reported in recent publications [20,46–48]. The aim was to compare the thermodynamic behavior of systems displaying widely different interactive characteristics. As anticipated, non-equilibrium lattice fluid models are combined with Fourier-transform infrared (FTIR) spectroscopy and gravimetric analyses. Information gathered from the experiments was used to identify the involved interactions to be accounted for in the modeling and to validate model predictions. A rather clear physical picture emerged on self- and cross-interactions, providing quantitative and qualitative indications on the involvement of the moieties present on the polymer backbone.

2. Materials and Methods

2.1. Materials

Totally amorphous PEI (see Scheme 1) was a commercial-grade product, kindly supplied by Goodfellow Co., PA, USA, in the form of a 50.0-μm-thick film. To obtain film thicknesses suitable for FTIR spectroscopy, the original product was first dissolved in chloroform (15% *w*/*w* concentration) and then cast onto a tempered glass support. The solution was spread using a calibrated Gardner knife, which allows one to control the film thickness in the range of 10–70 μm. The cast film was dried 1 h at room temperature and 1 h at 80 °C to allow most of the solvent to evaporate, and at 120 °C under vacuum over night. At the end of the drying protocol, the film was removed from the glass substrate by immersion in distilled water at 80 °C.

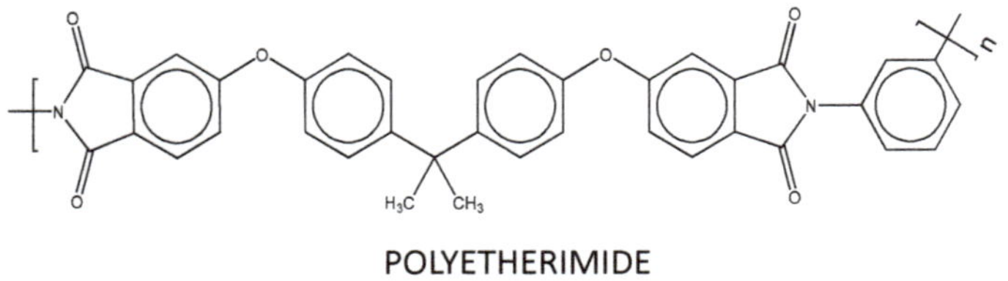

O O O O N N O O H3C CH3 n

POLYETHERIMIDE

Scheme 1. Chemical structure of PEI.

Thinner films (1.0–3.0 μm) were prepared via a two-step, spin-coating process performed with a Chemat KW-4A apparatus from Chemat Technologies Inc (Northridge, CA, USA). Spinning conditions were 12 s at 700 rpm for the first step and 20 s at 1500 rpm for the second step. The spin-coated films were dried in the same conditions as for the thicker films, and freestanding samples were removed in distilled water at room temperature.

Methanol used for sorption experiments was purchased from Sigma-Aldrich (Milano, Italy) with purity higher than 99.6%. Deionized water was used for water vapor sorption experiments. Before use, both water and methanol were degassed through several freezing–thawing cycles. CO_2 with a purity of 99.5% was purchased from Linde, Germany.

2.2. Gravimetric Measurements

The apparatus used to determine the sorption isotherms in PEI was based on the measurement of the elongation of a calibrated quartz spring microbalance (see Figure 1). The microbalance system recorded the weight gain of a sample exposed to a controlled environment (temperature and pressure of gaseous CO_2 or of water and methanol vapor) The quartz spring was placed in a water-jacketed glass cell that was connected, through service lines, to a CO_2 cylinder or to a water or methanol reservoir, to a turbo-molecular vacuum pump, and to pressure transducers. The polymer sample hung from the hook of the quartz spring. Changes in weight of the sample determined by sorption of the penetrant brought about the elongation of the spring that was monitored using a traveling camera, equipped with a macro objective, screwed on a computer-controlled piezoelectric slide. The microbalance provided a sensitivity of 1.0 μg with an accuracy of ±2.0 μg.

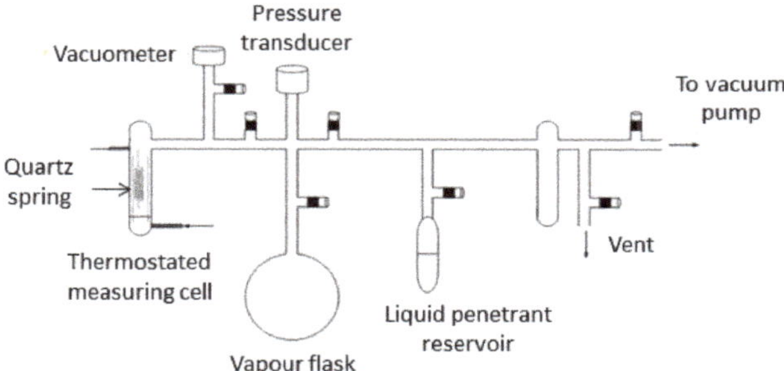

Figure 1. Schematic representation of quartz spring sorption apparatus in the configuration for vapour sorption. In the case of gas sorption, a penetrant reservoir is substituted by service lines to the gas cylinder. The polyetherimide (PEI) film sample hangs from the quartz spring located in the measuring cell.

Gravimetric sorption isotherms were collected at 18, 27, and 35 °C in the case of CO_2, at 30, 45, 60, and 70 °C in the case of water, and at 30 °C in the case of methanol. Sorption tests were performed starting from neat polymer by increasing the pressure of penetrant in a stepwise manner, waiting for the attainment of a constant weight at each step. Full details about the experimental procedure are given elsewhere [20–22].

2.3. In Situ FTIR Spectroscopy

Spectroscopic analysis of sorption processes was performed by time-resolved acquisition of the spectra of polymer films exposed to gaseous carbon dioxide or water and methanol vapors, at a constant pressure. These measurements were carried out in the transmission mode, and sorption kinetics was monitored up to the attainment of equilibrium.

The spectroscopic set-up consisted of a diffusion cell placed in the sample compartment of a suitably modified FTIR spectrometer. In the case of CO_2 sorption, the tests were performed with flowing gas. A schematic view of the apparatus is reported in Figure 2. The cell was connected to a gas line where the flux was regulated by a mass-flow controller and the downstream pressure was controlled by a solenoid valve. The cell temperature could be controlled between −190 °C and 350 °C. Differential sorption tests at 35 °C were performed by increasing stepwise the CO_2 pressure within the 40–150 Torr range.

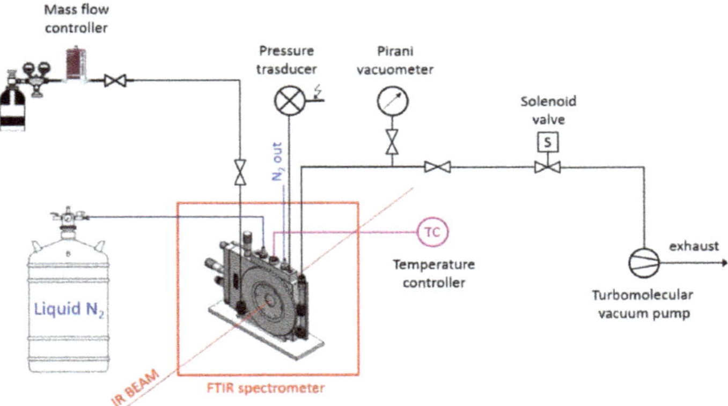

Figure 2. Schematic representation of experimental apparatus for Fourier-transform infrared (FTIR) spectroscopy measurements of CO_2 sorption. The PEI film sample is located inside the test cell, normally to the IR beam of the FTIR spectrometer.

In the case of water and methanol vapors, the tests were performed in static conditions. A custom-designed vacuum-tight FTIR cell was used to acquire the time-resolved FTIR spectra during the sorption experiments (see Figure 3). The cell, whose temperature could be controlled between −20 °C and 120 °C, positioned in the sample compartment of the spectrometer, was connected through service lines, to a water or methanol reservoir, a turbo-molecular vacuum pump, and pressure transducers. Full details of the experimental set-up are reported in References [49,50]. Before each sorption measurement, the sample was dried under vacuum overnight at the test temperature in the same apparatus used for the test. Differential sorption tests at 30 °C were performed by increasing stepwise the relative pressure of the vapor, p/p_0 (where p is the pressure of the vapor and p_0 is the vapor pressure at the test temperature), within the 0.0–0.6 range.

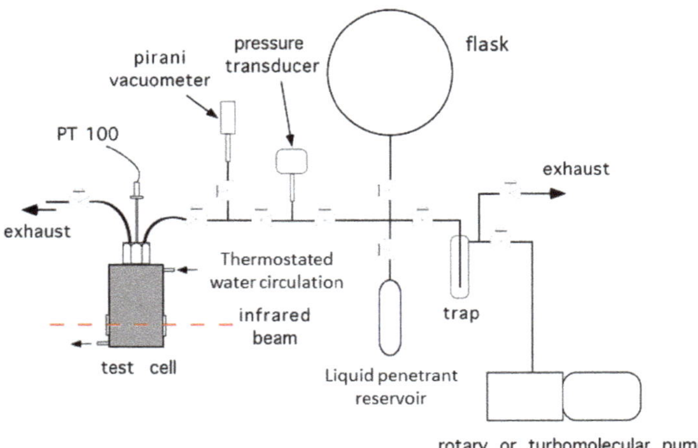

Figure 3. Schematic representation of experimental apparatus for FTIR measurements of H_2O and methanol vapor sorption. The PEI film sample is located inside the test cell, normally to the IR beam of the FTIR spectrometer.

Spectra were collected using a Spectrum 100 FTIR spectrometer (PerkinElmer, Norwalk, CT, USA), equipped with a Ge/KBr beam splitter and a wide-band deuterated triglycine sulfate (DTGS) detector.

The spectral resolution was set to 2 cm^{-1}, with an optical path difference (OPD) velocity of 0.5 cm/s. Collection time for a single spectrum was 2.0 s. Spectra were collected and stored for further processing in the single-beam mode. A dedicated software package was employed to control automated data acquisition (*Timebase* from Perkin-Elmer).

The cell without sample, at the test conditions, was used as background to obtain full absorbance spectra (i.e., polymer plus sorbed compound). The spectra representative of the different penetrants were isolated by difference spectroscopy (DC), which allowed us to eliminate the interference of the polymer substrate.

$$A_d(v) = A_s(v) - kA_r(v),$$

where v is the frequency, and $A_d(v)$, $A_s(v)$, and $A_r(v)$ are, respectively, the difference spectrum (the sorbed penetrant), the sample spectrum (the polymer containing the sorbed penetrant), and the reference spectrum (the dry polymer). The subtraction factor, k, allows to compensate for changes in the optical path (sample thickness) resulting from possible swelling. In the present cases, it was verified by spectroscopic means that a negligible volume change occurred; hence, a k value of 1 was consistently taken.

In the transmission FTIR measurements, sample thickness could be used to adjust the sensitivity, provided that the polymer substrate made no or limited interference in the frequency range where the relevant signals of the penetrant spectrum were located. Obviously, the sensitivity increment was achieved at the expense of the time to equilibration, and a trade-off was established between sensitivity and experiment duration. For this reason, thick samples (40–60 μm) were used to monitor the spectrum of the penetrant, whose concentration was very low, thus achieving an optimum signal-to-noise ratio. Conversely, to investigate the effect of the penetrant on the polymer spectrum, thin films were used (1–3 μm) that allowed keeping the intense carbonyl bands of PEI in the range of linearity of the absorbance vs. concentration curve.

3. Theoretical Background: Modeling of Sorption Thermodynamics

Sorption thermodynamics of low-molecular-weight compounds in glassy polymers was addressed using different approaches, some of which account for the molecular details associated with this process. In fact, there are various sorption modes, occurring simultaneously, that should be considered in modeling sorption equilibria: bulk dissolution and specific interactions (including self and cross hydrogen bonding, Lewis acid/Lewis base interactions, and penetrant clustering). Another relevant issue, when dealing with glassy polymers, is their non-equilibrium nature. In this section, we provide an overview of some significant lattice fluid modeling approaches proposed in the literature to interpret sorption thermodynamics in rubbery and glassy polymers. The reader is referred to Appendix A for the mathematical details of some theoretical approaches relevant in the present context.

In the last four decades, EoS-based theoretical models were introduced to address equilibrium thermodynamics of rubbery polymer–penetrant mixtures, possibly accounting for the occurrence of specific interactions (e.g., hydrogen bonding (HB) and Lewis acid/Lewis base interactions). Sorption thermodynamics in glassy polymers, endowed or not with specific interactions, was successfully interpreted by properly extending the equilibrium models, originally developed to describe sorption thermodynamics in rubbery polymers, to the case of polymers in an out-of-equilibrium glassy state.

The compressible lattice fluid (LF) model proposed by Sanchez–Lacombe (SL) [51–53] is one of the early examples of EoS-based approaches that were developed to describe sorption in rubbery polymers. Such a model assumes that a random mixing occurs within the polymer–penetrant mixture. The lattice fluid framework for fluid mixtures takes each component (i.e., polymer and low-molecular-weight penetrant in the present context) as consisting of a series of chemically bonded -mers, each one accommodated in a cell of the lattice. Since the lattice is compressible, not all the cells are occupied by the -mers, some of them being possibly empty (holes). Chemical bonds are established between first neighbor -mers belonging to the same molecule, but contacts are also established with other non-bonded -mers located in adjacent first-neighbor cells. The coordination number, z, which is a parameter of the model lattice, rules the number of contacts between first-neighbor cells. Contacts established between -mers of the same type are named homogeneous; otherwise, they are said to be heterogeneous. Sorption equilibrium in binary systems is ruled by the equivalence of chemical potentials of each component in the two phases in contact, i.e., the polymer–penetrant mixture and the gaseous phase. The SL theory provides the expressions for the chemical potentials of each of the mixture components, as well as the expression of the equations of state of the mixture and of each pure component. Since it is generally assumed that no polymer is present in the gaseous phase, the equivalence is only imposed to the chemical potentials of the penetrant in the two phases in contact. Densities of the two phases in contact are provided by the EoS of each phase. Parameters of the model are the three scaling parameters of the EoS for the two pure components and the value of the binary interaction parameter that is related to the energy associated to formation of heterogeneous contacts occurring between the -mers of the two components in a binary mixture and the energy associated to the formation of homogeneous contacts between the -mers of each component.

Later, lattice fluid theories were further developed to describe sorption of low-molecular-weight compounds in rubbery polymers for systems endowed with HB interactions. These approaches were developed by properly modifying available LF models that account only for mean field interactions to include the effect of occurrence of self and cross hydrogen bonding. Worthy of mention is the model proposed by Panayiotou and Sanchez (PS) [54] that, starting from the original Sanchez–Lacombe theory, introduced additional terms accounting for the formation of HB interactions. A key assumption in this development was the factorization of the configurational partition function in two separate contributions, respectively associated with mean field interactions and to specific interactions. The PS model, in fact, combines the SL mean field contribution with an HB contribution formulated on the basis of a combinatorial approach first proposed by Veytsman [55,56]. This model requires, in addition to the parameters already mentioned in the case of SL theory, the HB interaction parameters associated with each type of proton donor–proton acceptor adduct that can be formed in the system.

A remarkable intrinsic simplification of the SL model, as well as of the related PS model, is that the evaluation of the mean field contribution is, in both cases, based on a random arrangement of r-mers and holes in the lattice. This assumption is that a more inappropriate arrangement suggests a greater number of non-athermal contacts occurring between different kinds of r-mers [57]. Several approaches were then proposed to account for possible non-random distribution of contacts by following the pioneering work of Guggenheim [58]. As a first step, systems without specific interactions were addressed. Also in this case, the assumption was that the partition function can be factorized, in an ideal random contribution and in a non-random contribution "correction term". The latter was formulated considering the establishment of each contact as a reversible chemical reaction (the so-called "quasi-chemical approximation"). The original development proposed by Guggenheim was based on a lattice fluid system without holes, but Panayiotou and Vera (PV) improved this theory by introducing a compressible LF model (i.e., with the presence of hole sites) [59]. In the PV model, only contacts between -mers of the mixture components were non-random, while the hole site distribution was still random. Later, You et al. [60] and Taimoori and Panayiotou [61] further upgraded this approach introducing the non-randomness of all the possible couples of contacts, still adopting a non-random quasi-chemical approximation.

A further step consisted in the formulation of an LF-EoS theory accounting both for the occurrence of specific interactions and for the non-randomness of distribution of contacts within the lattice. In fact, Yeom et al. [62] and Panayiotou et al. [2,24,25,63,64] modified the PV approach to include the contribution of HB interactions. Again, the key assumption was the factorization of the configurational partition function in different terms: a mean field random term, a non-randomness "correction term", and a term accounting for specific interactions. In the present context, the non-random model is of particular interest, which also accounts for cross and self HB interactions, proposed by the group of Panayiotou in References [2,24,25]. We refer to this theory hereafter as the "non-random lattice fluid hydrogen bonding" (NRHB) model. The theory can be used to calculate the chemical potential of the polymer and of the penetrant, both in the polymer mixture and in the vapor or gaseous phase in equilibrium with it, as well as the EoS of both phases. The types of parameters required to perform calculations are similar to those required in the case of PS model. Examples of good correlations of the experimental sorption isotherms using the NRHB model were provided by Tsivintzelis at al. [65] for the case of mixtures of poly(ethylene glycol), poly(propylene glycol), poly(vinyl alcohol), and poly(vinyl acetate) with several solvents, including water.

The theories discussed so far are well suited to describe the sorption behavior of rubbery polymers. When dealing with modeling of sorption thermodynamics in a glassy polymer, one has to tackle an additional complexity, i.e., the non-equilibrium nature of the glassy state. Rational non-equilibrium thermodynamics provides an adequate theoretical framework to address this issue. In fact, using a proper reformulation, it is possible to extend the equilibrium theories developed for mixture thermodynamics—adequate for describing sorption thermodynamics in rubbery polymers—to the case of sorption in glassy polymers. This procedure is based on thermodynamics endowed with internal state variables. Order parameters that mark the departure, at a certain pressure and temperature, from the equilibrium conditions, are selected to play the role of internal state variables. In fact, in addition to the external state variables, which are the only ones needed to describe the state of a system at equilibrium (e.g., pressure, temperature, and concentration), a set of order parameters are used as internal state variables to provide a proper interpretation of the state of non-equilibrium glassy polymer–penetrant mixtures. This line of thought was introduced by Sarti and Doghieri [26,27], which adopted the density of the polymer in the mixture as the only order parameter and internal state variable. These authors introduced a procedure that enables the extension to non-equilibrium glassy systems of several equilibrium statistical thermodynamics mixture theories (in Appendix A, some further details are provided). This approach, referred to as "non-equilibrium thermodynamics for glassy polymers" (NETGP) [66], was demonstrated to be very effective in describing the thermodynamics of numerous binary and ternary glassy polymer–penetrant mixtures. A relevant difficulty that one encounters when applying this theory is that the expression for evolution kinetics of the internal state variable has to be known. To avoid dealing with complex evolution kinetics of non-equilibrium polymer density, a "simplified" version of this approach considers the polymer as being "frozen" in a kinetically locked pseudo-equilibrium (PE) state. This assumption is legitimate for glassy polymers at temperatures well below their glass transition temperature and when the concentration of penetrant within the polymer phase is low. To deal with cases in which such conditions do not occur, the theory was properly formulated to take also into account the structural evolution of the glassy system during penetrant sorption. The NETGP approach requires the knowledge of the same parameters as for the equilibrium theory from which it is derived. The phase equilibrium is still dictated by the equivalence of chemical potentials of each component in the two phases. However, differently from the case of sorption in rubbery polymers, no equation of state can be written for the polymer phase. In fact, in the case of glassy polymers, the value of polymer density cannot be calculated using an equilibrium EoS, but is determined by its intrinsic evolution kinetics; its value, which is necessary to perform model calculations, should be known from independent experimental data. In particular, in the present context, we refer to the so called "non-equilibrium lattice fluid" (NELF) model [26,27] as the NETGP extension of the equilibrium Sanchez–Lacombe theory to treat sorption in glassy polymers not endowed with specific interactions (see Appendix A, for details).

To deal with sorption thermodynamics for glassy polymer–penetrant systems exhibiting specific interactions, extensions of equilibrium models able to cope with self- and cross-interactions were then proposed. Referring to the case of EoS-based models, of particular relevance are the extensions of "statistical associating fluid theory" (SAFT) [67] and NRHB [68] approaches. Our group developed and applied an extension of the NRHB model to non-equilibrium glassy polymers to account for HB interactions in water/glassy polymer mixtures [22]. We refer to this model here as NETGP-NRHB (see Appendix A for details). Again, this model requires the same parameters as for the equilibrium NRHB model. In addition, the value of polymer density should also be known, since it cannot be provided by an EoS.

4. Results and Discussion

In this section, we discuss the results of experimental analysis and of theoretical modeling of the PEI–CO_2, PEI–water, and PEI–methanol systems. Results of in situ FTIR spectroscopy are presented along with their elaboration by means of two-dimensional correlation spectroscopy (2D-COS) [23,69–71]. Indications emerging from FTIR measurements were used to tailor the structure of sorption thermodynamics model, which was then used to fit gravimetric sorption isotherms.

4.1. PEI–CO_2 System

4.1.1. FTIR Analysis: Absorbance, Difference, and Two-Dimensional Correlation Spectra

As reported in Section 2.3, to achieve an optimum signal-to-noise ratio, thick samples were used to monitor the spectrum of the penetrant, whose concentration was very low. Conversely, to analyze the effect of the sorbed penetrant on the polymer spectrum, thin films were used in order to keep the relevant bands of PEI in the range of linearity of the absorbance vs. concentration curve. In Figure 4A, the spectra of a fully dried film (red trace) and of the same film equilibrated at 35 °C under a CO_2 pressure of 150 Torr (blue trace), collected on a 37.7-μm-thick film, are compared. This was the kind of sample used to perform the subtraction spectroscopy analysis on the spectral region populated by signals associated to penetrant molecules. In Figure 4B, the same comparison for the spectra collected on a much thinner film, with a thickness of 2.6 μm, is shown. This was the kind of sample used to perform the subtraction spectroscopy analysis in the spectral region with signals characteristic of the polymer, in order to detect possible perturbations related to the presence of sorbed CO_2. In particular, attention was focused on the intense carbonyl bands of PEI, and the thickness was selected to keep the associated absorbance vs. concentration curve in the range of linearity. It is worth noting that, in the case of the "thin" sample, the weakness of the penetrant signals, even at the maximum CO_2 pressure, did not allow a reliable band-shape analysis. Furthermore, the sorption kinetics was so fast that a time-resolved measurement and the associated two-dimensional correlation analysis (vide infra) were unfeasible.

It was immediately apparent, from the comparison of the two traces reported in Figure 4A, that sorbed carbon dioxide produced two signals, centered at 2336 and 655 cm^{-1}. The high-frequency band was due to the antisymmetric stretching mode (ν_3) at 2336 cm^{-1} plus a satellite peak at 2324 cm^{-1}. The latter component was a non-fundamental transition, in particular a hot band enhanced by Fermi resonance with the main stretching mode [72]. It is to be stressed that the peak at 2324 cm^{-1} was not due to a second CO_2 species, but to a further signal produced by the same species that generated the main absorption at 2336 cm^{-1}.

The ν_3 mode was well suited for quantitative analysis and for achieving information at the molecular level. The carbon dioxide bending mode (ν_2) found at around 655 cm^{-1} was less useful, being much weaker and superimposed onto intense polymer bands. The suppression of the polymer matrix interference by difference spectroscopy [50,69] allows one to isolate the spectrum of absorbed CO_2. The integrated absorbance of the 2336 cm^{-1} band was collected at each pressure as a function of sorption time up to attainment of sorption equilibrium.

In Figure 5A, the ν_3 band of CO_2 after sorption equilibrium at different pressures is reported. The correlation between integrated absorbance and gravimetric data, collected at 35 °C, at sorption equilibrium at different CO_2 pressures, is represented in Figure 5B. The behavior was linear through the origin, thus verifying the Lambert–Beer law, which allows one to convert the absorbance data into absolute concentration values. In the explored pressure range (40–160 Torr), the sorption isotherm was linear (see Figure 5C). The band-shape of the ν_3 mode was exactly coincident at all pressures (see

Figure 5A) indicating that, in the explored range, the molecular interactions formed with the polymer substrate did not depend on CO_2 concentration.

The time-resolved spectra were subjected to 2D-COS analysis, which is very effective for investigating molecular interactions [69]. This is a perturbative technique applied to systems that are initially at equilibrium; these are then subjected to an external stimulus and the spectral response of the system is treated by a correlation analysis. The covariance of two correlated signals (peak absorbance, in the present case) is measured as a function of a third common variable related to the perturbing function (time, in the present case). This procedure spreads the spectral data over a second frequency axis and allows unambiguous assignments through the correlation of bands. It produces synchronous and asynchronous maps. In the latter, the correlation intensity for signals evolving at the same rate vanishes [71], thus providing a resolution enhancement. Moreover, valuable dynamic information can be also gathered.

In Figure 6A,B, the synchronous spectrum in the 2250–2420 cm^{-1} range, obtained from the time-resolved spectra collected during the sorption experiment at 150 Torr and 308 K, and the power spectrum, i.e., the autocorrelation profile taken across the main diagonal, are reported. The synchronous spectrum (Figure 6A) displays a highly characteristic cross-shape; in the power spectrum (Figure 6B), a strong auto-peak was detected at 2336 cm^{-1}, corresponding to the main component in the frequency spectrum, plus a weak, fully resolved feature at 2324 cm^{-1}. The off-diagonal wings present in the synchronous map (Figure 6A) extend by about 60 cm^{-1} and reflect the presence of a further component, increasing on sorption or decreasing on desorption concurrently with the main peaks (at 2336 and 2324 cm^{-1}). The elongated shape of the off-diagonal feature indicated that this component was significantly broader than the main peaks.

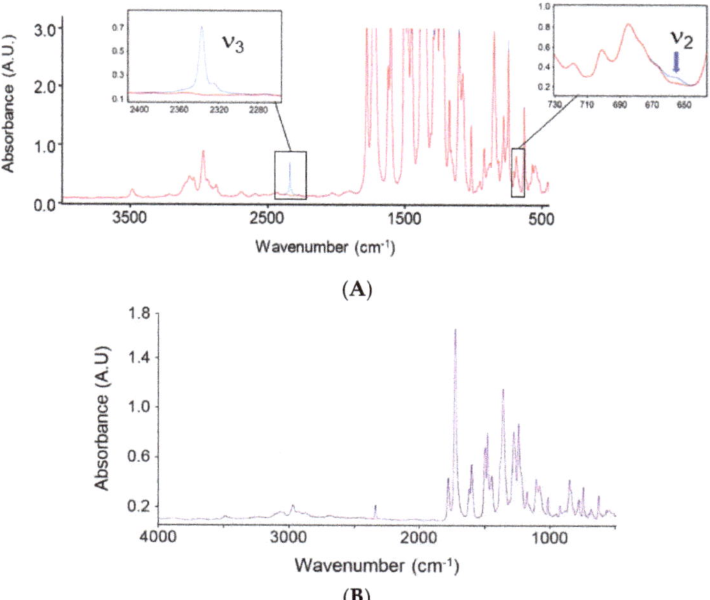

Figure 4. (**A**) Absorbance spectra of fully dried PEI (red trace) and PEI equilibrated at 35 °C at a CO_2 pressure of 150 Torr (blue trace). The insets highlight the bands of sorbed CO_2. The spectra were collected on a 37.7-µm-thick film. (**B**) Same as (**A**); measurement performed on a 2.6-µm-thick film.

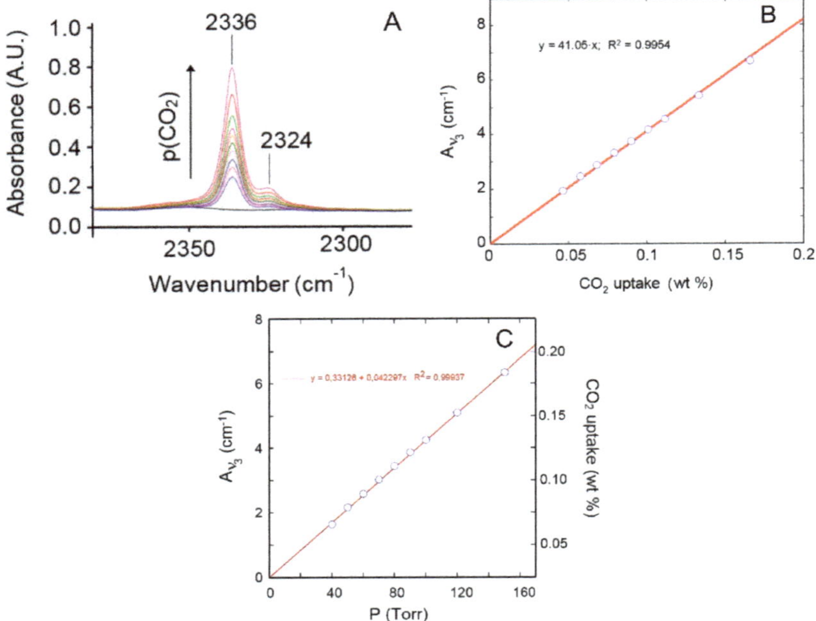

Figure 5. (**A**) The ν_3 band of CO_2 after sorption equilibrium at different pressures. (**B**) Absorbance of the ν_3 band as a function of sorbed CO_2 as evaluated gravimetrically at 35 °C. (**C**) FTIR isotherm at 35 °C. The right axis was obtained from the calibration curve of Figure 5B.

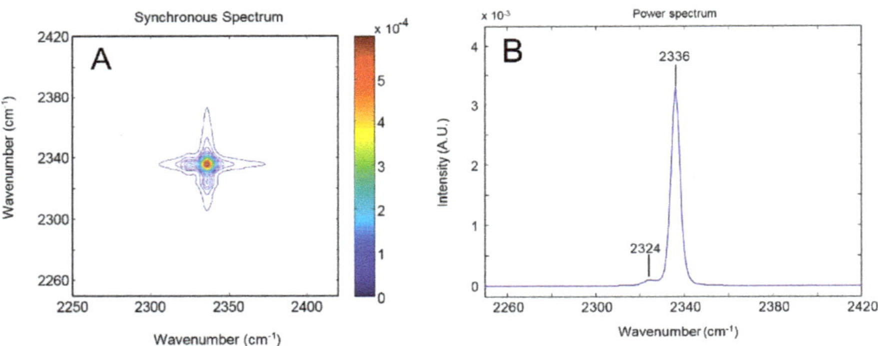

Figure 6. (**A**) Synchronous spectrum obtained from the time-resolved spectra collected during the sorption experiment performed at 150 Torr and at 35 °C. (**B**) Power spectrum calculated therefrom.

The asynchronous spectrum was featureless (see Figure 7A,B). Only noise was present despite the significant intensity of the band and its complex structure. This result was relevant; it demonstrated that all the components in the ν_3 profile evolved synchronously, which, in turn, signified that the species they originated from had comparable dynamics or, alternatively, that a single molecular species produced all the observed components. Albeit no conclusive evidence is yet available, we are inclined toward the second hypothesis, in view of previous studies on solvated low-molecular-weight compounds [73,74].

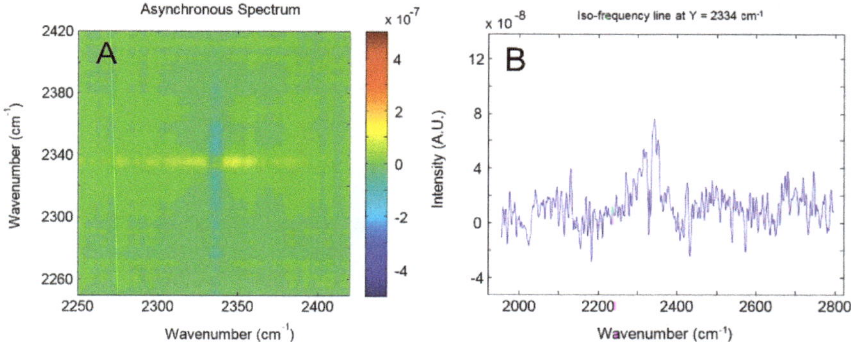

Figure 7. (**A**) Asynchronous spectrum obtained from the time-resolved spectra collected during the sorption experiment performed at 150 Torr and at 35 °C. (**B**) Iso-frequency profile at 2334 cm^{-1}.

Similar band profiles were reported in the literature (i.e., a broad Gaussian component superimposed onto a sharper peak of larger intensity) and were interpreted assuming the Gaussian part as being a remnant of the gas-phase spectrum, while the sharp, Lorentz-like component was interpreted as the purely vibrational transition active in the condensed phase (but not in the gas-phase) [73,74]. According to this interpretation, the two-component band-shape was the consequence of probe dynamics within the molecular environment (essentially free rotation in the early stages of the relaxation process, 0.2–1.0 ps), which produced the Gaussian part, and a random rotational diffusion regime at later stages (the so-called Debye regime), giving rise to the Lorentzian part. The band-shape models based on small-molecule mobility invariably tend to overlook the role played by molecular interactions. This is partially due to the difficulty in embracing the effects of specific contacts in a random collisional framework. In the present case, the specific interaction between CO_2 and PEI emerged clearly from the vibrational analysis (vide infra) and was taken into account when interpreting the band-shape and the 2D-COS results. Our view is that the probe molecules sorbed in PEI were characterized by very short free-rotation regimes (the Gaussian component is barely detectable in the frequency spectrum); the main component was due to a single molecular species of CO_2 forming a specific type of interaction with the polymer substrate. The uniqueness of the interaction is suggested by the sharp and highly symmetrical nature of the main ν_3 component, coupled with the absence of asynchronous features in the 2D-COS map. Work is in progress to substantiate this hypothesis and to put the above considerations on more quantitative grounds.

To complete the characterization of the probe-to-substrate interaction, we had to recognize the functional group(s) of the PEI backbone that was (were) involved in the formation of the adduct with the penetrant. In fact, there were several possible candidates, such as the imide carbonyls, the ether oxygens and, to a lesser extent, the aromatic rings. The analysis was performed by detecting the perturbation brought about by the probe to the spectrum of the substrate in terms of peak shifts and/or band-shape distortion, effects that were subsequently interpreted in terms of geometry and electron-density distribution of the molecular aggregate. The reversibility of the observed effects was also assessed, i.e., the obtainment of the unperturbed spectrum when the probe was fully desorbed. Measurements of this kind require films in the thickness range of 1.0–3.0 μm, in order to maintain the whole spectrum within the range of absorbance linearity. In the present case, these films were prepared ad hoc via a spin-coating process. Since the interaction was weak, the effects were expected to be very subtle; in these circumstances, difference spectroscopy was demonstrated to be a viable approach [46]. In Figure 8, the difference spectra in the frequency range 1820–1660 cm^{-1} are reported, obtained by subtracting the spectrum of the fully dried sample from the spectra of the polymer sample equilibrated with gaseous CO_2 at a pressure of 150 Torr and at the different temperatures. It is explicitly noted that the reference and the sample spectra to be subtracted were collected at the same temperature to avoid any spurious perturbation of the vibrational response. Lowering the temperature is expected to narrow the peaks while increasing the separation among the components, which should enhance the shift effect. An increase in the number of interacting CO_2 molecules is also possible at lower temperatures, as a consequence of the reduced kinetic energy of the probe.

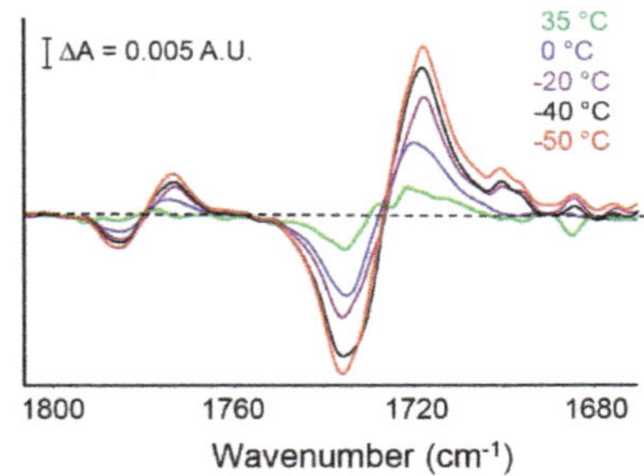

Figure 8. Difference spectra in the PEI carbonyl range. The subtraction involves the film equilibrated at the indicated temperatures as a sample spectrum and the fully dried sample as a reference. Both spectra were collected at the indicated temperatures. Equilibrium sorption measurements were performed at 150 Torr.

The difference spectra in Figure 8 display the typical first-derivative features associated with red-shifts, that is, a negative lobe preceding the positive. Thus, a lowering of the peak frequency was produced in the CO_2-containing samples. The effect was evident for both peaks occurring in the 1800–1680 cm^{-1} interval, according to the nature of the corresponding vibrational modes, i.e., the in-phase, ν_{ip}(C=O), and out-of-phase stretching, ν_{oop}(C=O), of the imide carbonyls. The two-lobe profiles were symmetric, i.e., the positive and the negative components had comparable intensities and grew progressively as the temperature decreased. Furthermore, the effect was fully reversible, as demonstrated by comparing the reference spectrum with that of the sample equilibrated at p = 150 Torr and subsequently desorbed. These results confirm that the observed shifts, albeit very subtle (maximum shift of the main carbonyl component = 0.25 cm^{-1}), actually originated from the probe/substrate interactions and that these interactions were weak. For comparison, we recall that the red-shift caused by the interaction of the same carbonyl groups with water molecules (H-bonding) amounted to 0.8 cm^{-1} [46]. All difference spectra were featureless below 1250 cm^{-1}. The absorbance spectrum of PEI displays two intense bands at 1238 and 1215 cm^{-1}, involving significant contributions from the C–O–C stretching modes [46]. The fact that these bands (as well as aromatic peaks) remained unaffected indicates that both the ether oxygens and the aromatic rings did not "see" the presence of the CO_2 molecules. The effects discussed above clarify the interaction mechanism: the imide carbonyls were selectively involved in a way that produced a lowering of the C=O force constant. The carbon atom of the carbon dioxide molecule, which brought a partial positive charge as a consequence of the two oxygens to which it was bound, formed a weak Lewis acid/Lewis base interaction with the imide carbonyls, as schematically represented in Figure 9. The above conclusions are in full agreement with the 2D-COS analysis.

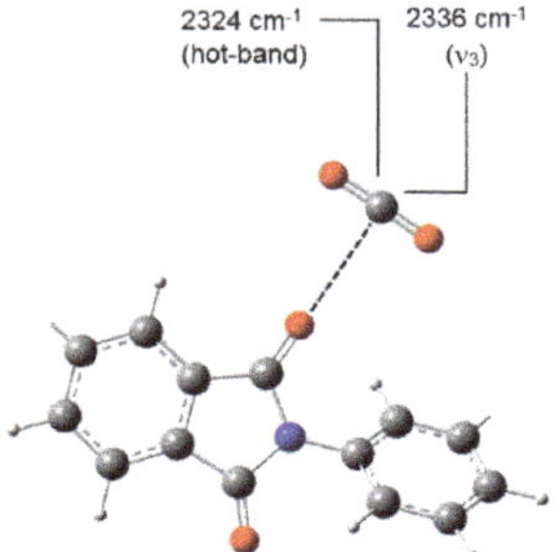

Figure 9. Schematic representation of the CO_2/C=O interaction.

4.1.2. Modeling Sorption Thermodynamics

The FTIR analysis presented in the previous section indicates that carbon dioxide molecules absorbed within PEI tended to establish weak Lewis acid/Lewis base interactions with carbonyls along the polymer backbone. The interactional energy was close to a mean field value; hence, for the sake of a macroscopic thermodynamics description of the system, it was pointless to introduce distinct specific interaction terms in the model (as for the NETGP-NRHB model). Based on these premises, the NELF approach (see Appendix A, for relevant equations) was used, which provided a good interpretation of gravimetric sorption isotherms. In Figure 10, the experimental gravimetric sorption isotherms, along with the curves obtained by a concurrent fitting of the data with the NELF model at three temperatures (18, 27, and 35 °C), are reported. Only one fitting parameter, i.e., the Sanchez–Lacombe binary interaction parameter, χ_{12} (see Appendix A, for its definition), was used. A value of $\chi_{12} = 0.0313 \pm 0.003$ was estimated, thus pointing to a limited deviation from the geometric mixing rule for characteristic pressure (see Equation (A24)). The scaling parameters for the SL EoS of pure PEI were calculated by fitting pressure, volume, and temperature (PVT) data available from a previous investigation [20] with the SL-EoS model, while those of pure CO_2 were taken from Reference [26]. The (non-equilibrium) density values of neat PEI (i.e., $\rho_{2,\infty}$ as defined in Appendix A) at each temperature (assumed here to be time-independent) used in the model were taken from the literature [20]. All parameters are reported in Tables 1 and 2.

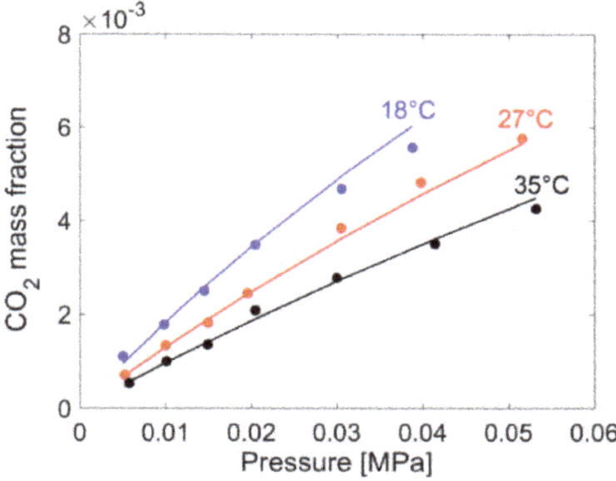

Figure 10. Experimental gravimetric sorption isotherms of CO_2 in PEI at 18, 27, and 35 °C. Continuous lines represent the results of concurrent data fitting performed with the non-equilibrium lattice fluid (NELF) model.

Table 1. Non-equilibrium polyetherimide (PEI) density [47].

Temperature (°C)	Density (g/cm^3)
18	1.2700 ± 0.0006
27	1.2680 ± 0.0006
30	1.2673 ± 0.0006
35	1.2663 ± 0.0006
45	1.2641 ± 0.0006
60	1.2610 ± 0.0006
70	1.2589 ± 0.0006

Table 2. Scaling parameters of Sanchez–Lacombe equation of state (EoS) for PEI and CO_2 [26,47].

Substance	T * (K)	P * (MPa)	ρ * (g/cm^3)
PEI	893	545	1.354
CO_2	300	630	1.515

4.2. PEI–H_2O System

4.2.1. FTIR Analysis: Absorbance, Difference, and Two-Dimensional Correlation Spectra

For the analysis of water absorbed in PEI, we considered the normal modes of the water molecule in the ν(OH) frequency range (3800–3200 cm^{-1}). Other regions of potential interest for the spectroscopic analysis were not available for the case at hand due the exceedingly low intensity of the signals. This system was discussed in a recent publication [46], and we present here the most relevant results.

In Figure 11A, the difference spectra in the OH stretching region (ν(OH)) are reported for the sample equilibrated with water vapor at different p/p_0 values. The band was representative of sorbed water and exhibited a profile with two maxima at 3655 and 3562 cm^{-1}. The well-resolved multicomponent band-shape clearly indicated the occurrences of different water species in contrast to the situation observed in the case of CO_2, where a single component pattern was detected.

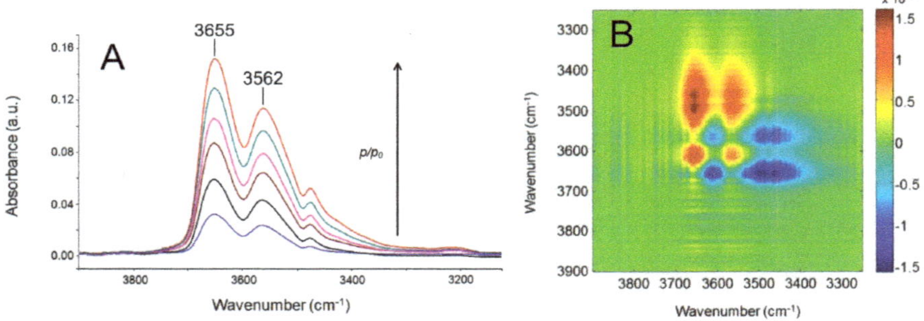

Figure 11. (**A**) The ν(OH) band of H_2O sorbed in PEI after sorption equilibrium at different p/p_0 at 30 °C. (**B**) Asynchronous spectrum obtained from the time-resolved spectra collected during the sorption experiment performed at p/p_0 = 0.6 and 30 °C.

To improve the resolution and to deepen the spectral interpretation, 2D-COS was also carried out in this case; the time-resolved spectra collected during a water sorption experiment performed at a p/p_0 = 0.6 were used for this purpose. The resulting asynchronous spectrum is reported in Figure 11B. The rich pattern contrasts with the featureless map detected for the CO_2–PEI system (compare Figure 11B with Figure 7A), thus confirming that asynchronous 2D-COS spectra are a powerful and sensitive tool to highlight complex molecular interaction scenarios. In particular, in the case of the H_2O–PEI system, it emerged that there existed two couples of signals, suggesting the presence of two distinct water species. In detail, the two sharp peaks at 3655–3562 cm^{-1} were assignedto the asymmetric and the symmetric stretching modes (ν_{as} and ν_s), respectively, of isolated water molecules interacting via H-bonding with the PEI backbone (cross-associated or first-shell water molecules). The second doublet at 3611–3486 cm^{-1} was related to water molecules self-interacting with the first shell species (self-associated or second-shell water molecules).

To identify the active sites of polymer backbone involved in the specific interaction with first-shell water molecules, an analysis was performed on thin samples (thickness in the range 1.0–3.0 μm). The red-shift of the ν_{ip}(C=O) and ν_{oop}(C=O) modes demonstrated the involvement of the imide carbonyls as proton acceptors in H-bonding. The peaks of the aryl ether groups remained unperturbed, which demonstrated that their interaction with water was negligible, if present. In addition, it was found that the largely prevalent stoichiometry of the first-shell adduct was 2:1, i.e., a single water molecule bridged two carbonyls: –C=O—H–O–H—O=C–. The two water species identified spectroscopically are schematically represented in Figure 12, where the peak frequencies they produced are also indicated.

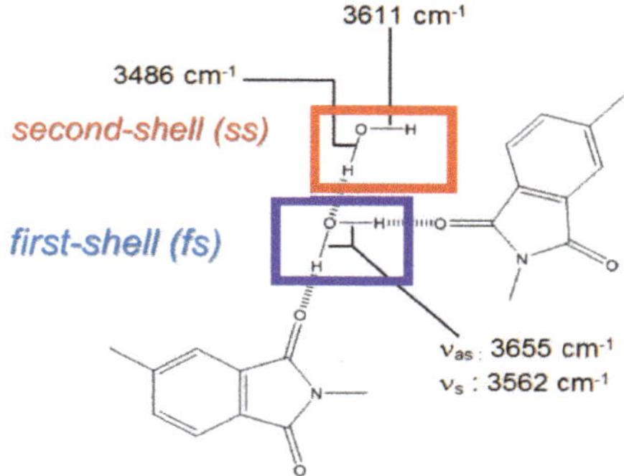

Figure 12. Schematic representation of the water species identified by the spectroscopic analysis.

Molecular dynamics (MD) calculations performed on this system [46] indicated that, at least in the low-to-intermediate relative pressure range of water vapor, these bridges formed by first-shell water were intramolecular, that is, formed by interacting with two contiguous carbonyl groups present on the same macromolecule. In Figure 13, a snapshot of an MD calculation is reported, showing the typical conformation of first-shell water molecules.

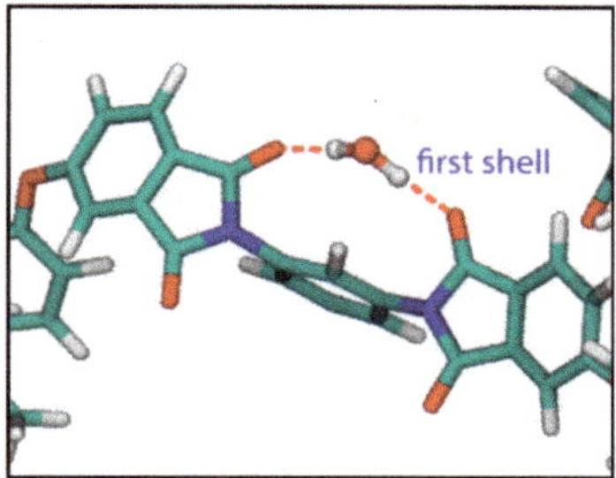

Figure 13. Snapshot of a typical first-shell (fs) configuration with an interacting water molecule bridging two consecutive carbonyl oxygens on the same PEI chain via hydrogen bonds.

Performing a quantitative analysis of the difference spectra of water, based on the information available from gravimetric measurements at different relative humidity, it was possible to obtain a quantitative estimation of the amount of the two water species identified spectroscopically. Details on the calculation procedure are reported in a previous publication [46].

4.2.2. Modeling Sorption Thermodynamics

In view of the significant HB interactions characterizing the PEI–H_2O system, the interpretation of sorption thermodynamics of water in glassy PEI was approached using the NETGP-NRHB model (see Appendix A for the relevant equations). Information gathered from the vibrational spectroscopy investigation, illustrated in Section 4.2.1, was exploited to tailor the H-bonding interaction contributions in the NETGP-NRHB model. In fact, only one proton acceptor was assumed to be present on the polymer backbone (i.e., the carbonyl of the imide group) and water molecules were assumed to be involved both in cross and self HB interactions. Moreover, the presence of two proton donors and two proton acceptors was assumed for each water molecule.

Also in this case, as for the PEI–CO_2 system, the non-equilibrium density of the polymer, at the different temperatures, was assumed to be time-independent and its value was taken to be equal to that of pure PEI. These density values, the values of the NRHB EoS scaling parameters for pure PEI and pure water, and the values of the self HB parameters for pure water were retrieved from the literature [20,75] and are reported in Tables 1 and 3. Experimental gravimetric water sorption isotherms in PEI were available at four temperatures (30, 45, 60, and 70 °C) [47]. For the meanings of scaling and HB interaction parameters appearing in Table 3, the reader is referred to Appendix A.

Table 3. Non-random lattice fluid hydrogen bonding (NRHB) scaling parameters and hydrogen-bonding (HB) interaction parameter for pure PEI and H_2O [20,75]. The meanings of symbols can be found in Appendix A.

Substance	ε_s^* (J/mol)	ε_h^* (J/(molK))	$v_{sp,0}^*$ (cm^3/g)	E_{11}^{0w}(J/mol)	S_{11}^{0w}(J/(molK))	s	V_{11}^{0w}(cm^3/mol)
PEI	6775 ± 250	5.503 ± 0.15	0.7228 ± 0.08	-	-	0.743 [a]	-
H_2O [b]	5336.5	−6.506	0.9703	−16,100	−14.7	0.861	0

[a] Calculated using UNIFAC group contribution calculation scheme [76,77]. [b] No confidence intervals were provided for water parameters in Reference [75].

Concurrent fitting of the four experimental gravimetric isotherms was performed using the NETGP-NRHB model, and the results are shown in Figure 14. The internal energy of formation of water–polymer cross HB, E_{12}^{0wp}, the entropy of formation of water–polymer cross HB, S_{12}^{0wp}, and the mean field binary interaction parameter, K_{12}, were used as fitting parameters (their meanings are discussed in Appendix A). In accordance with the previous relevant literature [65], the volume of formation of water–polymer cross HB, V_{12}^{0wp}, was taken to be equal to zero. In Table 4, the values of the model parameters as determined by a best fitting procedure along with their 95% confidence intervals are reported, determined using the Jacobian method implemented in the *nlinfit* and *nlparci* routines of the Matlab® software. As evident in Figure 14, the model provided a very good fitting of the experimental sorption isotherms. Calculations performed to carry out isotherm data fitting using the NETGP-NRHB model also provided a prediction for the amount of self HB (i.e., 1–1, involving only water molecules) and of cross HB (i.e., 1–2, between water molecules and carbonyl groups of PEI) interactions established in the PEI–water mixture equilibrated with water vapor at the different p/p_0 investigated. In particular, calculations provided the number of moles of HB self-interactions (1–1) and of moles of HB cross-interactions (1–2) per gram of amorphous PEI, indicated as n_{11}/m_2 and n_{12}/m_2, respectively. These values predicted using the NETGP-NRHB model, at a temperature of 30 °C, are compared in Figure 15 with the results obtained from FTIR spectroscopy and with the outcomes of MD simulations [46]. The agreement is satisfactory in the whole range of water concentrations.

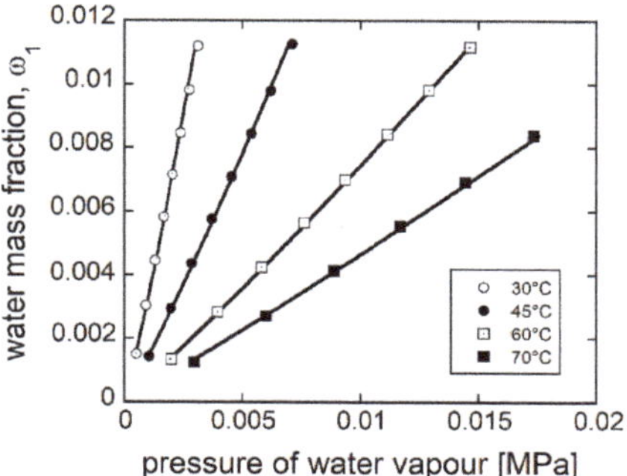

Figure 14. Experimental gravimetric sorption isotherms of H_2O in PEI at 30, 45, 60, and 70 °C. Dotted lines represent the results of concurrent data fitting performed with the non-equilibrium thermodynamics for glassy polymers/non-random lattice fluid hydrogen bonding (NETGP-NRHB) model.

Table 4. Non-equilibrium thermodynamics for glassy polymers (NETGP)-NRHB fitting parameters for PEI–water mixture (note that the value of V_{12}^{0wp} was set at 0 according to Reference [24]).

k_{12}	E_{12}^{0wp} (J/mol)	S_{12}^{0wp} (J/(mol·K))	V_{12}^{0wp} (cm^3/mol)
0.121 ± 0.005	−13,264 ± 200	−6.107 ± 0.100	0

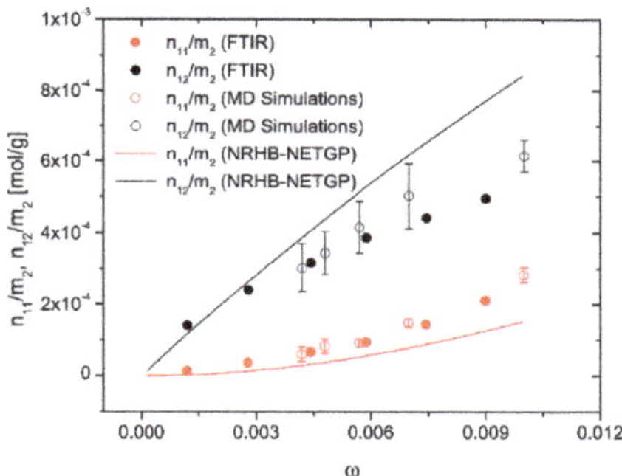

Figure 15. Comparison of the predictions of the NETGP-NRHB model for the number of self and cross HBs in the PEI–H_2O system at 30 °C with the outcomes from FTIR spectroscopy and of molecular dynamics (MD) simulations.

4.3. PEI–CH_3OH System

FTIR Analysis: Absorbance, Difference, and Two-Dimensional Correlation Spectra

For the methanol–PEI system at 30 °C, the ν(OH) band-shape (3650–3200 cm^{-1} range) was simpler than for water–PEI (compare Figures 16A and 11A) [47,48]. A single maximum was observed at 3575 cm^{-1}, superimposed to a broader component at lower frequencies. The occurrence of two signals only was confirmed by the asynchronous map (Figure 16B) displaying a single cross-correlation centered at 3441–3575 cm^{-1}. The other features appearing in the map were relative to the correlations with the ν(CH) signals and were not relevant for the molecular interaction analysis. Perturbation analysis of the PEI spectrum indicated that the carbonyl groups of the polyimide acted as proton acceptors in the H-bonding interaction with methanol, while the ν(C–O–C) band was unperturbed in the presence of methanol, thus indicating that the involvement of ether linkages in H-bonding with the penetrant could be ruled out.

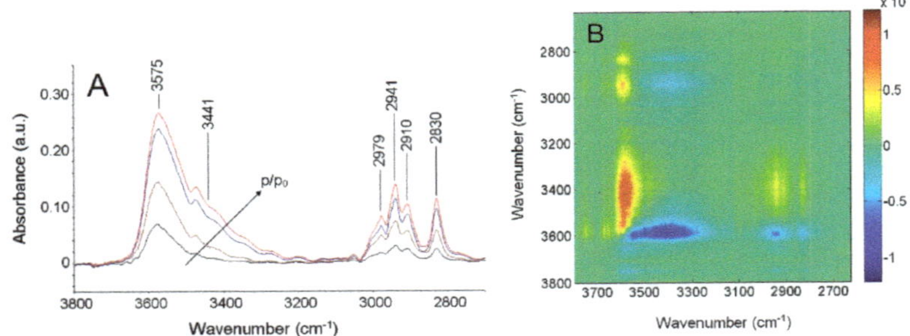

Figure 16. (**A**) The ν(OH) (3650–3200 cm^{-1} range) and ν(OH) (3650–3200 cm^{-1} range) of CH_3OH sorbed in PEI after sorption equilibrium at different p/p_0 at 30 °C. (**B**) Asynchronous spectrum obtained from the time-resolved spectra collected during the sorption experiment performed at $p/p_0 = 0.6$ and 30 °C.

In the light of all the experimental evidence, the peak at 3575 cm^{-1} was assigned to the OH group of methanol directly to the imide carbonyl, which represents the first-shell layer of the sorbed penetrant. The peak at 3441 cm^{-1} was due to self-associated methanol molecules (dimers), and in particular to the OH group bonded to oxygen of the first-shell species. A schematic representation the H-bonding aggregates established in the CH_3OH–PEI system is depicted in Figure 17.

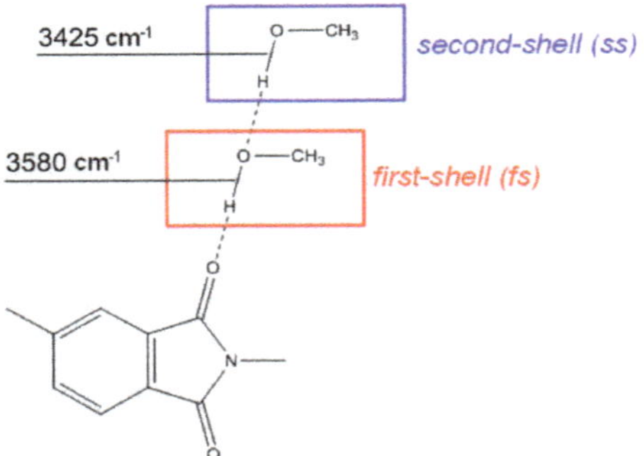

Figure 17. Schematic representation of the methanol species identified by the spectroscopic analysis.

As for the case of water, it was possible to quantify the amount of each of the two methanol species, as a function of the relative pressure of the methanol vapor phase at sorption equilibrium, by combining the gravimetric sorption isotherm, reported in Figure 18, with the results available from vibrational spectroscopy (see Figure 19) [47,48].

It was noted that, in the investigated range of relative pressure, the concentration of first-shell methanol, C_{fs}, was significantly higher than that of second-shell methanol, C_{ss}, thus indicating that, in this range, the monomer (i.e., a single molecule of methanol linked via HB to a carbonyl group), whose concentration was equal to $C_{fs} - C_{ss}$, was the largely prevailing species while self-associated aggregates were in a smaller amount and were likely to be dimers.

Fitting of the sorption isotherm was attempted both with the NELF and the NETGP-NRHB models assuming a time-independent value of polymer density, but the results were not satisfactory, likely due to a non-negligible structural evolution of PEI during methanol sorption. Since information on the actual value of polymer density was not available, we were not able to properly interpret the results with the illustrated class of lattice fluid models.

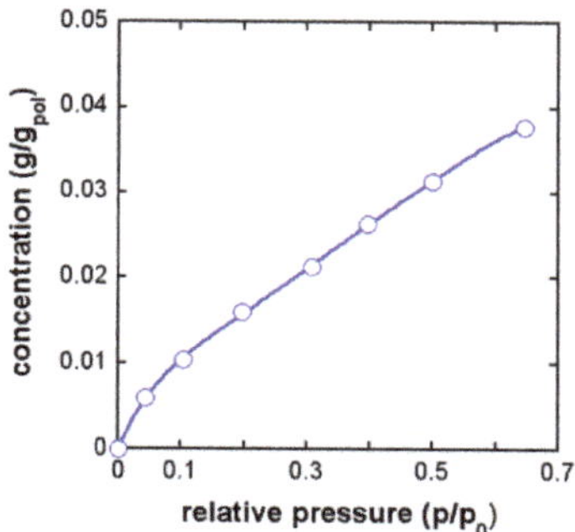

Figure 18. Experimental gravimetric sorption isotherms of CH_3OH in PEI at 30 °C. The continuous line does not represent any model and is intended to guide the eye.

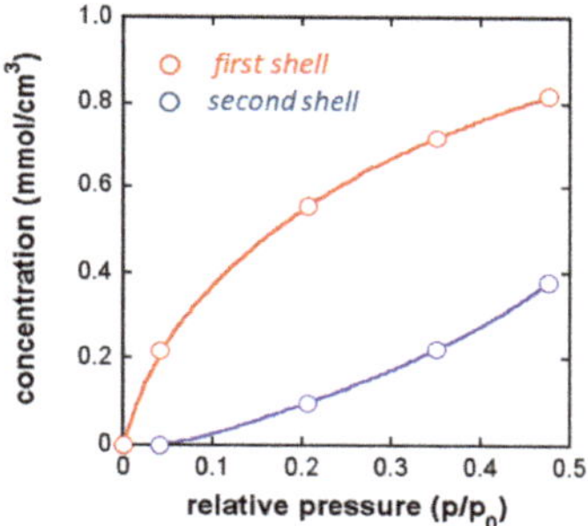

Figure 19. Amount of first- and second-shell methanol in PEI at equilibrium with pure methanol vapor phase at different relative pressures and at 30 °C. Continuous lines do not represent any model and are intended to guide the eye.

5. Conclusions

A molecular insight into sorption thermodynamics of low-molecular-weight penetrants in a glassy PEI was gained in the case of three penetrants with different interactive characteristics. To this aim, gravimetric analysis and FTIR spectroscopy were combined with macroscopic thermodynamics modeling.

For the case of sorption of carbon dioxide and water, non-equilibrium compressible lattice fluid theories were successfully used to interpret the experimental results. For the analysis of sorption data of carbon dioxide, an extension to non-equilibrium glassy polymers of the equilibrium Sanchez–Lacombe theory was adopted—the so-called NELF model. Conversely, for the case of water sorption, data were interpreted using an extension to non-equilibrium glassy polymers of the original NRHB approach—the so-called NETGP-NRHB model—to cope with specific H-bonding interactions occurring in the PEI–H_2O system.

In both cases, information gathered from vibrational spectroscopy was useful to choose the proper model and to tailor its structure. In fact, in the case of carbon dioxide, in view of the weak interactions established with groups present on the polymer backbone, the NELF approach, which does not explicitly account for specific interactions, was well suited to describe sorption thermodynamics. On the other hand, in the case of water, FTIR spectroscopy revealed the occurrence of two water species, one self-interacting via H-bonding and the other cross-interacting with carbonyl groups of PEI. This identification of the interactions to be accounted for in the sorption thermodynamics interpretation allowed us to tailor the structure of the H-bonding terms in the NETGP-NRHB model. In this case, FTIR analysis also provided quantitative information on self and cross HB interactions established within the system, which was in remarkably close agreement with predictions of the adopted theoretical approach.

In the case of methanol, we were not able to identify a suitable theoretical approach, likely due to the effect of the penetrant on the density of the polymer substrate in the time frame of the experiments. In this case, we limited ourselves to the spectroscopic identification of the molecular aggregates and the population analysis.

Author Contributions: Conceptualization, G.M. and P.M.; methodology, G.S. and P.L.M.; software, G.S.; formal analysis, G.M. and P.M.; investigation, G.S. and P.L.M.; data curation, G.S. and P.L.M.; writing—original draft preparation, G.M. and P.M.; writing—review and editing, G.M. and P.M.

Funding: This research received no external funding.

Conflicts of Interest: The authors declare no conflicts of interest.

Appendix A

A.1. The Equilibrium NRHB Model

The NRHB model [2,24,25] is an equation of state model for fluids used to describe the equilibrium thermodynamics of pure components and mixtures. The model is based on a non-random compressible lattice characterized by the presence of empty sites and accounting for hydrogen-bonding interactions. Each molecule of type i occupies r_i sites (or cells) of the lattice; this value is independent of the composition of the system. The total number of molecules in a system, N, is arranged in a lattice consisting of N_r sites, including N_0 empty sites. The lattice is characterized by a coordination number, z, that is generally assumed to be equal to 10. The system is made of N_1 molecules of type 1, N_2 molecules of type 2, and N_t molecules of type t, and is endowed with hydrogen-bonding specific interactions involving m types of proton donors and n types of proton acceptors. The model is based on a Gibbs partition function, Ψ, for the system, characterized by a tractable mathematical structure based on the assumption that the canonical partition function, Q, can be factorized in two contributions—one related to intermolecular interactions, apart from specific ones, and one associated to specific interactions. This corresponds to the decoupling of physical (van der Waals) and specific (hydrogen bonding) interactions. The canonical partition function is expressed as

$$Q = Q_P(T, N_0, N_1, N_2, \ldots, N_t) \cdot Q_H(T, N_0, N_1, N_2, \ldots, N_t), \tag{A1}$$

where the subscript P stands for "physical" and the subscript H stands for "hydrogen bonding".

The physical contribution to the canonical partition function is, in turn, expressed as

$$Q_P(T, N_0, N_1, N_2, \ldots, N_t) = \sum_{\underline{N}_{ij}} Q_P\left(T, N_0, N_1, N_2, \ldots, N_t, \underline{N}_{ij}\right), \tag{A2}$$

where $\underline{N}_{ij}$ represents a vector of variables, whose generic N_{ij} component represents the number of lattice fluid contacts between a segment of a molecule of type i and a segment of a molecule of type j, with i ranging from 0 to t and $j \geq i$. It is worth noting that not all the possible N_{ij} variables are present in this vector, since they are not all independent of each other. In fact, the following material balance equation holds:

$$2N_{ii} + \sum_{i \neq j}^{t} N_{ij} = N_i z q_i, \tag{A3}$$

where zq_i is the number of external lattice contacts per molecule for component i. As a consequence, the independent subset of variables is that made of N_{ij} with i ranging from 0 to t and $j > i$.

The hydrogen-bonding contribution to the canonical partition function is expressed as

$$Q_H(T, N_0, N_1, N_2, \ldots, N_t) = \sum_{\underline{N}_{\alpha\beta}^{HB}} Q_H\left(T, N_0, N_1, N_2, \ldots, N_t, \underline{N}_{\alpha\beta}^{H}\right), \quad \text{(A4)}$$

where $\underline{N}_{\alpha\beta}^{H}$ represents a vector of variables, whose generic $N_{\alpha\beta}^{H}$ component represents the total number of hydrogen-bonding interactions between proton donor groups of type α and proton acceptor groups of type β.

The Gibbs partition function, Ψ, can be then calculated from Q as follows:

$$\Psi(T, P, N_1, N_2, \ldots, N_t) = \sum_{N_0=0}^{\infty}\left[Q_P(T, N_0, N_1, N_2, \ldots, N_t)\cdot Q_H(T, N_0, N_1, N_2, \ldots, N_t)\cdot exp\left(-\frac{PV}{kT}\right)\right] \quad \text{(A5)}$$

where P and V are the pressure and the volume, respectively, of the mixture. It is noted that in the NRHB model, the "physical" canonical partition function, Q_P, is in turn factorized in two terms: a random term, Q_{PR}, and a non-random correction, Q_{PNR}, which accounts for the non-random distribution of contacts. Full details can be found in Reference [25].

The Gibbs energy can be then readily obtained as

$$G = -kTln[\Psi(T, P, N_1, N_2, \ldots, N_t)] \quad \text{(A6)}$$

According to a standard procedure of statistical thermodynamics, the sum in Equation (A5) can be approximated by its maximum term. This is mathematically equivalent to equating G to the logarithm of the generic term of the Gibbs partition function and finding its minimum with respect to the variables $\underline{N}_{ij}$, $\underline{N}_{\alpha\beta}^{H}$ and N_0.

This minimization procedures result in the following expressions [24,25]:

$$\frac{{\Gamma_{ij}}^2}{\Gamma_{ii}\,\Gamma_{jj}} = \exp\left(\frac{\Delta\varepsilon_{ij}}{RT}\right); \quad \text{(A7a)}$$

$$\frac{\nu_{\alpha\beta}}{\nu_{\alpha 0}\nu_{0\beta}} = \tilde{\rho}\exp\left(\frac{-G_{\alpha\beta}^{H}}{RT}\right); \quad \text{(A7b)}$$

$$\tilde{P} + \tilde{T}\left[\ln(1-\tilde{\rho}) - \tilde{\rho}\left(\sum_{i=1}^{t}\varphi_i\frac{l_i}{r_i} - \nu_H\right) - \frac{z}{2}\ln\left(1 - \tilde{\rho} + \frac{q}{r}\tilde{\rho}\right) + \frac{z}{2}ln\Gamma_{00}\right] = 0. \quad \text{(A7c)}$$

In Equation (A7a), the Γij's are non-random factors defined as N_{ij}/N_{ij}^{0}, where N_{ij}^{0} represents the number of lattice fluid contacts between a segment of a species of type i and a segment of a species of type j in the case of random arrangement of contacts in the lattice and $\Delta\varepsilon_{ij}$ is defined by the expression

$$\Delta\varepsilon_{ij} = \varepsilon_i + \varepsilon_j - 2\left(1 - k_{ij}\right)\sqrt{\varepsilon_i\varepsilon_j}, \quad \text{(A8)}$$

where ε_i is the interaction energy per i–i contact within the lattice and k_{ij} is the mean field binary interaction parameter measuring the deviation from geometric mixing rule for characteristic energies.

In Equation (A7b), $\nu_{\alpha\beta}$ represents the average number per molecular segment of hydrogen bonding established between a proton donor of type α and a proton acceptor of type β; $\nu_{\alpha 0}$ and $\nu_{0\beta}$ are the average number of unbonded proton donors of type α and the average number of unbonded proton acceptors of type β, respectively, per molecular segment; and $G_{\alpha\beta}^{H}$ is the molar Gibbs energy of formation of hydrogen bonding between proton donor group of type α and proton acceptor group of type β. The term $G_{\alpha\beta}^{H}$ is equal to $E_{\alpha\beta}^{H} - TS_{\alpha\beta}^{H} + PV_{\alpha\beta}^{H}$, where $E_{\alpha\beta}^{H}$, $S_{\alpha\beta}^{H}$, and $V_{\alpha\beta}^{H}$ are the molar internal energy, entropy, and volume of formation of hydrogen bonding, respectively, between proton donor group of type α and proton acceptor group of type β. The value of $V_{\alpha\beta}^{H}$ is generally taken as being equal to zero [65].

In Equation (A7c), representing the equation of state of the mixture, $\widetilde{P}$, $\widetilde{T}$, and $\widetilde{\rho}$ are the scaled pressure, the scaled temperature, and scaled density of the mixture, respectively; φ_i is the "close-packed" volumetric fraction of component i, zq is the averaged number of external lattice contacts per molecule in the mixture, r is the average number of cells occupied per molecule in the mixture, ν_H is the total number of hydrogen bonds per -mer in the mixture, and l_j is a parameter that can be calculated using the following expression [25]:

$$l_j = \frac{z}{2}(r_j - q_j) - (r_j - 1). \tag{A9}$$

The resulting general expression of the chemical potential for the component i in the polymer mixture is given as follows [25]:

$$\begin{aligned} \frac{\mu_{i,POL}}{RT} = \ln\frac{\varphi_i}{\delta_i r_i} & -r_i \sum_{j=1}^{t} \frac{\varphi_j l_j}{r_j} + ln\widetilde{\rho} + r_i(\widetilde{v}-1)\ln(1-\widetilde{\rho}) \\ & -\frac{z}{2} r_i \left[\widetilde{v} - 1 + \frac{q_i}{r_i}\right] \ln\left[1 - \widetilde{\rho} + \frac{q}{r}\widetilde{\rho}\right] + \frac{z q_i}{2}\left[\ln\Gamma_{ii} + \frac{r_i}{q_i}(\widetilde{v}-1)\ln\Gamma_{00}\right] \\ & + r_i \frac{\widetilde{P}\widetilde{v}}{\widetilde{T}} - \frac{q_i}{\widetilde{T}_i} + \frac{\mu_{i,\,POL}^{H}}{RT}, \end{aligned} \tag{A10}$$

where the hydrogen-bonding contribution to the penetrant chemical potential, $\mu_{i,\,POL}^{H}$, is given as follows [9]:

$$\frac{\mu_{i,\,POL}^{H}}{RT} = r_i \nu_H - \sum_{\alpha=1}^{m} d_{\alpha}^{i} \ln\left(\frac{\nu_d^{\alpha}}{\nu_{\alpha 0}}\right) - \sum_{\beta=1}^{n} a_{\beta}^{i} \ln\left(\frac{\nu_a^{\beta}}{\nu_{0\beta}}\right). \tag{A11}$$

In the previous expressions $\widetilde{v} = 1/\widetilde{\rho}$, δ_i is the number of configurations available to a molecule of component i in the "close-packed" state and is a characteristic quantity for the penetrant that accounts for the flexibility and the symmetry of the molecule (it cancels out in all equilibrium calculations of interest here), $\widetilde{T}_i$ is the scaled temperature of pure component i, d_{α}^{i} is the number of proton donors of type α on the molecule of type i, a_{β}^{i} is the number of proton acceptors of type β on the molecule of type i, and ν_d^{α} and ν_a^{β} are the total number of proton donors of type α and the average number of proton acceptors of type β, respectively, per -mer in the mixture.

The expression of equilibrium chemical potential for component i in the mixture can be obtained by coupling Equations (A7a)–(A7c) with Equation (A10). Phase equilibrium between a gas mixture and a polymer mixture is dictated by the equivalence of the equilbrium chemical potentials of each component in the two phases. In the case of penetrant sorption in a fluid or solid polymer, it is generally assumed that no polymer is present in the gaseous phase, and that the equilibrium is dictated only by the equivalence of chemical potentials of penetrants in the two phases.

The physical canonical partition function carries three scaling parameters for each component: (a) the enthalpic, $\varepsilon_h{}^*$, and the entropic, ε_s, contribution to the mean field interaction energy per mole of segment, and (b) $v_{sp,0}^*$, that is, the "close-packed" specific volume of the pure component at a reference temperature (298.15 K). For the case of a pure component i, the mean interaction energy per mole of segment, ε_{ii}^*, is expressed as

$$\varepsilon_{ii}^* = \varepsilon_h^* + (T - 298.15)\varepsilon_s^*. \tag{A12}$$

The so-called "close-packed" volume of a component (i.e., when no empty sites are present), can be calculated using the following equation (where v_{sp}^* is expressed in $cm^3 \cdot g^{-1}$):

$$v_{sp}^* = v_{sp,0}^* + (T - 298.15)v_{sp,1}^* - 0.135 \times 10^{-3} P, \tag{A13}$$

where P is the pressure (in MPa). The pressure term in Equation (A13) is taken to be equal to zero for low-molecular-weight compounds. $v^*_{sp,1}$ is a parameter that takes the same value for a given homologous series, and its values can be retrieved from the literature [25].

The value of r_i is expressed as a function of the lattice cell volume, v^*, of the "close-packed" volume, v^*_{sp} and of the molecular weight MW_i of the component i.

$$r_i = \frac{MW_i v^*_{sp}}{v^*}, \tag{A14}$$

where v^* is taken to be equal to 9.75 $cm^3 \cdot mol^{-1}$ for all fluids [25].

In the case of mixtures, proper mixing rules are introduced. Limiting the attention to binary mixtures, an average number of lattice cells occupied by a molecule in the mixture can be calculated as

$$r = \sum_{i=1}^{2} r_i x_i, \tag{A15}$$

where x_i represents the molar fraction of component i. The average lattice fluid intersegmental interaction energy per mole of average segment in the mixture is calculated as

$$\varepsilon^* = \sum_{i=1}^{2}\sum_{j=1}^{2} \theta_i \theta_j \varepsilon^*_{ij}, \tag{A16a}$$

where

$$\varepsilon^*_{12} = (1 - k_{12})\sqrt{\varepsilon^*_{11}\varepsilon^*_{22}}, \tag{A16b}$$

and θ are the concentration-dependent surface fraction of component i in the mixture.

The reduced temperature of the mixture, $\widetilde{T}$, the reduced temperature of component i, $\widetilde{T}_i$, reduced pressure of the mixture, $\widetilde{P}$, and reduced density of the mixture, $\widetilde{\rho}$, can be expressed as

$$\widetilde{T} = \frac{RT}{\varepsilon^*};\ \widetilde{T}_i = \frac{RT}{\varepsilon^*_{ii}} \widetilde{P} = \frac{v^* P}{\varepsilon^*};\ \widetilde{\rho} = \frac{\rho}{\rho^*}, \tag{A17}$$

where $\rho^* = \sum_{i=1}^{2} \frac{x_i MW_i}{r v^*}$

A.2. Extension of Equilibrium Theories to Non-Equilibrium Glassy Polymers

Equilibrium models, such as the NRHB theory, are not suited to describing mixture thermodynamics of low-molecular-weight compounds and non-equilibrium glassy polymers. To extend equilibrium approaches to the case of non-equilibrium glassy polymers, Doghieri and Sarti elaborated a procedure denominated the "non-equilibrium thermodynamics for glassy polymers (NETGP)" approach [26,27,69], which is rooted in thermodynamics endowed with internal state variables. In this section, we illustrate briefly the particular case of the application of this extension procedure to the Sanchez–Lacombe model [54] for binary polymer–penetrant mixtures. This was the first example of application of the NETGP procedure [26,27] that was then applied to other equation of state equilibrium models. In the present context, we refer to this theory as the "non-equilibrium lattice fluid" (NELF) model. Throughout this section, the subscript 1 refers to the penetrant and the subscript 2 refers to the polymer. The Sanchez–Lacombe theory is significantly simpler than the NRHB approach and is based on a random compressible lattice fluid model which accounts neither for the non-randomicity of contacts nor for the establishment of specific interactions among the components of the mixture.

In their extension of the Sanchez–Lacombe model, Sarti and Doghieri assumed that, for the non-equilibrium glassy polymer/penetrant mixture, the thermodynamic state is identified by the following set of variables: temperature, pressure, number of moles of penetrant, n_1, and density of the polymer in the mixture, ρ_2 [26,27]. The density of the polymer is an order parameter that rules the departure from equilibrium and takes on the role of an internal state variable of the system, whose rate of change over time is itself a function of the state of the system.

$$\frac{d\rho_2}{dt} = f(T, P, n_1, \rho_2). \tag{A18}$$

In the implementation of this extension procedure, it is frequently feasible to assume $f(T, p, n_1, \rho_2) \cong 0$. In fact, in the case of polymer systems at temperatures well below their glass transition temperature, the density relaxation may occur on a time span much longer than that required to reach phase equilibrium, thus implying that ρ_2 is "frozen" at a roughly constant non-equilibrium value, referred to as $\rho_{2,\infty}$. This is legitimate for low concentrations of penetrant within the polymer. In such cases $\rho_{2,\infty}$ is taken as being equal to the starting value of dry polymer. We actually deal, in this case, with a pseudo phase equilibrium, in view of the non-equilibrium nature of the glassy polymer.

It can be demonstrated [27] that the phase pseudo-equilibrium between a pure penetrant gaseous phase and the "frozen" polymer–penetrant mixture is still dictated by the equivalence of chemical potentials. Assuming that no polymer is present in the gaseous phase, in a binary case, this condition applies only to the penetrant

$$\mu_{1,POL}\left(T, p, n_1^{PE}, \rho_{2,\infty}\right) = \mu_{1,GAS}(T, p), \tag{A19}$$

where $\mu_{1,POL}$ and $\mu_{1,GAS}$ are the penetrant chemical potential in the polymer phase and in the gaseous phase, respectively. In Equation (A19), $\mu_{1,GAS}$ is consistently provided by the Sanchez–Lacombe theory written for pure penetrant and is coupled with related equation of state expression. The symbol n_1^{PE} highlights that we refer to a pseudo-equilibrium polymer phase.

In Equation (A19), $\mu_{1,POL}$ is obtained by deriving the Gibbs energy as a function of the number of moles of penetrant.

$$\mu_{1,POL} = \left.\frac{\partial G}{\partial n_1}\right|_{T,p,\rho_2}. \tag{A20}$$

The expression for G is provided by an expression similar to Equation (6) where the partition function Ψ is evaluated, using the expression provided by Sanchez–Lacombe theory, considering only the generic term of the summation, evaluated at a single value of ρ_2. Actually, a change of variable is introduced, expressing this generic term as a function of ρ_2 using the following relationship that relates ρ_2 to N_0:

$$\rho_2 = \frac{m_2}{V} = \frac{m_2}{rNv^* + N_0 v^*}, \tag{A21}$$

where m_2 is the mass of polymer in the polymer–penetrant mixture (that is fixed). In performing the derivative in Equation (A20), the generic term for G is evaluated at $\rho_2 = \rho_{2,\infty}$. By this procedure, Sarti et al. [26] obtained the following expression for $\mu_{1,POL}$ according to the NELF approach:

$$\frac{\mu_{1,POL}}{RT} = 1 + \ln(\widetilde{\rho}\varphi_1) - \left[r_1^0 + \frac{(r_1 - r_1^0)}{\widetilde{\rho}}\right] ln(1 - \widetilde{\rho}) - r_1 - \frac{\widetilde{\rho}\left[r_1^0 v_1^* (P_1^* + P^* - \varphi_2^2 \Delta P^*)\right]}{RT}. \tag{A22}$$

It is worth noting that, unlike NRHB, in this model, the number of -mers occupied in the lattice by component i takes on different values for the component in the mixture (r_i, which is a function of concentration) and for the pure component (r_i^0). This last value is obtained by the scaling parameters of the model for the pure component i. P_i^* is the scaling parameter for the pressure in the Sanchez and Lacombe theory named "characteristic pressure" of pure component i, while P^* is obtained by a mixing rule as

$$P^* = \varphi_1 P_1^* + \varphi_2 P_2^* - \varphi_1 \varphi_2 \, \Delta P^*, \tag{A23}$$

with

$$\Delta P^* = P_1^* + P_2^* - 2\chi_{12}\sqrt{P_1^* P_2^*}. \tag{A24}$$

In Equation (A24), χ_{12} represents the Sanchez–Lacombe binary interaction parameter. It is worth noting that, unlike the case of equilibrium expression for chemical potential obtained by Sanchez–Lacombe theory, the expression (A22) for non-equilibrium chemical potential does not need to be coupled with the minimization condition for G as a function of N_0.

In Equation (A22), we have

$$\widetilde{\rho} = \frac{\rho_{2,\infty}}{\omega_2 \rho^*}, \tag{A25}$$

where ω_i is the mass fraction of component i and ρ is the "close-packed" density of the mixture, which is a function of concentration according to

$$\rho^* = -\frac{\rho_1^* \rho_2^*}{\left(-\omega_1 \rho_2^* - \rho_1^* + \omega_1 \rho_1^*\right)}, \tag{A26}$$

where ρ_i^* is the scaling parameter for the density in the Sanchez–Lacombe theory for the component i.

The parameters of the NELF model are the binary interaction parameter, χ_{12} and the scaling parameters of each pure component, P_i^*, T_i^*, and ρ_i^*. T_i^* is the characteristic temperature of component i. The reduced pressure, temperature, and density of each component are consequently defined as

$$\widetilde{P}_i = \frac{P}{P_i^*}; \quad \widetilde{T}_i = \frac{T}{T_i^*}; \; \widetilde{\rho}_i = \frac{\rho}{\rho_i^*}. \tag{A27}$$

For the mixture, the corresponding scaling parameters are P^*, T^*, and ρ^*, which can be obtained, by proper mixing rules, from the corresponding pure component quantities.

The binary interaction parameter can be retrieved by fitting sorption isotherms with the NELF model for chemical potential. Conversely, the pure component scaling parameters are generally obtained by fitting equilibrium PVT data in the case of polymers and liquid–vapor equilibrium data in the case of low-molecular-weight compounds. Obviously, the value of $\rho_{2,\infty}$ is also to be known and can be retrieved by standard methods for measuring the density of polymers (e.g., by helium picnometry).

A procedure similar to that applied by Sarti and Doghieri to develop the NELF approach was applied by our group to extend the NRHB model to the case of non-equilibrium glassy polymers [72], developing the so-called NETGP-NRHB model. The procedure is more complex since, in this case, the number of internal state variables is incremented. In fact, in addition to ρ_2, there are also the variables $\underline{N}_{\alpha\beta}^H$ and the variables $\underline{N}_{ij}$ with i ranging from 0 to t and $j > i$.

As already discussed, at equilibrium, the values of these variables are dictated by Gibbs energy minimization conditions. Conversely, in non-equilibrium conditions, the values of these variables are ruled by equations for their evolution kinetics that, in turn, are a function of the state of the system.

To simplify an otherwise complex matter, we assumed in developing the NETGP-NRHB approach that, in analogy to the NELF model, the evolution kinetics of ρ_2 is extremely slow. Conversely, we assumed that the evolution kinetics of all the other internal state variables is "instantaneous". As a consequence, the values of variables $\underline{N}_{\alpha\beta}^H$ and $\underline{N}_{ij}$ are those that the system would exhibit at equilibrium for the specific value of (non-equilibrium) $\rho_2 = \rho_{2,\infty}$ and for the current value of concentration at the system pressure and temperature. Consequently, in a binary system, the expression for chemical potential is the one obtained from

$$\mu_{1,POL} = \left.\frac{\partial G}{\partial n_1}\right|_{T,\, p, n_2, \rho_2,\, \underline{N}_{ij}, \underline{N}_{\alpha\beta}^H}, \tag{A28}$$

where the values of $\underline{N}_{\alpha\beta}^H$ and $\underline{N}_{ij}$ are obtained by the minimization conditions (A7a) and (A7b).

In Equation (A28), G is provided by Equation (6) where the partition function ψ is evaluated, using the expression provided by NRHB theory, considering only the generic term of the summation corresponding to $\rho_2 = \rho_{2,\infty}$ and to the values of $\underline{N}_{\alpha\beta}^{H}$ and $\underline{N}_{ij}$ obtained by the minimization conditions.

Having made the hypothesis of "instantaneous equilibrium" for the variables $\underline{N}_{\alpha\beta}^{H}$ and $\underline{N}_{ij}$, the constitutive class for non-equilibrium system actually becomes the same as for NELF and, for a binary system, consists of the "external" variables T, P, n_1, and n_2 and of the "internal" variable ρ_2. We can then use the same arguments invoked for the NELF model to demonstrate that, at the phase pseudo-equilibrium between glassy polymer–penetrant phase and pure penetrant gaseous phase, the following equation holds:

$$\mu_{1,POL}\left(T, P, n_1^{PE}\right) = \left.\frac{\partial G}{\partial n_1}\right|_{T,\ p, n_2, \rho_2,\ \underline{N}_{ij}, \underline{N}_{\alpha\beta}^{H}} = \mu_{1,GAS}(T,P). \quad \text{(A29)}$$

In Equation (A29), $\mu_{1,GAS}$ is consistently provided by the equilibrium NRHB theory written for pure penetrant (see Section A.1) and is coupled with the related equation of state expression. The following expression was finally obtained [22] for the penetrant chemical potential in the polymer–penetrant mixture:

$$\begin{aligned}\frac{\mu_{1,POL}}{RT} = \ln\frac{\varphi_1}{\delta_1 r_1} \quad & -r_1 \sum_{j=1}^{2} \frac{\varphi_j l_j}{r_j} + ln\tilde{\rho} + r_1(\tilde{v}-1)\ln(1-\tilde{\rho}) - \frac{z}{2} r_1\left[\tilde{v} - 1 + \frac{q_1}{r_1}\right]\ln\left[1-\tilde{\rho}+\frac{q}{r}\tilde{\rho}\right] \\ & + \frac{zq_1}{2}\left[\ln\Gamma_{11} + \frac{r_1}{q_1}(\tilde{v}-1)\ln\Gamma_{00}\right] - \frac{q_1}{\tilde{T}_1} \\ & + \tilde{T}\left[\ln(1-\tilde{\rho}) - \tilde{\rho}\left(\sum_{j=1}^{2}\frac{\varphi_j l_j}{r_j}\right) - \frac{z}{2}\ln\left(1-\tilde{\rho}+\frac{q}{r}\tilde{\rho}\right) + \frac{z}{2}\ln\Gamma_{00}\right] \\ & \cdot \frac{r x_2 \cdot \left.\frac{\partial \tilde{v}}{\partial x_1}\right|_{P,T,\rho_2,\underline{N}_{ij},\ \underline{N}_{\alpha\beta}^{H}}}{\tilde{T}} + r_1 v_H - \sum_{i}^{m} d_i^1 \ln\left(\frac{v_d^i}{v_{i0}}\right) - \sum_{j}^{n} a_j^1 \ln\left(\frac{v_a^j}{v_{0j}}\right) \\ & + v_H \left.\frac{\partial ln\tilde{v}}{\partial x_1}\right|_{P,T,\rho_2,\underline{N}_{ij},\underline{N}_{\alpha\beta}^{H}} x_2 r.\end{aligned} \quad \text{(A30)}$$

This expression, coupled with Equations (A7a) and (A7b), provides the values of non-equilibrium chemical potential of penetrant in the polymer–penetrant mixture, according to the NETGP-NRHB model.

The parameters of the NETGP-NRHB model are the NRHB EoS parameters for pure components, i.e., ε_h^{*}, ν_s, and $v_{sp,0}^{*}$, and the hydrogen bonding parameters, i.e., the $E_{\alpha\beta}^{H}$ and $S_{\alpha\beta}^{H}$ for each proton donor–proton acceptor couple. The values of EoS parameters and of the hydrogen-bonding parameters for pure component interactions (self hydrogen-bonding interactions) can be retrieved from equilibrium PVT data in the case of polymers and liquid–vapor equilibrium data in the case of penetrants. The hydrogen-bonding parameters for interactions between polymer and penetrant (cross hydrogen bonding interactions), as well as the binary interaction parameter, k_{12}, are retrieved from fitting of sorption isotherms with the NETGP-NRHB model. Regarding q_i, its value can be obtained as $q_i = s_i \times r_i$, where s_i is the surface-to-volume ratio for a molecule of component i and can be estimated using the UNIFAC group contribution model [77].

It is explicitly noted that, once fitting of the experimental sorption isotherms is carried out, the model calculations also provide a prediction for the values of the set of variables $\underline{N}_{\alpha\beta}^{H}$ and $\underline{N}_{ij}$ for the mixture.

References

1. Kontogeorgis, G.M.; Folas, G.K. *Thermodynamic Models for Industrial Applications: From Classical and Advanced Mixing Rules to Association Theories*, 1st ed.; John Wiley & Sons, Ltd.: Chichester, UK, 2010; ISBN 978-0-470-69726-9.

2. Panayiotou, C.G.; Birdi, K.S. *Handbook of Surface and Colloid Chemistry*, 3rd ed.; CRC Press Taylor and Francis Group: New York, NY, USA, 2009.
3. Iwamoto, R.; Kusanagi, H. Hydration Structure and Mobility of the Water Contained in Poly(ethylene terephthalate). *J. Phys. Chem. A* **2015**, *119*, 2885–2894. [CrossRef] [PubMed]
4. Sammon, C.; Mura, C.; Yarwood, J.; Everall, N.; Swart, R.; Hodge, D. FTIR−ATR Studies of the Structure and Dynamics of Water Molecules in Polymeric Matrixes. A Comparison of PET and PVC. *J. Phys. Chem. B* **1998**, *102*, 3402–3411. [CrossRef]
5. Davis, E.M.; Elabd, Y.A. Water Clustering in Glassy Polymers. *J. Phys. Chem. B* **2013**, *117*, 10629–10640. [CrossRef] [PubMed]
6. Taylor, L.S.; Langkilde, F.W.; Zografi, G. Fourier transform Raman spectroscopic study of the interaction of water vapor with amorphous polymers. *J. Pharm. Sci.* **2001**, *90*, 888–901. [CrossRef] [PubMed]
7. De Angelis, M.G.; Sarti, G.C. Solubility of Gases and Liquids in Glassy Polymers. *Annu. Rev. Chem. Biomol.* **2011**, *2*, 97–120. [CrossRef] [PubMed]
8. Belova, N.A.; Safronov, A.P.; Yampolskii, Y.P. Inverse Gas Chromatography and the Thermodynamics of Sorption in Polymers. *Polym. Sci. Ser. A* **2012**, *54*, 859–873. [CrossRef]
9. Galizia, M.; Chi, W.S.; Smith, Z.P.; Merkel, T.C.; Baker, R.W.; Freeman, B.D. Polymers and Mixed Matrix Membranes for Gas and Vapor Separation: A Review and Prospective Opportunities. *Macromolecules* **2017**, *50*, 7809–7843. [CrossRef]
10. Glass, J.E. Adsorption characteristics of water-soluble polymers. I. Poly(vinyl alcohol) and poly(vinylpyrrolidone) at the aqueous-air interface. *J. Phys. Chem.* **1968**, *72*, 4450–4458. [CrossRef]
11. Zhao, Z.-J.; Wang, Q.; Zhang, L.; Wu, T. Structured Water and Water—Polymer Interactions in Hydrogels of Molecularly Imprinted Polymers. *J. Phys. Chem. B* **2008**, *112*, 7515–7521. [CrossRef]
12. Chiessi, E.; Cavalieri, F.; Paradossi, G. Water and Polymer Dynamics in Chemically Cross-Linked Hydrogels of Poly(vinyl alcohol): A Molecular Dynamics Simulation Study. *J. Phys. Chem. B* **2007**, *111*, 2820–2827. [CrossRef]
13. Eslami, H.; Müller-Plathe, F. Molecular Dynamics Simulation of Water Influence on Local Structure of Nanoconfined Polyamide-6,6. *J. Phys. Chem. B* **2011**, *115*, 9720–9731. [CrossRef] [PubMed]
14. Mensitieri, G.; Iannone, M. Modelling accelerated ageing in polymer composites. In *Ageing of Composites*, 1st ed.; Martin, R., Ed.; Woodhead Publishing, Ltd.: Cambridge, UK; CRC Press: Boca Raton, FL, USA, 2008; pp. 224–281. ISBN 9781845693527.
15. Li, D.; Yao, J.; Wang, H. Thin Films and Membranes with Hierarchical Porosity. In *Encyclopedia of Membrane Science and Technology*, 1st ed.; Hoek, E.M.V., Tarabara, V.V., Eds.; Wiley-Blackwell: Hoboken, NJ, USA, 2013; ISBN 978-0-470-90687-3.
16. Bernstein, R.; Kaufman, Y.; Freger, V. Membrane Characterization. In *Encyclopedia of Membrane Science and Technology*, 1st ed.; Hoek, E.M.V., Tarabara, V.V., Eds.; Wiley-Blackwell: Hoboken, NJ, USA, 2013; ISBN 978-0-470-90687-3.
17. Brunetti, A.; Barbieri, G.; Drioli, E. Gas Separation, Applications. In *Encyclopedia of Membrane Science and Technology*, 1st ed.; Hoek, E.M.V., Tarabara, V.V., Eds.; Wiley-Blackwell: Hoboken, NJ, USA, 2013; ISBN 978-0-470-90687-3.
18. Koros, W.J.; Burgess, S.K.; Chen, Z. Polymer Transport Properties. In *Encyclopedia of Polymer Science and Technology*, 4th ed.; John Wiley & Sons: Hoboken, NJ, USA, 2015; ISBN 9781118633892.
19. Koros, W.J.; Fleming, G.K.; Jordan, S.M.; Kim, T.H.; Hoehn, H.H. Polymeric membrane materials for solution-diffusion based permeation separations. *Prog. Polym. Sci.* **1988**, *13*, 339–401. [CrossRef]
20. Scherillo, G.; Petretta, M.; Galizia, M.; La Manna, P.; Musto, P.; Mensitieri, G. Thermodynamics of water sorption in high performance glassy thermoplastic polymers. *Front. Chem.* **2014**, *2*, 25. [CrossRef] [PubMed]
21. Scherillo, G.; Galizia, M.; Musto, P.; Mensitieri, G. Water Sorption Thermodynamics in Glassy and Rubbery Polymers: Modeling the Interactional Issues Emerging from FTIR Spectroscopy. *Ind. Eng. Chem. Res.* **2013**, *52*, 8674–8691. [CrossRef]
22. Scherillo, G.; Sanguigno, L.; Galizia, M.; Lavorgna, M.; Musto, P.; Mensitieri, G. Non-equilibrium compressible lattice theories accounting for hydrogen bonding interactions: Modelling water sorption thermodynamics in fluorinated polyimides. *Fluid Phase Equilibr.* **2012**, *334*, 166–188. [CrossRef]
23. Musto, P. Two-Dimensional FTIR Spectroscopy Studies on the Thermal-Oxidative Degradation of Epoxy and Epoxy−Bis(maleimide) Networks. *Macromolecules* **2003**, *36*, 3210–3221. [CrossRef]

24. Panayiotou, C.; Pantoula, M.; Stefanis, E.; Tsivintzelis, I.; Economou, I.G. Nonrandom Hydrogen-Bonding Model of Fluids and Their Mixtures. 1. Pure Fluids. *Ind. Eng. Chem. Res.* **2004**, *43*, 6592–6606. [CrossRef]
25. Panayiotou, C.; Tsivintzelis, I.; Economou, I.G. Nonrandom Hydrogen-Bonding Model of Fluids and Their Mixtures. 2. Multicomponent Mixtures. *Ind. Eng. Chem. Res.* **2007**, *46*, 2628–2636. [CrossRef]
26. Doghieri, F.; Sarti, G.C. Nonequilibrium Lattice Fluids: A Predictive Model for the Solubility in Glassy Polymers. *Macromolecules* **1996**, *29*, 7885–7896. [CrossRef]
27. Sarti, G.C.; Doghieri, F. Predictions of the solubility of gases in glassy polymers based on the NELF model. *Chem. Eng. Sci.* **1998**, *53*, 3435–3447. [CrossRef]
28. Robeson, L.M. Polymer membranes for gas separation. *Curr. Opin. Solid. State Mater. Sci.* **1999**, *4*, 549–555. [CrossRef]
29. Seo, J.; Jang, W.; Han, H. Thermal, optical, and water sorption properties in composite films of poly(ether imide) and bismaleimides: Effect of chemical structure. *J. Appl. Polym. Sci.* **2009**, *113*, 777–783. [CrossRef]
30. Karimi, M.; Albrecht, W.; Heuchel, M.; Kish, M.H.; Frahn, J.; Weigel, T.; Hofmann, D.; Modarress, H.; Lendlein, A. Determination of water/polymer interaction parameter for membrane-forming systems by sorption measurement and a fitting technique. *J. Membr. Sci.* **2005**, *265*, 1–12. [CrossRef]
31. Merdas, I.; Thominette, F.; Verdu, J. Humid ageing of polyetherimide. Chemical and physical interactions with water. In *Polyimides and Other High Temperature Polymer: Synthesis, Characterization and Applications*; Mittal, K.L., Ed.; VSP: Utrecht, The Netherlands, 2003; Volume 2, pp. 255–266. ISBN 90-6764-378-5.
32. Okamoto, K.-I.; Tanihara, N.; Watanabe, H.; Tanaka, K.; Kita, H.; Nakamura, A.; Kusuki, Y.; Nakagawa, K. Sorption and diffusion of water vapor in polyimide films. *J. Polym. Sci. Pol. Phys.* **1992**, *30*, 1223–1231. [CrossRef]
33. Wang, Y.; Jiang, L.; Matsuura, T.; Chung, T.S.; Goh, S.H. Investigation of the fundamental differences between polyamide-imide (PAI) and polyetherimide (PEI) membranes for isopropanol dehydration via pervaporation. *J. Membr. Sci.* **2008**, *318*, 217–226. [CrossRef]
34. Wang, Y.; Chung, T.S.; Neo, B.W.; Gruender, M. Processing and engineering of pervaporation dehydration of ethylene glycol via dual-layer polybenzimidazole (PBI)/polyetherimide (PEI) membranes. *J. Membr. Sci.* **2011**, *378*, 339–350. [CrossRef]
35. Zhao, W.; Shi, B. Removal of Volatile Organic Compounds from Water by Pervaporation Using Polyetherimide-Polyethersulfone Blend Hollow Fiber Membranes. *Sep. Sci. Technol.* **2009**, *44*, 1737–1752. [CrossRef]
36. Qiu, W.; Kosuri, M.; Zhou, F.; Koros, W.J. Dehydration of ethanol–water mixtures using asymmetric hollow fiber membranes from commercial polyimides. *J. Membr. Sci.* **2009**, *327*, 96–103. [CrossRef]
37. Huang, R.Y.M. (Ed.) *Pervaporation Membrane Separation Processes*; Membrane Science and Technology Series; Elsevier Science Publishers: Amsterdam, The Netherlands, 1991; Volume 1, ISBN O-444-88227-8.
38. Semenova, S.I. Separation properties of polyimides. In *Polyimide Membranes-Applications, Fabrications, and Properties*; Ohya, H., Kudryavsev, V.V., Semenova, S.I., Eds.; Kodansha: Tokyo, Japan, 1996; pp. 103–178. ISBN 4-06-208189-X.
39. Feng, X.; Huang, R.Y.M. Pervaporation and performance of asymmetric polyetherimide membranes for isopropanol dehydration by pervaporation. *J. Membr. Sci.* **1996**, *109*, 165–172. [CrossRef]
40. Yanagishita, H.; Maejima, C.; Kitamoto, D.; Nakane, T. Preparation of asymmetric polyimide membrane for water/ethanol separation in pervaporation by the phase inversion process. *J. Membr. Sci.* **1994**, *86*, 231–240. [CrossRef]
41. Huang, R.Y.M.; Feng, X. Dehydration of isopropanol by pervaporation using aromatic polyetherimide membranes. *Sep. Sci. Technol.* **1993**, *28*, 2035–2048. [CrossRef]
42. Yanagishita, H.; Kitamoto, D.; Nakane, T. Separation of alcohol aqueous solution by pervaporation using asymmetric polyimide membrane. *High Perform. Polym.* **1995**, *7*, 275–281. [CrossRef]
43. Okamoto, K.; Tanihara, N.; Watanabe, H.; Tanaka, K.; Kita, H.; Nakamura, A.; Kusuki, Y.; Nakagawa, K. Vapor permeation and pervaporation separation of water-ethanol through polyimide membranes. *J. Membr. Sci.* **1992**, *68*, 53–63. [CrossRef]
44. Chen, W.J.; Martin, C.R. Highly methanol selective membranes for the pervaporation separation of methyl-t-butyl ether/methanol mixtures. *J. Membr. Sci.* **1995**, *104*, 101–108. [CrossRef]
45. Nakagawa, K.; Asakura, Y.; Nakanishi, S.; Hoshino, H.; Kouda, H.; Kusuki, Y. Separation of vapor mixtures of water and alcohol by aromatic polyimide hollow fibers. *Kobunshi Ronbunshu* **1989**, *46*, 405–411. [CrossRef]

46. De Nicola, A.; Correa, A.; Milano, G.; La Manna, P.; Musto, P.; Mensitieri, G.; Scherillo, G. Local Structure and Dynamics of Water Absorbed in Polyetherimide: A Hydrogen Bonding Anatomy. *J. Phys. Chem. B* **2017**, *121*, 3162–3176. [CrossRef] [PubMed]
47. Musto, P.; Galizia, M.; La Manna, P.; Pannico, M.; Mensitieri, G. Diffusion and Molecular Interactions in a Methanol/Polyimide System Probed by Coupling time-resolved FTIR Spectroscopy with Gravimetric Measurements. *Front. Chem.* **2014**, *2*, 1–9. [CrossRef]
48. Galizia, M.; La Manna, P.; Pannico, M.; Mensitieri, G.; Musto, P. Methanol diffusion in polyimides: A molecular description. *Polymer* **2014**, *55*, 1028–1039. [CrossRef]
49. Cotugno, S.; Larobina, D.; Mensitieri, G.; Musto, P.; Ragosta, G. A novel spectroscopic approach to investigate transport processes in polymers: The case of water–epoxy system. *Polymer* **2001**, *42*, 6431–6438. [CrossRef]
50. Mensitieri, G.; Cotugno, S.; Musto, P.; Ragosta, G.; Nicolais, L. Transport of water in high Tg Polymers: A comparison between interacting and non-interacting systems. In *Polyimides and Other High Temperature Polymers: Synthesis, Characterization and Applications*; Mittal, K.L., Ed.; VSP: Utrecht, The Netherlands, 2003; Volume 2, pp. 267–285. ISBN 90-6764-378-5.
51. Sanchez, I.C.; Lacombe, R.H. An Elementary Molecular Theory of Classical Fluids. Pure Fluids. *J. Phys. Chem.* **1976**, *80*, 2352–2362. [CrossRef]
52. Sanchez, I.C.; Lacombe, R.H. Statistical Thermodynamics of Fluid Mixtures. *J. Phys. Chem.* **1976**, *80*, 2568–2580.
53. Sanchez, I.C.; Lacombe, R.H. Statistical Thermodynamics of Polymer Solutions. *Macromolecules* **1978**, *11*, 1145–1156. [CrossRef]
54. Panayiotou, C.; Sanchez, I.C. Hydrogen Bonding in Fluids—An Equation-of-State Approach. *J. Phys. Chem.* **1991**, *95*, 10090–10097. [CrossRef]
55. Veytsman, B.A. Are Lattice Models Valid for Fluids with Hydrogen Bonds? *J. Phys. Chem.* **1990**, *94*, 8499–8500. [CrossRef]
56. Veytsman, B.A. Equation of State for Hydrogen-Bonded Systems. *J. Phys. Chem.* **1998**, *B 102*, 7515–7517. [CrossRef]
57. Prausnitz, J.M.; Lichtenthaler, R.N.; Gomes de Azevedo, E. *Molecular Thermodynamics of Fluid-Phase Equilibria*, 3rd ed.; Prentice Hall PTR: Upper Saddle River, NJ, USA, 1998; ISBN 978-0139777455.
58. Guggenheim, E.A. *Mixtures: The Theory of the Equilibrium Properties of Some Simple Classes of Mixtures, Solutions and Alloys*; Oxford University Press: Oxford, UK, 1952.
59. Panayiotou, C.; Vera, J.H. Statistical Thermodynamics of r-Mer Fluids and Their Mixtures. *Polym. J.* **1982**, *14*, 681–694. [CrossRef]
60. You, S.S.; Yoo, K.P.; Lee, C.S. An Approximate Nonrandom Lattice Theory of Fluids: General Derivation and Application to Pure Fluids. *Fluid Phase Equilibr.* **1994**, *93*, 193–213. [CrossRef]
61. Taimoori, M.; Panayiotou, C. The non-random distribution of free volume in fluids: Non-polar systems. *Fluid Phase Equilibr.* **2001**, *192*, 155–169. [CrossRef]
62. Yeom, M.S.; Yoo, K.P.; Park, B.H.; Lee, C.S. A Nonrandom Lattice Fluid Hydrogen Bonding Theory for Phase Equilibria of Associating Systems. *Fluid Phase Equilibr.* **1999**, *158*, 143–149. [CrossRef]
63. Vlachou, T.; Prinos, I.; Vera, J.H.; Panayiotou, C. Nonrandom Distribution of Free Volume in Fluids and their Mixtures: Hydrogen-Bonded Systems. *Ind. Eng. Chem. Res.* **2002**, *41*, 1057–1063. [CrossRef]
64. Panayiotou, C. The QCHB Model of Fluids and their Mixtures. *J. Chem. Thermodyn.* **2003**, *35*, 349–381. [CrossRef]
65. Tsivintzelis, I.; Kontogeorgis, G.M. Modeling the Vapor–Liquid Equilibria of Polymer–Solvent Mixtures: Systems with Complex Hydrogen Bonding Behavior. *Fluid Phase Equilib.* **2009**, *280*, 100–109. [CrossRef]
66. Doghieri, F.; Quinzi, M.; Rethwisch, D.G.; Sarti, G.C. Predicting Gas Solubility in Glassy Polymers through Nonequilibrium EOS. In *Advanced Materials for Membrane Separations*; Pinnau, I., Freeman, B.D., Eds.; ACS Symposium ser. N. 876; American Chemical Society (ACS): Washington, DC, USA, 2004; pp. 74–90, ISBN 0-8412-3833-2.
67. De Angelis, M.G.; Doghieri, F.; Sarti, G.C.; Freeman, B.D. Modeling gas sorption in amorphous Teflon through the non equilibrium thermodynamics for glassy polymers (NETGP) approach. *Desalination* **2006**, *193*, 82–89. [CrossRef]
68. Pantoula, M.; von Schnitzler, J.; Eggers, R.; Panayiotou, C. Sorption and swelling in glassy polymer/carbon dioxide systems Part II—Swelling. *J. Supercrit. Fluid* **2007**, *39*, 426–434. [CrossRef]

69. Musto, P.; Ragosta, G.; Mensitieri, G.; Lavorgna, M. On the Molecular Mechanism of H_2O Diffusion into Polyimides: A Vibrational Spectroscopy Investigation. *Macromolecules* **2007**, *40*, 9614–9627. [CrossRef]
70. Czarnecki, M.A. Two-Dimensional Correlation Spectroscopy: Effect of Normalization of the Dynamic Spectra. *Appl. Spectrosc.* **1999**, *53*, 1392–1397. [CrossRef]
71. Noda, I.; Ozaki, Y. *Two-Dimensional Correlation Spectroscopy: Applications in Vibrational and Optical Spectroscopy*; John Wiley & Sons: Chichester, UK, 2005; ISBN 978-0-471-62391-5.
72. Musto, P.; La Manna, P.; Pannico, M.; Mensitieri, G.; Gargiulo, N.; Caputo, D. Molecular interactions of CO_2 with the CuBTC metal organic framework: An FTIR study based on two-dimensional correlation spectroscopy. *J. Mol. Struct.* **2018**, *1166*, 326–333. [CrossRef]
73. Rothshild, W.G. Mobility of Small Molecules in Viscous Media. *Macromolecules* **1968**, *1*, 43–47.
74. Stolov, A.A.; Morozov, A.I.; Remizov, A.B. Infrared spectra and spinning diffusion of methyl bromide in solutions. *Vib. Spectosc.* **1997**, *15*, 211–218. [CrossRef]
75. Tsivintzelis, I.; Grenner, A.; Economou, I.G.; Kontogeorgis, G.M. Evaluation of the Nonrandom Hydrogen Bonding (NRHB) Theory and the Simplified Perturbed-Chain Statistical Associating Fluid Theory (sPC-SAFT). 2. Liquid-Liquid Equilibria and Prediction of Monomer Fraction in Hydrogen Bonding Systems. *Ind. Eng. Chem. Res.* **2008**, *47*, 5651–5659. [CrossRef]
76. Fredenslund, A.; Jones, R.L.; Prausnitz, J.M. Group-contribution estimation of activity coefficients in nonideal liquid mixtures. *AIChE J.* **1975**, *21*, 1086–1099. [CrossRef]
77. Sorensen, M.J. Group Contribution Estimation Methods. In *Models for Thermodynamic and Phase Equilibria Calculations*; Sandler, S.I., Ed.; Marcel Dekker: New York, NY, USA, 1994; ISBN 978-0824791308.

 *membranes*

Article

Experimental Mixed-Gas Permeability, Sorption and Diffusion of CO_2-CH_4 Mixtures in 6FDA-mPDA Polyimide Membrane: Unveiling the Effect of Competitive Sorption on Permeability Selectivity

Giuseppe Genduso *, Bader S. Ghanem and Ingo Pinnau *

Functional Polymer Membranes Group, Advanced Membranes and Porous Materials Center, Division of Physical Science and Engineering, King Abdullah University of Science and Technology, Thuwal 23955-6900, Saudi Arabia; bader.ghanem@kaust.edu.sa

* Correspondence: giuseppe.genduso@kaust.edu.sa (G.G.); ingo.pinnau@kaust.edu.sa (I.P.)

Received: 8 November 2018; Accepted: 20 December 2018; Published: 8 January 2019

Abstract: The nonideal behavior of polymeric membranes during separation of gas mixtures can be quantified via the solution-diffusion theory from experimental mixed-gas solubility and permeability coefficients. In this study, CO_2-CH_4 mixtures were sorbed at 35 °C in 4,4′-(hexafluoroisopropylidene)diphthalic dianhydride (6FDA)-m-phenylenediamine (mPDA)—a polyimide of remarkable performance. The existence of a linear trend for all data of mixed-gas CO_2 versus CH_4 solubility coefficients—regardless of mixture concentration—was observed for 6FDA-mPDA and other polymeric films; the slope of this trend was identified as the ratio of gas solubilities at infinite dilution. The CO_2/CH_4 mixed-gas solubility selectivity of 6FDA-mPDA and previously reported polymers was higher than the equimolar pure-gas value and increased with pressure from the infinite dilution value. The analysis of CO_2-CH_4 mixed-gas concentration-averaged effective diffusion coefficients of equimolar feeds showed that CO_2 diffusivity was not affected by CH_4. Our data indicate that the decrease of CO_2/CH_4 mixed-gas diffusion, and permeability selectivity from the pure-gas values, resulted from an increase in the methane diffusion coefficient in mixtures. This effect was the result of an alteration of the size sieving properties of 6FDA-mPDA as a consequence of CO_2 presence in the 6FDA-mPDA film matrix.

Keywords: mixed-gas sorption; mixed-gas diffusion; mixed-gas permeation; competitive sorption; gas separation membranes; 6FDA-mPDA polyimide

---

1. Introduction

Conventional distillation or absorption and adsorption systems are reliable and, for existing plants, economically feasible technologies. However, membrane units could potentially replace these traditional unit operations, ensuring better economics and lower environmental impact [1]. To fully exploit the potential of membrane-based gas-separations, polymeric materials of high permeability and permeability selectivity are required. Moreover, these materials must be mechanically strong for formation of integral asymmetric and thin-film composite membranes and should also ensure stable separation performances over time. Currently, most experimental studies reported in the literature are based on pure-gas permeation properties of membrane materials (e.g., cellulose acetate, polysulfone, polyphenylene oxide, polyimides, polymers of intrinsic microporosity (PIM), and others) [2–6]. However, when gas permeation is performed under mixed-gas conditions, permeability and permeability selectivity can deviate significantly from the pure-gas values. For example, the pure-gas CO_2/CH_4 permeability selectivity of PIM-1 is ~15 (in the range of 0–10 atm CO_2 pressure), but drops to ~8 in a CO_2/CH_4 mixture at 10 atm partial CO_2 pressure [4], that is, only ~50% of the ideal value. Because the overall intention of CO_2-CH_4 separation studies is to design membrane materials for industrial applications, correct identification of the reasons for the deviation from ideality of permeability selectivity is crucial. Therefore, one should describe gas transport at both thermodynamic (sorption) and kinetic (diffusion) levels. The solution-diffusion theory that is commonly applied to describe the transport of fluids through dense polymer membranes affirms that permeability (P_i) can be directly determined as the combination of solubility (S_i) and diffusion (D_i) coefficients of a specific gas in a membrane [7]. Hence, to accurately describe the transport of gases through membranes, one should preferably couple mixed-gas permeation data, measured by a well-established experimental technique [8], with mixed-gas sorption or diffusion data. Fraga et al. [9], for example, designed and implemented a new time lag apparatus for the direct measurement of mixed-gas permeability and diffusion coefficients in Pebax 2533, HyflonAD60X, and PIM-EA-TB—the same instrument was used by Monteleone et al. [10] to test mixed-gas diffusion coefficients of a spirobifluorene-based polymer of intrinsic microporosity (PIM-SBF-1). On the basis of our experimental capabilities, we estimated mixed-gas diffusion coefficient values by applying the solution-diffusion theory and combining experimental mixed-gas permeability and mixed-gas sorption values. Particularly, a newly developed sorption system was used for the mixed-gas sorption studies reported here [11]. To the best of our knowledge, so far, four types of mixed-gas sorption units have been described in the literature [11–15] and utilized to test a total of only 11 polymers [11–23]. Furthermore, the number of studies covering the diffusion of gas mixtures in polymers—retrieved from experiments of mixed-gas sorption and permeation—is even smaller. Some previous studies include the following: CO_2-C_2H_6 in XLPEO [24], *n*-C_4H_{10}-CH_4 in poly [1-trimethylsilyl-1-propyne] (PTMSP) [25], CO_2-CH_4 in 6FDA-TADPO [22], CO_2-CH_4 in PEO-based multi-block copolymer [26], *n*-C_4H_{10}-CH_4 in polydimethylsiloxane (PDMS) [27], and CO_2-CH_4 in PDMS [11]. In our previous work [11], we showed that co-permeation of CO_2 in mixtures with CH_4 in rubbery polydimethylsiloxane (PDMS) increased mixed-gas diffusion of CH_4.

Similar to other commercial glassy polymers (e.g., polysulfones, cellulose acetates), polyimides have attracted the attention of the academic and industrial community for the following reasons: (i) strong thermal and mechanical properties (high glass transition and thermal decomposition temperatures, high ultimate tensile strength/elongation at break, and Young's modulus) and (ii) excellent combination of pure-gas permeability and permeability selectivity [5]. In particular, polyimides based on the commercially available monomer 4,4′-(hexafluoroisopropylidene)diphthalic dianhydride (6FDA) are typically solution processable and, therefore, suitable for hollow fibers spinning [5,28]. The polycondensation reaction of 6FDA with m-phenylenediamine (mPDA)—a commercially available monomer—yields the high-performance 6FDA-mPDA polyimide [29,30], shown in Figure 1. 6FDA-mPDA displays an interesting combination of pure-gas CO_2 permeability of ~14 Barrer and CO_2/CH_4 permeability selectivity of ~70 measured at 2 atm and 35 °C [27]. Therefore, this membrane material is a particularly attractive polymer to perform a case study of mixed-gas sorption and diffusion of CO_2-CH_4 mixtures in polyimides. Such a study may also clarify whether the deviation of the CO_2 mixed-gas permeability from the pure-gas values can be ascribed to *competitive sorption* phenomena—as commonly assumed in the literature [31–33]—or to phenomena related to CO_2-induced dilation [34].

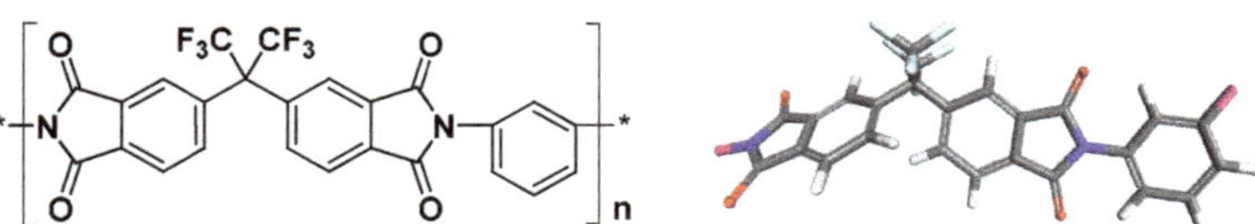

Figure 1. Chemical structure and 3D representation of the repeat unit of 4,4′-(hexafluoroisopropylidene)diphthalic dianhydride (6FDA)-m-phenylenediamine (mPDA) polyimide.

In this work, we aimed to provide a full experimental description of sorption and diffusion of CO_2-CH_4 mixtures in isotropic 6FDA-mPDA films at 35 °C. First, we show the results of CO_2-CH_4 mixed-gas sorption experiments, and then perform an analysis of mixture effects on the solubility coefficients and solubility selectivity at various equilibrium pressures. To analyze mixed-gas solubility data, isothermal surfaces of gas uptake were estimated via the following: (i) linear interpolation and (ii) the extension of the dual-mode sorption model [35] to mixtures supported by an empirical expression for better data fitting. In the second part of this work, experimental mixed-gas permeability values of 6FDA-mPDA obtained with equimolar CO_2-CH_4 mixtures previously reported by our group [29] were divided by experimental mixed-gas solubilities to estimate CO_2-CH_4 mixed-gas concentration-averaged effective diffusion coefficient data. These data show the interaction between CO_2 and the polymeric matrix of 6FDA-mPDA and the effect of CO_2 diffusion on CH_4 diffusion. Moreover, using our experimental data, we were able to clarify the impact of competitive sorption and CO_2-sorption related phenomena on transport and separation of CO_2-CH_4 mixtures in 6FDA-mPDA.

2. Materials and Methods

2.1. Materials

6FDA-mPDA was synthesized according to the procedure reported elsewhere [29]. The polymer had a weight-averaged molecular weight (Mw) of 141,000 g/mol and a polydispersity index (Mw/Mn) of 1.2. Isotropic polyimide films made by solution casting from chloroform were air-dried, soaked in methanol for 12 h, and then dried at 120 °C under vacuum for 24 h. Complete solvent removal was confirmed by thermal gravimetric analysis (TGA). The geometric density of 6FDA-mPDA was determined at room temperature (22 °C) from membrane area (via image scanning), thickness (547-400S micrometer, Mitutoyo, Japan), and weight measurements (XPE204, Mettler Toledo, Columbia, SC, USA). Three pieces of 6FDA-mPDA film of 50 microns and three pieces of 240 microns were measured with fresh samples (i.e., 0 days aging). The geometric density of 6FDA-mPDA was 1.42 ± 0.02 g/cm^3. Moreover, the density of 240-µm thick 6FDA-mPDA films aged for >3 months was identical to that of the fresh film samples. This value was slightly lower than other values reported in the literature—that is, 1.464 g/cm^3 [36], 1.46 g/cm^3 [37,38], 1.456 ± 0.014 g/cm^3 [28], and 1.45 ± 0.01 g/cm^3 [39], which were measured via the Archimedes' principle procedure.

Certified gas mixtures of 11 mol% and 90 mol% CO_2 in CH_4 were purchased from Air Liquide; gas mixtures of 37 mol% and 51 mol% CO_2 in CH_4 were purchased from AHG Specialty Gas Center (Jeddah, Saudi Arabia).

2.2. Methods

2.2.1. Barometric System

The design and operation of the system used for barometric pure- and mixed-gas sorption experiments were introduced in detail elsewhere [11] and are shown here in Figure 2. In brief, gases are introduced from volume V_B to V_A, which contains the polymer sample (V_P). Volume V_C is connected to V_A, V_B, the gas chromatograph (GC, Agilent 490 Micro GC Natural Gas Analyzer, Santa Clara, USA), gas cylinders (custom-made mixtures and carrier gas), and a vacuum pump. V_C allows a certain operational flexibility to this mixed-gas system; for example, by addition from V_C, gases can be mixed in V_B. V_C can receive gas samples from both V_B and V_A for GC analysis. Furthermore, when using custom-made mixtures, the valve between V_B and V_C can be left open to increase the volume of the feed chamber so that more gas can be expanded to V_A. P_A and P_B transducers of 35 atm range were used for both volumes V_A and V_B and were exchanged with transducers of 50 and 100 atm range, respectively, to explore high partial pressures (>15 atm partial gas pressure).

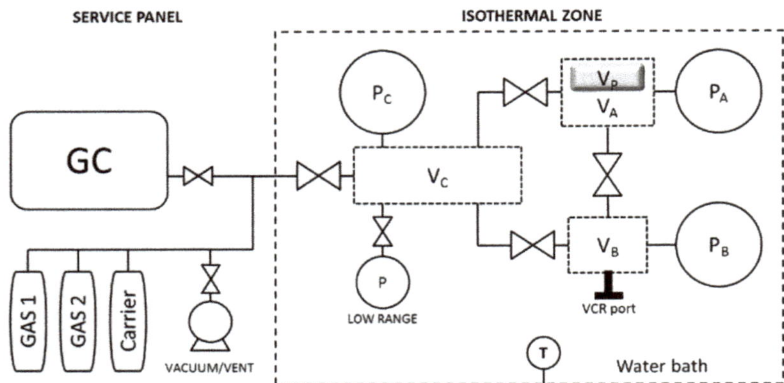

Figure 2. Schematic of the barometric mixed-gas unit used for sorption experiments discussed in this work (adapted from the literature [11]). GC—gas chromatograph.

Active volumes in a barometric sorption system are the feed and the sample chamber because they are used to calculate gas uptakes by mass balance. Dead-volumes in tubes, transducers, and valves do not act actively during sorption experiments. The minimization of these non-active volumes (e.g., by insertion of metallic rods in all tubes [11]) maximizes the difference between amount of *i*-gas in V_A at time zero (i.e., immediately after gas is expanded from V_B to V_A) and the same amount at equilibrium; therefore, volume optimization increments the sensitivity of the mixed-gas sorption system and it was crucial to explore the low solubility coefficient of 6FDA-mPDA toward methane at high partial pressures (see later in this work). All volumes of the pressure-decay system were calibrated via a gas expansion procedure ($V_A = 9.00 \pm 0.03$ cm^3; $V_B = 7.77 \pm 0.03$ cm^3; $Vc = 10.44 \pm 0.04$ cm^3), already well described in the literature [9–14], in which a known reservoir volume, which was previously calibrated via water filling at room temperature, is added to V_A (or eventually to V_B). To the best data accuracy, all volume values employed reflected the actual situation of the system—that is, any maintenance-operation on the system was always followed by a leak-test and a re-calibration procedure, and were always very close to the first calibration values.

2.2.2. Barometric Pure-Gas Sorption

CH_4 and CO_2 barometric pure-gas sorption experiments were performed in the system shown in Figure 2. A fresh film sample of 1.16 g was loaded and degassed for 24 h at 35 °C. Gas was admitted in V_B and was expanded after pressure equilibration to V_A. Data were acquired continuously at a rate of one point every two seconds using custom-made software operating in LabVIEW (National Instruments™, Austin, USA) until the average pressure variation was approximately -2×10^{-7} atm/s during 100 min; above this pressure, variation value uptakes were constant. At equilibrium, more gas was added to V_B and expanded to V_A.

Gas uptakes were estimated as reported previously in the literature [12–15] from the difference between the amount gas first admitted to V_A at time zero and the amount of the same gas not sorbed by the polymer sample at equilibrium. Molar amounts were calculated via the equation of state of gases (corrected with compressibility factor) from pressure transducer readings. The Soave-Redlich-Kwong (S-R-K) equation of state [40,41] was used to estimate compressibility factors (S-R-K compressibility factors matched with values obtained via the virial equation of state [19]) and partial gas fugacities. The S-R-K parameters of CO_2 and CH_4 can be found elsewhere [15].

The well-established dual-mode sorption (DMS) model [42] was used to fit CH_4 and CO_2 pure-gas uptake data. Because DMS model interpolation of gas uptake from fitted isotherms is usually very accurate, the DMS model was also used for estimation of CO_2 and CH_4 solubilities at infinite dilution of 6FDA-mPDA and other polymers discussed in this work (see Supporting Information).

2.2.3. Gravimetric Gas Sorption

Pure-gas sorption experiments were performed via an Intelligent Gravimetric Analyzer (IGA) by Hiden Isochema (Warrington, UK). A fresh sample was loaded in this gravimetric system and degassed at 35 °C under high vacuum ($<10^{-7}$ mbar) for at least 24 h. When the sample weight was stable, sorption measurements were initiated; gas was introduced in the sample chamber at a rate of 0.1 atm/min to reach the desired equilibrium pressure. After equilibration, gas was added cumulatively to obtain a further pressure point.

2.2.4. Barometric Mixed-Gas Sorption and Data Analysis

To perform mixed-gas sorption experiments (see system in Figure 2), the constant feed concentration procedure described in our previous work [11] was applied. In the framework of this procedure, custom-made mixtures were added to V_B and, after pressure equilibration, expanded to V_A. It should be noted that during this expansion procedure, the valve between A and B volumes was opened and then immediately closed to avoid any interaction between the feed gas in V_B and the gas of V_A now sorbing into the polymer sample (hence, we can assume that at time zero, V_A and V_B have the same gas composition). At sorption equilibrium (i.e., average pressure variation approximately -2×10^{-7} atm/s during 100 minutes), two gas samples from V_B and V_A were sent to V_C for GC analysis. Then, both V_B and V_A were degassed for a time long enough (about the same time allowed for sorption) to remove any sorbed (detectable) gas from the polymer sample in V_A; the 6.8 atm range transducer—mounted on V_C (see "low range" in Figure 2)—was used to detect gas desorption. If no desorption could be detected, the same gas mixture was added at a higher pressure to V_B and then expanded to V_A to perform the next mixed-gas uptake experiment. Once an experimental series at fixed mixture composition reached the highest total feed pressure allowed by V_B (which depended on the maximum value of gas-cylinder pressure, P_B transducer pressure range, and V_B dimension), the next experimental series was carried out from a feed pressure of 7 atm. It should be noted that although we set the sequence of the concentration series in the direction of increasing CO_2 concentrations (i.e., mixtures of higher methane concentration were run first), the sample might have undergone CO_2-conditioning when going from the last pressure of a series to the first experiment of the new series. This sample conditioning might have introduced a certain over-estimation of a few gas uptake data points; however, we anticipate that this effect was relatively small because between two consecutive experimental data points, the difference in CO_2 partial pressure was always lower than 8 atm, and because of the extensive sample degassing mentioned above between each experiment. Mixed-gas uptakes were calculated via mass balance and GC composition data [12–15].

Unless otherwise stated (see Table S4), the same sample used for pure-gas sorption uptakes was employed during mixed-gas sorption experiments.

CO_2 and CH_4 mixed-gas sorption uptakes in 6FDA-mPDA in the form of three-dimensional data points were fitted via MATLAB® software (version R2016b, The MathWorks, Inc, Natick, MA, USA). All data were linearly interpolated or were fitted via the DMS model for mixtures [35] supported by an empirical equation for better data fitting. All details of this fitting analysis can be found in the Supporting Information of this work.

2.2.5. Pure- and Mixed-Gas Permeation and Diffusion Coefficients

Pure- and mixed-gas concentration-averaged effective diffusion coefficients for the case of 50:50 mol% CO_2/CH_4 feed concentration were calculated as previously done elsewhere [14,24,26,27,43] from $\overline{D}_i = P_i/S_i^{feed}$; where P_i is the pure- or mixed-gas permeability at permeate pressures approaching zero, these permeabilities were previously published by our research group for 6FDA-mPDA [29], and S_i^{feed} is the pure- or mixed-gas solubility coefficient of the i-gas at the feed pressure. Because our experimental mixed-gas solubility coefficients were obtained via *constant feed concentration* experiments (i.e., we could only control the concentration at the beginning of the experiment and not at equilibrium [11]), we used models to predict the S_i^{feed} values at the fixed 50:50 mol% CO_2/CH_4 equilibrium concentration. We predicted these values in two ways: (i) with a modified version of the dual-mode sorption model (details can be found in the Supporting Information), and (ii) with the use of the *'linearinterp'* model within the *'fit'* function of Matlab R2016b. The *'linearinterp'* model simply connects all data points of the 3D uptake diagrams with planes (see Figure S5), therefore, the quality of the prediction of S_i^{feed} strictly depends on the number of data points and on experimental accuracy. Both prediction methods for S_i^{feed} produced very similar values (see later in the Results and Discussion section).

3. Results and Discussion

3.1. Experimental Pure- and Mixed-Gas Sorption Data

To assess the effects of multicomponent gas sorption on the sorption capacity of the individual mixture components, pure-gas sorption experiments were first performed by barometric and gravimetric techniques. Pure-gas sorption isotherms obtained via the barometric system with a 6FDA-mPDA film were in excellent agreement with those determined gravimetrically (this comparison validated our instrument accuracy) using film and powder samples (Figure 3a,b), indicating that sorption did not depend on the physical state of the 6FDA-mPDA samples.

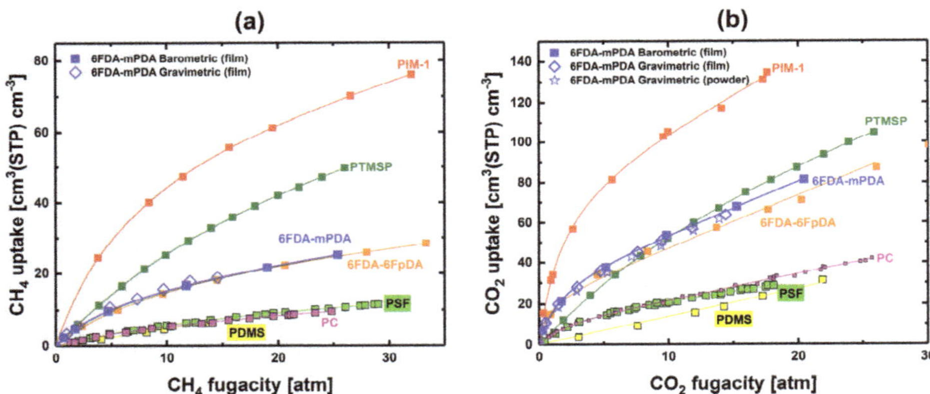

Figure 3. (**a**) CH_4 and (**b**) CO_2 pure-gas sorption isotherms at 35 °C vs. gas fugacity. Blue squares were obtained with 6FDA-mPDA films in our custom-built barometric system. Blue diamonds were obtained with 6FDA-mPDA films via gravimetric sorption. CO_2 pure-gas gravimetric sorption was also performed with a 6FDA-mPDA powder sample (blue stars). Sorption isotherms for polydimethylsiloxane (PDMS) [11], polysulfone (PSF) [45], polycarbonate (PC) [42,46], 6FDA-6FpDA polyimide [44], poly [1-trimethylsilyl-1-propyne] (PTMSP) [15], and polymers of intrinsic microporosity (PIM)-1 [20] are also shown. Interpolations were performed via the dual-mode sorption (DMS) model [42]—DMS parameters were determined in-house. PDMS uptakes were interpolated linearly.

Figure 3 shows that the 6FDA-mPDA polyimide follows the general pure-gas sorption behavior of another fluorine-containing polyimide, that is, the 6FDA-6FpDA [44] (Figure 3a,b). CO_2 and CH_4 isotherms of 6FDA-mPDA are located between the curves of low-free-volume glassy polymers, PSF (polysulfone) and PC (polycarbonate), and high-free-volume PIM-1. The CH_4 sorption uptake in high-free-volume glassy PTMSP is much higher than in 6FDA-mPDA, as shown in Figure 3a. However,

CO_2 uptakes are comparable for PTMSP and 6FDA-mPDA (Figure 3b). Hence, this example shows how gas/polymer affinity, as well as free volume, plays an important role in gas sorption.

The CO_2-CH_4 mixed-gas solubility coefficient data for 6FDA-mPDA as function of gas fugacity are shown in Figure 4 (all data are also listed in Table S4). The presence of CO_2 strongly influenced the solubility of CH_4 (Figure 4a), and, similarly but less markedly, CH_4 also affected CO_2 solubility (Figure 4b). The inserts in Figure 4 show the predictions of the extension to mixtures of the dual-mode sorption model (DMS-mix [35]) for CH_4 and CO_2, respectively—these inset graphs are intended to provide a reference framework and to guide the reader through the mixed-gas data. The qualitative agreement between predictions of the DMS model and experiments (Figure 4) suggests that competitive sorption in the glassy polymer may be the *main* reason of the deviation of the mixed-gas data from the pure-gas solubility coefficient trends. Other effects that are not accounted for by the DMS model and that impact solubility at high pressures are presented in the Supporting Information of this paper.

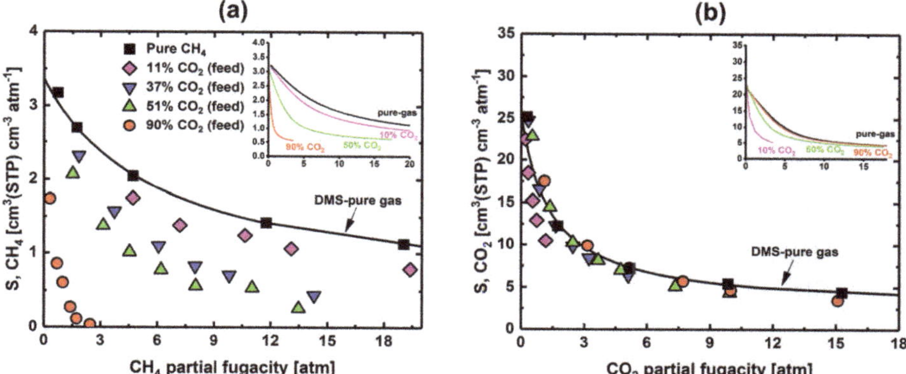

Figure 4. (**a**) Experimental CH_4 solubility coefficient vs. CH_4 fugacity and (**b**) CO_2 solubility coefficient vs. CO_2 fugacity of 6FDA-mPDA. The inset graphs report the solubility coefficient behavior of CH_4 and CO_2 at various equilibrium concentrations—these curves were predicted using the dual-mode sorption model extended to mixtures (DMS-mix) [35] and the pure-gas DMS sorption parameters (Table S1). Note that the feed mixture concentration is the parameter for the experimental data in (**a**,**b**), whereas the concentration at equilibrium is the parameter for the DMS-mix predictions (insert graphs); hence, the comparison between experimental data and predictions is qualitative (i.e., the DMS-mix curves guide the reader through the data).

Because 6FDA-mPDA has strong affinity to CO_2, and hence high sorption uptake, the experimental data exhibit small scattering. Conversely, solubility coefficients of methane were scattered at low CH_4 feed concentrations and at high total pressures, because the accuracy limit of the system was approached (methane solubility coefficients lower than ~0.5 cm^3(STP) cm^{-3} atm^{-1} require volumes optimization of the barometric pressure decay system, as discussed elsewhere [11]).

3.2. Solubility Selectivity Analysis

We further analyzed our experimental results of mixed-gas solubility for 6FDA-mPDA via a plot of CO_2 mixed-gas solubility coefficient versus CH_4 mixed-gas solubility coefficient (Figure 5a). Note that each mixed-gas sorption experiment produces two solubility coefficients: one for CO_2 and one for CH_4; hence, these two solubility coefficients generate a single data point in the plot of Figure 5a. Experimental mixed-gas solubility coefficient data at 35 °C are also shown from previous studies for PIM-1 [20], TZ-PIM-1 [23], PTMSP [15], and PPO [18]—data of AO-PIM-1 [23] and polynonene [23] are plotted in Figure S1. Interestingly, all experimental data could be fitted with a straight line regardless of mixture concentration.

The straight line fitted to the data in Figure 5a follows the following equation:

$$S_{CO_2} = \alpha^0_{CO_2/CH_4} \cdot S_{CH_4} + B \quad (1)$$

where S_{CO_2} and S_{CH_4} are the solubility coefficient of CO_2 and CH_4, respectively; $\alpha^0_{CO_2/CH_4}$ is the mixed-gas selectivity at infinite dilution, that is, the slope of the straight line; and B is the intercept. We rearranged this equation as follows:

$$\alpha^{mix,S}_{CO_2/CH_4} = \frac{S_{CO_2}}{S_{CH_4}} = \alpha^0_{CO_2/CH_4} + \frac{B}{S_{CH_4}} \quad (2)$$

where $\alpha^{mix,\,S}_{CO_2/CH_4}$ is the mixed-gas solubility selectivity coefficient of the membrane material.

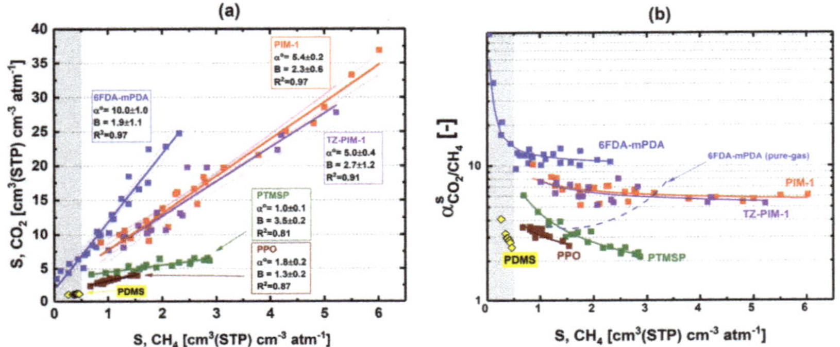

Figure 5. (**a**) CO_2 vs. CH_4 mixed-gas solubility coefficient of 6FDA-mPDA at 35 °C—solid lines were estimated via linear fitting of experimental data (the dotted curves delimit the confidence intervals of each linear interpolation); (**b**) data of CO_2/CH_4 mixed-gas solubility selectivity vs. CH_4 mixed-gas solubility coefficient of 6FDA-mPDA (CO_2/CH_4 mixed-gas solubility selectivity vs. CH_4 mixed-gas solubility coefficient mixed- (solid line) and pure-gas trends (dashed-line) are also shown for comparison). Mixed-gas solubility coefficient data from previous reports on PDMS [11], PIM-1 [20], TZ-PIM-1 [23], PTMSP [15], and PPO [18] are included in (**a**,**b**).

When S_{CH_4} is high, and $\frac{B}{S_{CH_4}} \ll \alpha^0_{CO_2/CH_4}, \alpha^{mix,\,S}_{CO_2/CH_4} = \alpha^0_{CO_2/CH_4}$ (Equation (2)); this condition may be found for very low equilibrium pressures. Hence, $\alpha^0_{CO_2/CH_4}$ corresponds to the ratio between the solubility coefficient of CO_2 and CH_4 at infinite dilution. For most of the polymers shown in Figure 5a, the pure-gas solubility selectivity at infinite dilution (estimated via the DMS model in the form of Equation S2) was in good agreement with the experimental values found via linear interpolation of CO_2 mixed-gas solubility coefficient versus CH_4 mixed-gas solubility coefficient data (i.e., $\alpha^0_{CO_2/CH_4}$); this comparison is discussed in the Supporting Information of this work. When S_{CH_4} is low (i.e., for high equilibrium pressures), a second limiting condition is found; in this case, the CO_2/CH_4 solubility selectivity diverges positively from the $\alpha^0_{CO_2/CH_4}$ value if $B > 0$.

The significance of B and $\alpha^0_{CO_2/CH_4}$ in Equations (1) and (2) can be better appreciated in Figure 5b, where the mixed-gas solubility selectivity of 6FDA-mPDA, PIM-1, TZ-PIM-1, PTMSP, PPO, and the predictions of Equation (2) are plotted against the solubility coefficient of CH_4. For all polymers, predictions by Equation (2) follow the experimental data, and as S_{CH_4} decreases, the mixed-gas solubility selectivity increases ($B > 0$)—especially in the grey region of Figure 5. Note that during linear fitting of CO_2 versus CH_4 mixed-gas solubilities of 6FDA-mPDA (Figure 5a), the data in the grey region had almost no influence on $\alpha^0_{CO_2/CH_4}$, while they could affect the value of B, but not the sign—this confirms the overall trends of solubility selectivity shown in Figure 5b. Interestingly, the mixed-gas data of rubbery PDMS [11] also follow the trend seen for 6FDA-mPDA.

In Figure 5b, the solubility selectivity increases with pressure ($B > 0$) from $\alpha^{o}_{CO_2/CH_4}$; hence, $\alpha^{o}_{CO_2/CH_4}$ appears to be a characteristic solubility selectivity value of glassy polymers. We plotted data of $\alpha^{o}_{CO_2/CH_4}$ versus CO_2 solubility coefficient at infinite dilution in the 2014 upper bound solubility plot (Figure 6) discussed by Lou et al. [47]. The solubility selectivity at infinite dilution of AO-PIM-1 was obtained from pure-gas uptake experiments (see Supporting Information). For both PTMSP and PPO, the solubility selectivity at infinite dilution estimated from Equation (2) and from the DMS model (i.e., from the ratio of DMS solubility coefficients at infinite dilution calculated via Equation S2) did not agree within the respective standard deviations and were both plotted in Figure 6. The values of $\alpha^{o}_{CO_2/CH_4}$ and CO_2 solubility coefficient at infinite dilution for 6FDA-TADPO were uncertain, because of the limited number of pure- and mixed-gas data reported [22,48]. Although the standard error is very large, the pure-gas 6FDA-TADPO solubility trend is qualitatively similar to 6FDA-mPDA (Figure 6). Similarly, PIM-1, TZ-PIM-1, and AO-PIM-1 points group in a confined region of the 2014 CO_2/CH_4 solubility upper bound plot.

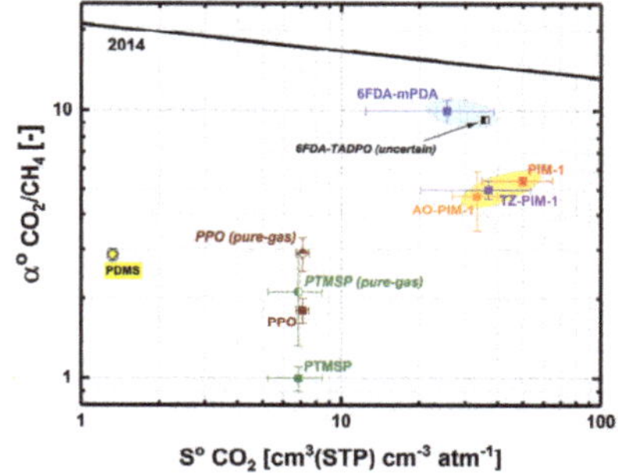

Figure 6. CO_2 solubility coefficient vs. CO_2/CH_4 solubility selectivity (at infinite dilution) data of all polymers tested for CO_2-CH_4 mixed-gas sorption at 35 °C [11,15,18,20,22,23,48]. CO_2/CH_4 solubility selectivities were obtained from linear interpolation of CO_2 mixed-gas solubility coefficient vs. CH_4 mixed-gas solubility coefficient data or from pure-gas uptake data via DMS model equations—see the discussion in the Supporting Information and data values in Table S2. CO_2 solubility coefficients were estimated from experimental data of pure-gas CO_2 uptake. The 2014 CO_2/CH_4 solubility upper bound was discussed elsewhere [47].

3.3. *Equimolar CO_2-CH_4 Mixed-Gas Diffusion*

To elucidate the phenomena that affect the separation performance of 6FDA-mPDA polyimide in mixed-gas conditions, we first show previously reported pure- and 50 mol% mixed-gas permeability data of 6FDA-mPDA from our group [29]—here, these data were corrected with fugacity coefficients and re-plotted in Figure 7. Secondly, we incorporate the results of our pure- and mixed-gas solubility experiments to clarify the contribution of mixed-gas solubility to permeability. Finally, CH_4 and CO_2 pure- and mixed-gas concentration-averaged effective diffusion coefficients are discussed.

After correction of partial pressures and driving forces with fugacity coefficients, permeability trends show that CO_2 mixed-gas permeability suffers from the presence of methane (a local minimum is found at about ~10 atm partial fugacity), whereas CH_4 mixed-gas permeability increases in mixed-gas conditions, particularly above ~10 atm partial fugacity (Figure 7a). Overall, the permeability selectivity of the mixture strongly diverges from the pure-gas trend (Figure 7b); at about 18 atm partial fugacity, almost 35% of the ideal permeability selectivity is lost. Frequently, it is assumed that competitive sorption is the cause for this loss of permeability selectivity. The data of mixed-gas solubilities at 50 mol% equilibrium concentration (Figure 8) show that competitive sorption strongly affects CH_4 mixed-gas solubility—that is, when partial pressures increase, the CH_4 mixed-gas solubility coefficients diverge from the pure-gas values. Because the effect of competitive sorption on the solubility coefficient of CO_2 is limited, we found that at 2 atm CO_2 partial pressure, the CO_2/CH_4 solubility selectivity jumps from a value of ~5 in the pure-gas state to ~10 in the mixture; moreover, as previously discussed, the CO_2/CH_4 mixed-gas solubility selectivity increases with partial pressures. Thus, the effects of gas mixture on solubility selectivity are beneficial during separation of CO_2 from CH_4 (equimolar feed) and cannot be held responsible for the loss of permeability selectivity from ideality (Figure 7b).

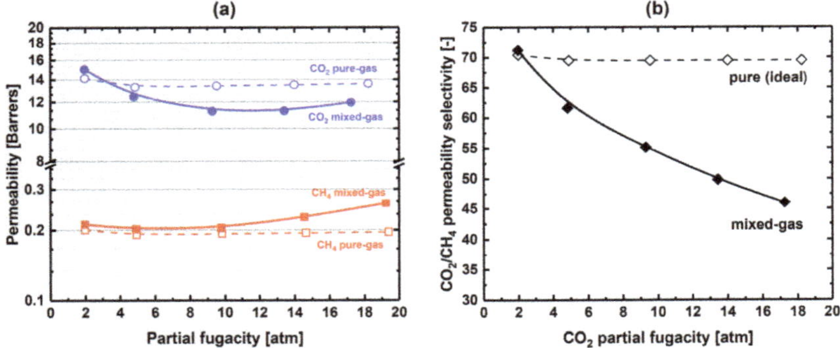

Figure 7. (**a**) CH_4 and CO_2 pure- and mixed-gas permeability data (6FDA-mPDA) vs. partial fugacities; (**b**) CO_2/CH_4 pure- and mixed-gas permeability selectivity data vs. CO_2 partial fugacities (feed was equimolar). CH_4 and CO_2 permeabilities based on partial pressures were previously reported by our group [29] and corrected with fugacity coefficients. Lines are drawn to guide the eye.

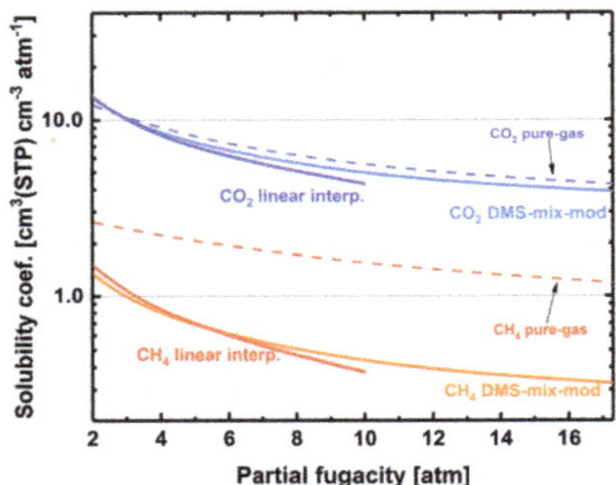

Figure 8. CH_4 and CO_2 pure- and mixed-gas solubility coefficients (6FDA-mPDA) vs. partial fugacities. The mixed-gas solubility coefficient curves for methane and carbon dioxide were obtained by linear interpolations of mixed-gas experimental data or by fitting with a modified version of the DMS model for gas mixtures (DMS-mix-mod)—more details can be found in the Supporting Information of this work (Figure S4 and Figure S5).

Hence, kinetic effects must be responsible for the deviation of mixed-gas permeability selectivity from the pure-gas values. Figure 9a shows the variation of pure- and mixed-gas concentration-averaged effective diffusion coefficients with partial fugacities. The mixed-gas diffusivities of CH_4 notably deviate from the pure-gas trend, whereas CO_2 diffuses in the same manner in pure- and mixed-gas environments with almost no disturbance by methane (Figure 9a). The CO_2/CH_4 mixed-gas diffusion selectivity drops from an average pure-gas value of ~18 to a mixed-gas value of ~5 (Figure 9b), simply because in the mixture, CH_4 diffusion is enhanced compared with that in the pure-gas environment because of the presence of CO_2 in the mixture. Similar effects on CO_2-CH_4 diffusion were observed for PDMS rubber [11]; in this case, sorption of CO_2 induced the decline of the mixed-gas diffusion and permeability selectivity relative to the pure-gas values. A similar case was discussed by Ribeiro et al. [24], who described how the increase of C_2H_6 mixed-gas diffusion coefficient and the decrease of the CO_2/C_2H_6 permeability selectivity could be ascribed to CO_2-induced "plasticization" of a XLPEO rubber film.

Finally, from the analysis of the data in Figure 7a, one may conclude that "plasticization" of 6FDA-mPDA takes place at partial fugacities greater than ~10 atm, where the gas permeability shows minima for both CO_2 and CH_4. However, CH_4 mixed-gas diffusion rises immediately at low feed pressure after the polymer matrix sorbs CO_2 (Figure 9a); in other words, the effect of CO_2 sorption on gas transport occurs over the entire range of partial pressures explored in this work, and the local minima seen in Figure 7a were produced by counteracting thermodynamic and kinetic contributions to transport.

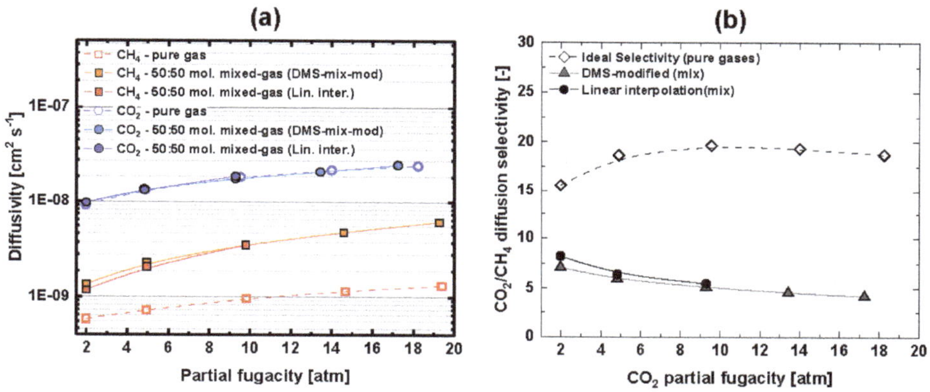

Figure 9. (**a**) CH_4 and CO_2 pure- and mixed-gas diffusion coefficients (6FDA-mPDA) vs. partial fugacity; (**b**) CO_2/CH_4 pure- and mixed-gas diffusion selectivity vs. CO_2 partial fugacity. Lines are drawn to guide the eye.

4. Conclusions

To quantify the deviation from ideality of CO_2-CH_4 mixed-gas permeability and CO_2/CH_4 mixed-gas permeability selectivity of 6FDA-mPDA at 35 °C, sorption and diffusion contributions to permeation were decoupled. Experimental data of mixed-gas solubility revealed a decrease of both CO_2 and, more markedly, CH_4 solubility due to mixture effects. We found that CO_2 versus CH_4 mixed-gas solubility coefficients of 6FDA-mPDA (and other glassy polymers previously studied) follow a linear trend regardless of equilibrium concentration. The slope of the trend line agrees well with the CO_2/CH_4 solubility selectivity at infinite dilution, and the intercept indicates the way in which solubility selectivity deviates at increasing pressures. We found the same behavior reviewing mixed-gas sorption data of glassy polymers reported in the literature. In all cases, the CO_2/CH_4 solubility selectivity increases with pressure from the value of solubility selectivity at infinite dilution.

Because the CO_2/CH_4 solubility selectivity of 6FDA-mPDA improved under mixed-gas conditions, the decline of CO_2/CH_4 mixed-gas permeability selectivity from the corresponding pure-gas permeability selectivities—typically observed during CO_2-CH_4 permeation in polymeric films—could not be attributed to competitive sorption (as frequently assumed in the literature). Hence, we studied the kinetic behavior of 6FDA-mPDA to elucidate the effect of gas mixture effects on concentration-averaged effective diffusion coefficients as estimated from experimental mixed-gas sorption and permeation data. We observed that after CO_2 was added to CH_4 in a mixture, even at a low concentration, the concentration-averaged effective diffusion coefficient of CH_4 deviated from the pure-gas values, whereas the concentration-averaged effective diffusion coefficient of CO_2 essentially followed the pure-gas trend; hence, the departure of CO_2/CH_4 permeability selectivity of

6FDA-mPDA from the pure-gas values can be explained by a depression of the size sieving capability of 6FDA-mPDA (i.e., it makes CH_4 diffusion faster than in the pure-gas environment) induced by the presence of CO_2 by sorption in the polymeric film matrix.

Supplementary Materials: The following are available online at http://www.mdpi.com/2077-0375/9/1/10/s1. Figure S1: Data of CO_2 mixed-gas solubility vs. CH_4 mixed-gas solubility coefficient of AO-PIM-1 and polynonene [2]. For both interpolations, the slope was fixed at the value of pure-gas solubility selectivity at infinite dilution. For both interpolations, the slope was fixed at the value of pure-gas solubility selectivity at infinite dilution. Figure S2: In red and in blue, two examples of the behavior of the *switch* function (Equation S5) used in this work to track the behavior of K_{Di} and b_i parameters. In this graph, $A_i^{pure} = 1$. Figure S3: Comparison between experimental uptakes and model predictions for (a) CH_4 and (b) CO_2. Black squares are predictions of the DMS-mix with pure-gas parameters. Red circles are predictions of the DMS-mix-mod. Figure S4: CH_4 (a) and CO_2 (b) mixed-gas uptakes in 6FDA-mPDA. Surfaces were obtained via the DMS-mix-mod fitting. The DMS-mix-mod allowed us to predict the solubility behavior beyond the region covered by experimental data. Figure S5: CH_4 (a) and CO_2 (b) mixed-gas uptakes in 6FDA-mPDA. Surfaces were obtained via linear interpolation. Table S1: Dual-mode model parameters of methane and carbon dioxide in 6FDA-mPDA for pure-gas sorption at 35 °C. Table S2: Comparison between solubility selectivities at infinite dilution retrieved from mixed-gas and pure-gas sorption data. Table S3: Parameters derived from DMS-mix-mod fitting. Table S4: Pure- and mixed-gas uptake data presented in this work.

Author Contributions: Conceptualization, I.P. and G.G.; methodology, I.P. and G.G.; investigation, G.G.; data acquisition, G.G.; writing—original draft preparation, G.G.; review and editing, I.P.; polymer synthesis, B.S.G.

Funding: This work was supported by funding (BAS/1/1323-01-01) from King Abdullah University of Science and Technology (KAUST).

Conflicts of Interest: The authors declare no conflict of interest.

References

1. Sholl, D.S.; Lively, R.P. Seven chemical separations to change the world. *Nature* **2016**, *532*, 435–438. [CrossRef] [PubMed]
2. Budd, P.M.; Msayib, K.J.; Tattershall, C.E.; Ghanem, B.S.; Reynolds, K.J.; McKeown, N.B.; Fritsch, D. Gas separation membranes from polymers of intrinsic microporosity. *J. Membr. Sci.* **2005**, *251*, 263–269. [CrossRef]
3. Swaidan, R.; Ma, X.; Litwiller, E.; Pinnau, I. High pressure pure- and mixed-gas separation of CO_2/CH_4 by thermally-rearranged and carbon molecular sieve membranes derived from a polyimide of intrinsic microporosity. *J. Membr. Sci.* **2013**, *447*, 387–394. [CrossRef]
4. Swaidan, R.; Ghanem, B.S.; Litwiller, E.; Pinnau, I. Pure- and mixed-gas CO_2/CH_4 separation properties of PIM-1 and an amidoxime-functionalized PIM-1. *J. Membr. Sci.* **2014**, *457*, 95–102. [CrossRef]
5. Sanders, D.F.; Smith, Z.P.; Guo, R.; Robeson, L.M.; McGrath, J.E.; Paul, D.R.; Freeman, B.D. Energy-efficient polymeric gas separation membranes for a sustainable future: A review. *Polymer* **2013**, *54*, 4729–4761. [CrossRef]
6. Puleo, A.C.; Paul, D.R.; Kelley, S.S. The effect of degree of acetylation on gas sorption and transport behavior in cellulose acetate. *J. Membr. Sci.* **1989**, *47*, 301–332. [CrossRef]
7. Wijmans, J.G.; Baker, R.W. The solution-diffusion model: A review. *J. Membr. Sci.* **1995**, *107*, 1–21. [CrossRef]
8. O'Brien, K.C.; Koros, W.J.; Barbari, T.A.; Sanders, E.S. A new technique for the measurement of multicomponent gas transport through polymeric films. *J. Membr. Sci.* **1986**, *29*, 229–238. [CrossRef]
9. Fraga, S.C.; Monteleone, M.; Lanč, M.; Esposito, E.; Fuoco, A.; Giorno, L.; Pilnáček, K.; Friess, K.; Carta, M.; McKeown, N.B.; et al. A novel time lag method for the analysis of mixed gas diffusion in polymeric membranes by on-line mass spectrometry: Method development and validation. *J. Membr. Sci.* **2018**, *561*, 39–58. [CrossRef]
10. Monteleone, M.; Esposito, E.; Fuoco, A.; Lanč, M.; Pilnáček, K.; Friess, K.; Bezzu, C.G.; Carta, M.; McKeown, N.B.; Jansen, J.C. A novel time lag method for the analysis of mixed gas diffusion in polymeric membranes by on-line mass spectrometry: Pressure dependence of transport parameters. *Membranes* **2018**, *8*, 73. [CrossRef]
11. Genduso, G.; Litwiller, E.; Ma, X.; Zampini, S.; Pinnau, I. Mixed-gas sorption in polymers via a new barometric test system: Sorption and diffusion of CO_2-CH_4 mixtures in polydimethylsiloxane (PDMS). Manuscript submitted for publication.

12. Koros, W.J.; Paul, D.R. Design considerations for measurement of gas sorption in polymers by pressure decay. *J. Polym. Sci. Polym. Phys. Ed.* **1976**, *14*, 1903–1907. [CrossRef]
13. Ribeiro, C.P.; Freeman, B.D. Carbon dioxide/ethane mixed-gas sorption and dilation in a cross-linked poly(ethylene oxide) copolymer. *Polymer* **2010**, *51*, 1156–1168. [CrossRef]
14. Raharjo, R.D.; Freeman, B.D.; Sanders, E.S. Pure and mixed gas CH_4 and n-C_4H_{10} sorption and dilation in poly(1-trimethylsilyl-1-propyne). *Polymer* **2007**, *48*, 6097–6114. [CrossRef]
15. Vopička, O.; De Angelis, M.G.; Sarti, G.C. Mixed gas sorption in glassy polymeric membranes: I. CO_2/CH_4 and n-C_4/CH_4 mixtures sorption in poly(1-trimethylsilyl-1-propyne) (PTMSP). *J. Membr. Sci.* **2014**, *449*, 97–108. [CrossRef]
16. Sanders, E.S.; Koros, W.J.; Hopfenberg, H.B.; Stannett, V.T. Mixed gas sorption in glassy polymers: Equipment design considerations and preliminary results. *J. Membr. Sci.* **1983**, *13*, 161–174. [CrossRef]
17. Sanders, E.S.; Koros, W.J.; Hopfenberg, H.B.; Stannett, V.T. Pure and mixed gas sorption of carbon dioxide and ethylene in poly(methyl methacrylate). *J. Membr. Sci.* **1984**, *18*, 53–74. [CrossRef]
18. Story, B.J.; Koros, W.J. Sorption of CO_2/CH_4 mixtures in poly(phenylene oxide) and a carboxylated derivative. *J. Appl. Polym. Sci.* **1991**, *42*, 2613–2626. [CrossRef]
19. Raharjo, R.D.; Freeman, B.D.; Sanders, E.S. Pure and mixed gas CH_4 and n-C_4H_{10} sorption and dilation in poly(dimethylsiloxane). *J. Membr. Sci.* **2007**, *292*, 45–61. [CrossRef]
20. Vopička, O.; De Angelis, M.G.; Du, N.; Li, N.; Guiver, M.D.; Sarti, G.C. Mixed gas sorption in glassy polymeric membranes: II. CO_2/CH_4 mixtures in a polymer of intrinsic microporosity (PIM-1). *J. Membr. Sci.* **2014**, *459*, 264–276. [CrossRef]
21. Gemeda, A.E.; De Angelis, M.G.; Du, N.; Li, N.; Guiver, M.D.; Sarti, G.C. Mixed gas sorption in glassy polymeric membranes. III. CO_2/CH_4 mixtures in a polymer of intrinsic microporosity (PIM-1): Effect of temperature. *J. Membr. Sci.* **2017**, *524*, 746–757. [CrossRef]
22. Kamaruddin, H.D.; Koros, W.J. Some observations about the application of Fick's first law for membrane separation of multicomponent mixtures. *J. Membr. Sci.* **1997**, *135*, 147–159. [CrossRef]
23. Gemeda, A.E. Solubility, Diffusivity and Permeability of Gases in Glassy Polymers. Ph.D. Thesis, Università di Bologna, Bologna, Italy, 2015.
24. Ribeiro, C.P.; Freeman, B.D.; Paul, D.R. Pure- and mixed-gas carbon dioxide/ethane permeability and diffusivity in a cross-linked poly(ethylene oxide) copolymer. *J. Membr. Sci.* **2011**, *377*, 110–123. [CrossRef]
25. Raharjo, R.D.; Freeman, B.D.; Paul, D.R.; Sanders, E.S. Pure and mixed gas CH_4 and n-C_4H_{10} permeability and diffusivity in poly(1-trimethylsilyl-1-propyne). *Polymer* **2007**, *48*, 7329–7344. [CrossRef]
26. Reijerkerk, S.R.; Nijmeijer, K.; Ribeiro, C.P.; Freeman, B.D.; Wessling, M. On the effects of plasticization in CO_2/light gas separation using polymeric solubility selective membranes. *J. Membr. Sci.* **2011**, *367*, 33–44. [CrossRef]
27. Raharjo, R.D.; Freeman, B.D.; Paul, D.R.; Sarti, G.C.; Sanders, E.S. Pure and mixed gas CH_4 and n-C_4H_{10} permeability and diffusivity in poly(dimethylsiloxane). *J. Membr. Sci.* **2007**, *306*, 75–92. [CrossRef]
28. Tanis, I.; Brown, D.; Neyertz, S.J.; Heck, R.; Mercier, R. A comparison of homopolymer and block copolymer structure in 6FDA-based polyimides. *Phys. Chem. Chem. Phys.* **2014**, *16*, 23044–23055. [CrossRef]
29. Alaslai, N.; Ghanem, B.; Alghunaimi, F.; Litwiller, E.; Pinnau, I. Pure- and mixed-gas permeation properties of highly selective and plasticization resistant hydroxyl-diamine-based 6FDA polyimides for CO_2/CH_4 separation. *J. Membr. Sci.* **2016**, *505*, 100–107. [CrossRef]
30. Qiu, W.; Zhang, K.; Li, F.S.; Zhang, K.; Koros, W.J. Gas separation performance of carbon molecular sieve membranes based on 6FDA-mPDA/DABA (3:2) polyimide. *ChemSusChem* **2014**, *7*, 1186–1194. [CrossRef]
31. Gleason, K.L.; Smith, Z.P.; Liu, Q.; Paul, D.R.; Freeman, B.D. Pure- and mixed-gas permeation of CO_2 and CH_4 in thermally rearranged polymers based on 3,3′-dihydroxy-4,4′-diamino-biphenyl (HAB) and 2,2′-bis-(3,4-dicarboxyphenyl) hexafluoropropane dianhydride (6FDA). *J. Membr. Sci.* **2015**, *475*, 204–214. [CrossRef]
32. Swaidan, R.; Ghanem, B.; Litwiller, E.; Pinnau, I. Physical aging, plasticization and their effects on gas permeation in "rigid" polymers of intrinsic microporosity. *Macromolecules* **2015**, *48*, 6553–6561. [CrossRef]
33. Swaidan, R.; Ghanem, B.; Al-Saeedi, M.; Litwiller, E.; Pinnau, I. Role of intrachain rigidity in the plasticization of intrinsically microporous triptycene-based polyimide membranes in mixed-gas CO_2/CH_4 separations. *Macromolecules* **2014**, *47*, 7453–7462. [CrossRef]

34. Staudt-Bickel, C.J.; Koros, W. Improvement of co$_2$/ch$_4$ separation characteristics of polyimides by chemical crosslinking. *J. Membr. Sci.* **1999**, *155*, 145–154. [CrossRef]
35. Koros, W.J. Model for sorption of mixed gases in glassy polymers. *J. Polym. Sci. Polym. Phys. Ed.* **1980**, *18*, 981–992. [CrossRef]
36. Comesaña-Gándara, B.; Hernández, A.; de la Campa, J.G.; de Abajo, J.; Lozano, A.E.; Lee, Y.M. Thermally rearranged polybenzoxazoles and poly(benzoxazole-co-imide)s from ortho-hydroxyamine monomers for high performance gas separation membranes. *J. Membr. Sci.* **2015**, *493*, 329–339. [CrossRef]
37. Cheng, S.-X.; Chung, T.-S.; Wang, R.; Vora, R.H. Gas-sorption properties of 6FDA–durene/1,4-phenylenediamine (pPDA) and 6FDA–durene/1,3-phenylenediamine (mPDA) copolyimides. *J. Appl. Polym. Sci.* **2003**, *90*, 2187–2193. [CrossRef]
38. Heck, R.; Qahtani, M.S.; Yahaya, G.O.; Tanis, I.; Brown, D.; Bahamdan, A.A.; Ameen, A.W.; Vaidya, M.M.; Ballaguet, J.P.R.; Alhajry, R.H.; et al. Block copolyimide membranes for pure- and mixed-gas separation. *Sep. Purif. Technol.* **2017**, *173*, 183–192. [CrossRef]
39. Joly, C.; Le Cerf, D.; Chappey, C.; Langevin, D.; Muller, G. Residual solvent effect on the permeation properties of fluorinated polyimide films. *Sep. Purif. Technol.* **1999**, *16*, 47–54. [CrossRef]
40. Redlich, O.; Kwong, J.N.S. On the thermodynamics of solutions. V. An equation of state. Fugacities of gaseous solutions. *Chem. Rev.* **1949**, *44*, 233–244. [CrossRef]
41. Soave, G. 20 years of Redlich-Kwong equation of state. *Fluid Phase Equilib.* **1993**, *82*, 345–359. [CrossRef]
42. Koros, W.J.; Chan, A.H.; Paul, D.R. Sorption and transport of various gases in polycarbonate. *J. Membr. Sci.* **1977**, *2*, 165–190. [CrossRef]
43. Sadrzadeh, M.; Shahidi, K.; Mohammadi, T. Effect of operating parameters on pure and mixed gas permeation properties of a synthesized composite PDMS/PA membrane. *J. Membr. Sci.* **2009**, *342*, 327–340. [CrossRef]
44. Coleman, M.R. Isomers of Fluorine-Containing Polyimides for Gas Separation Membranes. Ph.D. Thesis, The University of Texas at Austin, Austin, TX, USA, 1992.
45. Erb, A.J.; Paul, D.R. Gas sorption and transport in polysulfone. *J. Membr. Sci.* **1981**, *8*, 11–22. [CrossRef]
46. Koros, W.J.; Paul, D.R.; Rocha, A.A. Carbon dioxide sorption and transport in polycarbonate. *J. Polym. Sci. Polym. Phys. Ed.* **1976**, *14*, 687–702. [CrossRef]
47. Lou, Y.; Hao, P.; Lipscomb, G. NELF predictions of a solubility–solubility selectivity upper bound. *J. Membr. Sci.* **2014**, *455*, 247–253. [CrossRef]
48. Walker, D.R.B. Synthesis and Characterization of Polypyrrolones for Gas Separation Membranes. Ph.D. Thesis, The University of Texas at Austin, Austin, TX, USA, 1993.

Article

Temperature and Pressure Dependence of Gas Permeation in a Microporous Tröger's Base Polymer

Elsa Lasseuguette [1], Richard Malpass-Evans [2], Mariolino Carta [3], Neil B. McKeown [2] and Maria-Chiara Ferrari [1,*]

1 School of Engineering, University of Edinburgh, Robert Stevenson Road, Edinburgh EH9 3FB, UK; e.lasseuguette@ed.ac.uk
2 EastChem, School of Chemistry, University of Edinburgh, David Brewster Road, Edinburgh EH9 3FJ, UK; R.Malpass-Evans@ed.ac.uk (R.M.-E.); Neil.McKeown@ed.ac.uk (N.B.M.)
3 Department of Chemistry, College of Science, Grove Building, Singleton Park, Swansea University, Swansea SA2 8PP, UK; mariolino.carta@swansea.ac.uk
* Correspondence: m.ferrari@ed.ac.uk; Tel.: +44-131-650-5689

Received: 31 October 2018; Accepted: 5 December 2018; Published: 14 December 2018

Abstract: Gas transport properties of PIM-EA(H_2)-TB, a microporous Tröger's base polymer, were systematically studied over a range of pressure and temperature. Permeability coefficients of pure CO_2, N_2, CH_4 and H_2 were determined for upstream pressures up to 20 bar and temperatures up to 200 °C. PIM-EA(H_2)-TB exhibited high permeability coefficients in absence of plasticization phenomena. The permeability coefficient of N_2, CH_4 and H_2 increased with increasing temperature while CO_2 permeability decreased with increasing temperature as expected for a glassy polymer. The diffusion and solubility coefficients were also analysed individually and compared with other polymers of intrinsic microporosity. From these results, the activation energies of permeation, diffusion and sorption enthalpies were calculated using an Arrhenius equation.

Keywords: microporous polymer; gas permeability; activation energy; CO_2 capture

---

1. Introduction

Membranes are one of the most promising technologies to compete with conventional separation processes for gas separations including post- and pre-combustion carbon capture. Studies on the use of polymeric membranes in an Integrated Gasification Combined Cycle (IGCC) power plant [1–3] show their viability and their competitiveness with the currently more developed solvent-based technology. Process simulations [1] have shown an advantage for hydrogen selective materials for this application and new membrane materials are currently under development [3]. The performance of the materials in the relatively harsh conditions of the separation (50 bars and 200 °C) needs to be investigated before production can be scaled up [4].

The increase of gas pressure can have a negative impact on membrane performance, due to plasticization effects. For glassy polymers, many gases, such as O_2, N_2 and H_2, can permeate through the polymer without modifying the polymer's properties due to their relatively low solubility in the polymer. Therefore, with the pressure's increase, the gas permeability slightly decreases, as expected from the dual sorption—dual mobility model [5]. On the contrary, highly sorbing gases such as CO_2 can induce a swelling of the polymer matrix, that is, plasticization, leading to a large increase of the gas permeability with increasing pressure. In addition, the influence of temperature on gas separation performance has been investigated for a large number of polymers. Depending on the polymer, the membrane performance can be improved by an increase of temperature as shown by the Robeson plot in Figure 1. Most polymers, including ultrahigh-free volume polymers such as PTMSP and Teflon AF, present higher hydrogen selectivity over CO_2 at high temperature. For example, Merkel et al. [6]

reported H_2S, CO_2, H_2, N_2 and CO permeation as a function of temperature up to 240 °C. At room temperature, PTMSP appears to be more permeable to the more condensable gases, such as CO_2 and H_2S than to H_2. However, it becomes hydrogen selective at elevated temperatures.

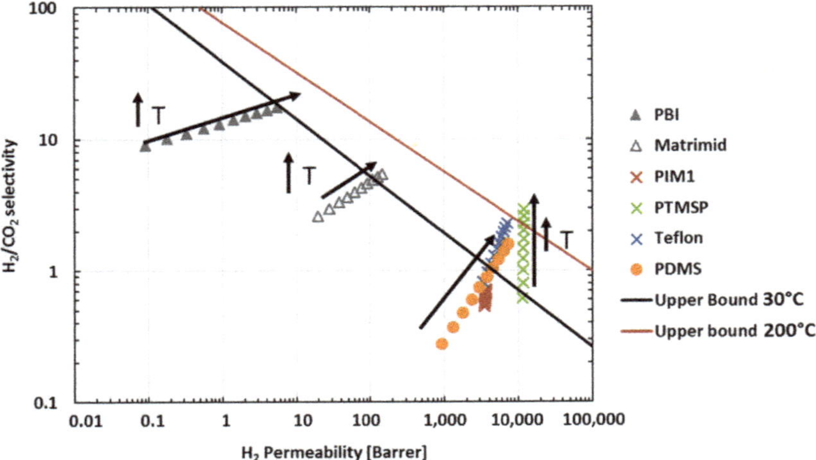

Figure 1. Influence of temperature on membrane performance (calculated from [7]) [6,8–11].

For the first Polymer of Intrinsic Microporosity, PIM-1, Budd et al. [8] showed that the CO_2 permeability coefficient decreased gradually as the temperature increased, whereas the H_2 permeability coefficient increased. Thus, PIM-1 also becomes slightly more H_2 selective at higher temperature. Recently, Fuoco et al. [12] studied the temperature dependence of gas permeation in triptycene-based ultrapermeable PIMs, such as PIM-TMN-Trip. With increasing temperature, the permeability coefficient increased for the bulkier penetrants (N_2 and CH_4), while for the faster penetrants (CO_2 and O_2) it decreased and for the very small penetrants (H_2 and He) it was constant. Therefore, PIM-TMN-Trip became more selective to H_2 at high temperature; these ultrapermeable polymers behave as microporous solids, in which the pore dimensions are rather large in comparison with the diffusing gas molecules. Such studies of the temperature and pressure dependence of transport properties are essential for understanding the behaviour of membranes over a wide range of conditions, in order to assist any consideration of industrial use.

Recently, a new type of PIM has been developed using a polymerization reaction based on the formation of the bridged bicyclic diamine called Tröger's base (TB: 6H,12H-5,11-methanodibenzo [b,f][1,5]diazocine), such as PIM-EA(Me_2)-TB [13] or PIM-EA(H_2)-TB [14,15] (Figure 2). PIM-EA(Me_2)-TB demonstrates at ambient temperature very fast gas permeability and good selectivity, surpassing the Robeson's upper bound in the case of O_2/N_2, H_2/N_2 and H_2/CH_4 gas pairs [16,17]. This is due to the large diffusivity selectivity that favours transport of gas molecules of smaller kinetic diameters (H_2, CO_2) over that of larger molecules (N_2, CH_4).

PIM-EA(H_2)-TB differs from PIM-EA(Me_2)-TB only by the absence of methyl groups at the bridgehead (9,10) position of the ethanoanthracene (EA) unit, which modifies its chain packing in the solid state. PIM-EA(H_2)-TB presents an inter-chain distance, d-space, of 7.7 Å and 32% free volume, whereas PIM-EA(Me)-TB has values of 11 Å and 30%, respectively [18]. With these differences, a higher separation performance for PIM-EA(H_2)-TB is expected. However, few papers have been published on this polymer. Bernardo et al. [15] developed thin film composite based on PIM-EA(H_2)-TB and they studied the impact of the residual casting solvent on the separation performance at 25 °C and 1 bar. In addition, Benito et al. [19] studied composite membranes based on a ultrathin layer of PIM-EA(H_2)-TB for CO_2/N_2 separation at 35 °C and 3 bar.

Here we report a novel study on the permeation properties of PIM-EA(H_2)-TB over a large temperature and pressure range for a series of gases (CO_2, H_2, N_2 and CH_4).

Figure 2. The chemical structure of PIM-EA(Me_2)-TB and PIM-EA(H_2)-TB.

2. Experimental Section

The detailed synthetic procedure for making PIM-EA(H_2)-TB and its structural characterization are reported elsewhere [15]. Robust flat films of thickness between 130 and 200 μm were cast from chloroform with their thickness determined using a digital micrometre (Mitutoyo, Kawasaki, Japan). The permeation properties were measured in a constant volume-variable pressure apparatus (Figure 3) using pure CO_2, N_2, CH_4 and H_2 (Table 1) with pressures up to 20 bar (10 bar for H_2) and temperatures up to 200 °C.

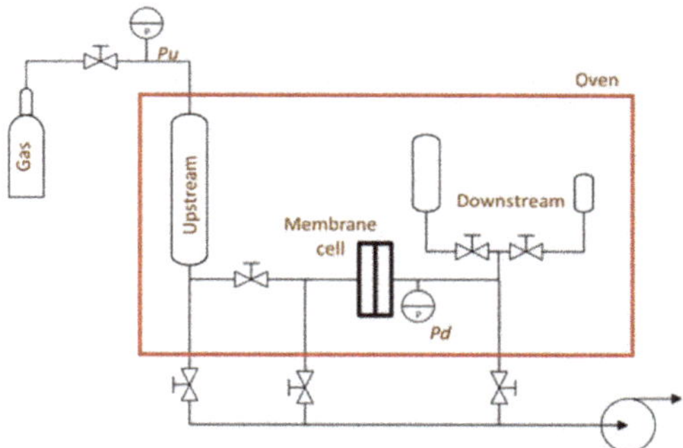

Figure 3. Constant volume-variable pressure apparatus.

Table 1. Kinetic diameter and critical temperature [20].

Gas	Kinetic Diameter (d) (Å)	Critical Temperature (T_c) (K)
H_2	2.89	33.2
N_2	3.64	126.2
CH_4	3.8	190.6
CO_2	3.3	304.2

For each measurement campaign (i.e., one gas and either variable T or variable P), the sample was carefully treated with methanol prior to the measurement in order to start from the same ageing history. The methanol treatment consists of soaking the sample in methanol for 2 h, drying it under ambient conditions for 20 min and under vacuum at 30 °C overnight. At the end of the campaign, the gas permeability at 30 °C and 1 bar was re-measured in order to check the absence of physical/chemical ageing. Moreover, each campaign's duration was short, carried out over a maximum of 3 days in order to limit physical ageing. By using this procedure, the physical ageing was minimised and had no apparent impact on the results for permeability and selectivity.

The permeability was obtained from the evolution of pressure of the downstream side (MKS Baratron 615A (Andover, MA, USA)). The permeability coefficient, P, was determined from the slope of the pressure vs. time curve under steady state condition. Before each experiment, the apparatus was vacuum-degassed and a leak rate determined from the pressure increase in the downstream part. Three different downstream volumes could be selected accordingly to the permeation rate of the gas.

The time lag, θ, was used to determine the diffusivity coefficient D (Equation (1)).

$$D = \frac{l^2}{6\theta} \quad (1)$$

The solubility coefficient, S, for the gas in the polymer was evaluated indirectly, assuming the validity of the diffusion-solution mechanism (Equation (2)):

$$S = \frac{P}{D} \quad (2)$$

The ideal selectivity between two gas species i and j is the ratio of the two single gas permeabilities (Equation (3)).

$$\alpha_{ij} = \frac{P(i)}{P(j)} \quad (3)$$

3. Results

3.1. Permeability

Permeation measurements on methanol treated films of PIM-EA(H_2)-TB were carried out using pure N_2, H_2, CO_2 and CH_4 at several pressures (1 to 20 bar) and temperatures (30 °C to 200 °C).

Table 2 reports the results from the time lag experiment at 30 °C and 1 bar.

Table 2. Gas permeabilities and ideal selectivities (CO_2/Gas, H_2/Gas) for MeOH treated film PIM-EA(H_2)-TB at 30 °C, 1 bar (Errors calculated by statistical analysis of repeated measurements from separately prepared films (between 3 and 5)).

30 °C, 1 bar		N_2	H_2	CO_2	CH_4
PIM-1 [8]	Permeability (Barrer)	252	2936	5303	440
	Selectivity CO_2/Gas	21	1.8	-	12
	Selectivity H_2/Gas	12	-	0.5	6.7
PIM-EA(Me_2)-TB [16]	Permeability (Barrer)	525	7760	7140	699
	Selectivity CO_2/Gas	13.6	0.9	-	10
	Selectivity H_2/Gas	14.8	-	1.1	11
PIM-EA(H_2)-TB (This study)	Permeability (Barrer) (± Error)	238 (± 3%)	5188 (± 1%)	5990 (± 1%)	372 (± 3%)
	Selectivity CO_2/Gas	25	1	-	16
	Selectivity H_2/Gas	22	-	1	14

PIM-EA(H_2)-TB presents high CO_2 and H_2 permeability coefficients and good ideal selectivity over N_2 and CH_4. The order of gas permeabilities for PIM-EA(H_2)-TB is $CO_2 > H_2 > CH_4 > N_2$, the same as that for PIM-1. CO_2, which is the most condensable gas, is the most permeable due to the predominant role of solubility in PIMs [8]. In comparison with PIM-EA(Me_2)-TB, the permeability coefficients obtained for PIM-EA(H_2)-TB are lower. This can be explained by the methyl groups increasing the distance between polymer chains of PIM-EA(Me_2)-TB, relative to PIM-EA(H_2)-TB, which ensures higher free volume and, hence, higher permeability [16] but reduces selectivity.

Figure 4 shows the Robeson plots for five gas pairs, H_2/CH_4, H_2/N_2, H_2/CO_2, CO_2/CH_4 and CO_2/N_2.

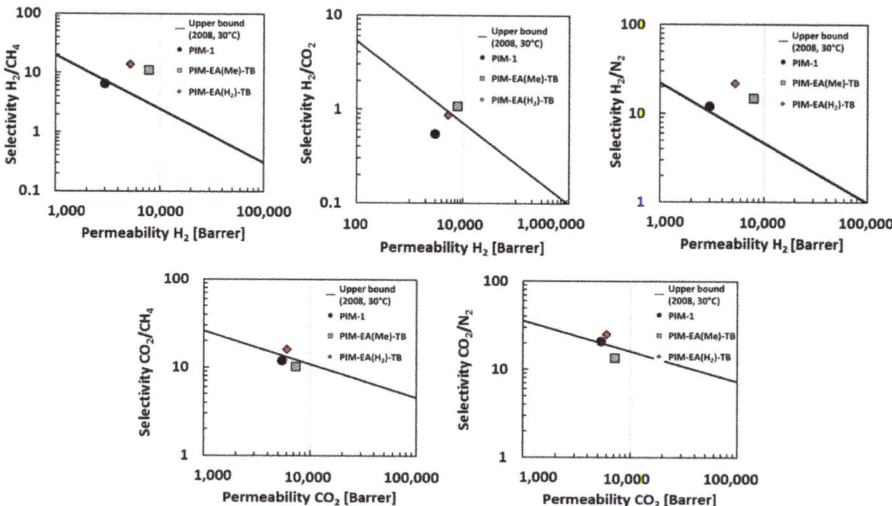

Figure 4. Robeson plots for H_2/CH_4, H_2/N_2, H_2/CO_2, CO_2/CH_4 and CO_2/N_2 for PIM-1 [8] (●), PIM-EA(Me$_2$)-TB [16] (■) and PIM-EA(H_2)-TB [our study] (◆) at 30 °C and 1 bar. The lines represents the 2008 upper bound for each gas pair [21].

As shown on Figure 4, the data for PIM-EA(H_2)-TB are located above the 2008 upper bound for all five gas pairs. For H_2/CH_4 and H_2/N_2, they are clearly higher than for PIM-1 and PIM-EA(Me)-TB. This demonstrates the potential of PIM-EA(H_2)-TB for industrial applications, such as carbon capture (CO_2/N_2 mixture), natural gas sweetening and biogas treatment (CO_2/CH_4 mixture) or hydrogen recovery (H_2/CH_4 mixture).

3.2. *Diffusivity and Solubility Coefficients*

The gas transport in PIM-EA(H_2)-TB was analysed using the solution-diffusion model (Equation (2)), to provide the diffusivity and sorption coefficients (Table 3).

Table 3. Diffusivity and solubility coefficients for MeOH treated film PIM-EA(H_2)-TB, at 30 °C, 1 bar (Errors calculated by statistical analysis of repeated measurements from separately prepared films).

30 °C, 1 bar	N_2	H_2	CO_2	CH_4
D (10^{-7} cm^2/s) (± Error)	9.7 (± 12%)	500.0 (± 9%)	8.2 (± 3%)	1.3 (± 11%)
S (cm^3(STP)/(cm^3·cmHg)) (± Error)	3×10^{-2} (± 15%)	9×10^{-3} (± 10%)	9×10^{-1} (± 4%)	3×10^{-1} (± 14%)

The diffusivity and solubility values of PIM-EA(H_2)-TB are similar to those of polymers from the same family (PIM-EA(Me)-TB) [16] with a very high value of CO_2 solubility coefficient. This affinity towards CO_2 may be enhanced by the presence of the amine groups in the TB moiety.

Diffusivity and solubility data are plotted in Figure 5 as correlations of log D versus d^2 and log S versus T_c, respectively, where d is the kinetic diameter and T_c is the critical temperature of the gases.

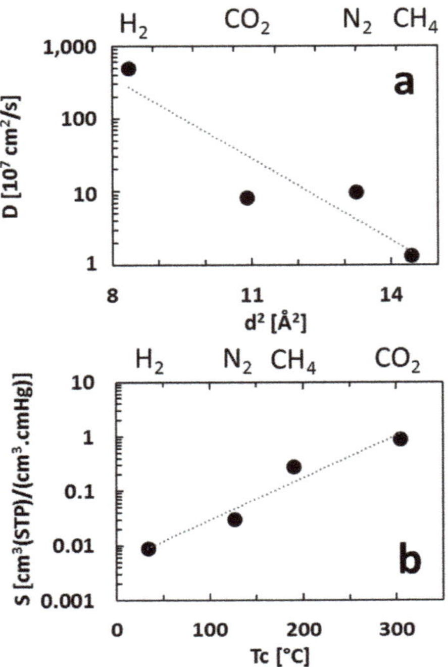

Figure 5. Diffusivity (**a**) and solubility (**b**) coefficients of PIM-EA(H_2)-TB for H_2, CO_2, CH_4 and N_2 at 30 °C and 1 bar.

Figure 5a shows that the diffusivity coefficient of PIM-EA(H_2)-TB decreases with increasing molecular size of the permeate. Larger molecules interact with more segments of the polymer chains than do smaller molecules and thus the mobility selectivity always favours the passage of smaller molecules over larger ones [20]. Moreover, this decrease is large due to the glassy state of the polymer where the highly rigid polymer chains of PIM-EA(H_2)-TB are essentially fixed and do not move readily to accommodate the transport of larger molecules. It is noteworthy that the value of diffusivity for CO_2 is slightly lower than for N_2. Generally, in polymers, the smaller molecule, that is, CO_2, is expected to diffuse faster than N_2, which is a larger molecule. This unusual inversion is found for polymer with high CO_2 affinity [13,17,22] and is caused by the specific interaction between CO_2 and amine groups slowing diffusion [23].

The sorption coefficient of the gas within PIM-EA(H_2)-TB increases with its critical temperature (i.e., its condensability) as is usually observed for polymers (Figure 5b).

3.3. *Effect of Pressure*

The permeability coefficients of each gas were measured as a function of upstream feed pressure. The measurements were carried out with H_2, CO_2, CH_4 and N_2 at 30 °C and pressures up to 20 bar (10 bar for H_2) (Figure 6).

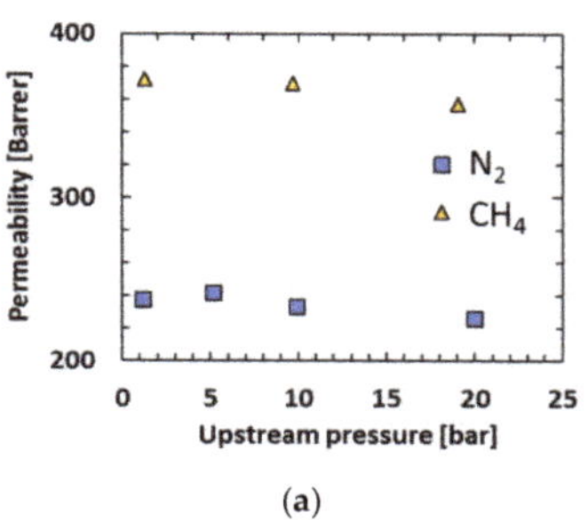

(a)

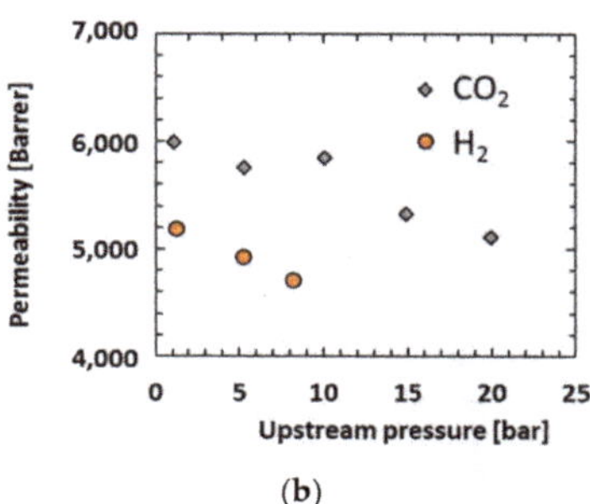

(b)

Figure 6. Permeability coefficients of PIM-EA(H_2)-TB for CH_4, N_2(**a**) and H_2, CO_2 (**b**) at 30 °C.

The permeability of nitrogen is constant with increasing pressure while CO_2 and CH_4 permeabilities decrease with increasing pressure, which is classical behaviour for glassy polymers [24] and is due to the filling of Langmuir sorption sites. At higher pressures, the contribution of the Langmuir region to the overall permeability is weaker and gas permeability approaches a constant value associated with simple dissolution (Henry's law) transport. In contrast to the majority of glassy polymers, PIM-EA(H_2)-TB does not exhibit the typical increase in CO_2 permeability associated with "plasticization" in the high pressure range for CO_2. A similar behaviour has been also noted for other polymers of intrinsic microporosity, such as PIM-1 or PIM-EA(Me)-TB [17,24,25]. However, the decrease in H_2 permeability is higher than expected [25].

Despite the decrease of permeability coefficients, the ideal selectivities of PIM-EA(H_2)-TB stay constant with the increase of the feed pressure (Table 4). However, it should be noted that ideal selectivity is usually not representative of behaviour at high pressure in mixed gas systems due to the interactions between different gases.

Table 4. Selectivity of PIM-EA(H_2)-TB for CH_4, N_2, H_2, CO_2 at 30 °C for different pressures.

Selectivity, 30 °C	H_2/CO_2	H_2/N_2	H_2/CH_4	CO_2/N_2	CO_2/CH_4	CH_4/N_2
1 bar	1	22	14	25	16	2
5 bar	1	20	-	24	-	-
10 bar	1	20	14	25	16	2
20 bar	-	-	-	23	14	2

3.4. *Effect of Temperature*

The temperature effect on gas permeability through PIM-EA(H_2)-TB was studied over a temperature range of 30–200 °C for pure gas at different pressures. The values of the permeability coefficients are summarised in the Table S1. Figure 7 shows the permeability coefficient of N_2, CO_2, H_2 and CH_4 as a function of the inverse absolute temperature at 1 bar.

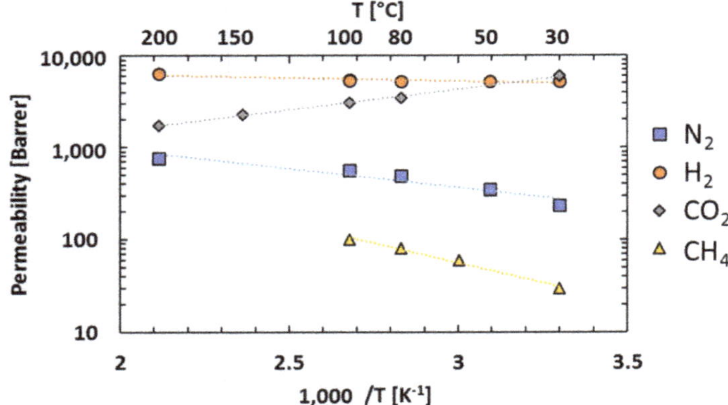

Figure 7. Permeability coefficients of N_2, CO_2, H_2 and CH_4 as a function of the inverse absolute temperature (at 1 bar) (The dotted lines represent the best curve-fits of the experimental data with Arrhenius equation).

The permeability coefficient of N_2, CH_4 and H_2 increases with increasing temperature while for CO_2 it decreases with increasing temperature. In order to explore the temperature dependence of the gas permeability, the data were correlated with the Arrhenius equation.

$$P = P_0 exp\left(-\frac{E_p}{RT}\right) \tag{4}$$

where P_0 is the pre-exponential factor ((cm^3(STP)·cm)/(cm^2·s·cmHg)), E_p is the activation energy of permeation (J/mol), T is the temperature (K) and R is the ideal gas constant (8.314 kJ/(mol·K)). E_p for the transport of each gas through PIM-EA(H_2)-TB were determined from the slopes ($-E_P/R$) of the best curve-fits through the permeation data in Figure 7. The E_p values at 1 bar are summarized in Table 5.

Table 5. Activation energy of gas permeation for PIM-EA(H_2)-TB, PIM-1, PIM-TMN-Trip and PTMSP.

Gas	E_P (kJ/mol)			
	PIM-EA(H_2)-TB (This Study)	**PIM-1** [26]	**PIM-TMN-Trip** [12]	**PTMSP** [26]
H_2	0.5	−0.4	−2.8	−2.1
N_2	8.6	14.3	4.4	−3.5
CH_4	13.1	19.4	9.5	−5.3
CO_2	−8.6	−1	−7.7	−11.7

PIM-EA(H_2)-TB presents high values for the activation energy of permeation for N_2 and CH_4, which means that the permeability coefficients depend strongly on the temperature. On the contrary, for the smaller gases, such as H_2, E_P is close to zero as the dependence on temperature is much weaker. For CO_2, the activation energy of permeation is negative. This behaviour is routinely observed for microporous solids, such as PIM-1, PIM-TMN-Trip and PTMSP, in which the pore dimensions are relatively large in comparison with the diffusing gas molecules [11].

Since the gas transport in a microporous membrane is based on a solution-diffusion mechanism, the impact of temperature on the permeation can be better understood when looking at diffusion and solubility individually. The activation energy of permeation can be represented as the sum of the activation energies of diffusion, E_D and sorption ΔH_s. Table 6 lists the activation energies of gas permeation and diffusion as well as the enthalpy of sorption of all the gases in PIM-EA(H_2)-TB. For all the gases at 1 bar, the activation energy of diffusion, E_D, is positive, which means that the diffusivity increases with the temperature, which is expected as the main effect of increasing the temperature is an increase of molecular vibrations which facilitates diffusion. In contrary, the sorption enthalpy, ΔH_s, is negative as expected since the sorption is an exothermic process.

Table 6. Activation energies for gas permeation (E_p), for diffusion (E_d) and for sorption (ΔH_s) of PIM-EA(H_2)-TB for N_2, CO_2, H_2 and CH_4 at 1 bar.

1 bar	E_P (kJ/mol)	E_D (kJ/mol)	ΔH_s (kJ/mol)
CO_2	−8.6	8.1	−16.7
N_2	8.6	18.5	−9.9
H_2	0.5	5.2	−4.6
CH_4	13.1	17.9	−4.8

For CH_4, N_2 and H_2, the absolute value of E_D is greater than ΔH_s and so the energy of activation E_p is positive, which means that diffusion rather than sorption dominates the response of permeation to temperature. For CO_2, the absolute value of E_D is smaller than ΔH_s, which induces a negative activation energy E_P. The CO_2 transport is mainly influenced by the gas solubility, which is characteristic of microporous polymer, with similar results being found for PIM-1 and PTMSP [5,8,11].

Based upon these effects, the increase of temperature improves H_2/CO_2 selectivity modestly moving the data for PIM-EA(H_2)-TB close to the 200 °C upper bound (Figure 8, however, even its enhanced high temperature selectivity (~2) is insufficient for viable pre-combustion application. In contrast, the selectivity for CO_2 or H_2 over N_2 or CH_4 decreases dramatically at higher temperatures suggesting that optimal performance is obtained at lower temperatures (Figure 9).

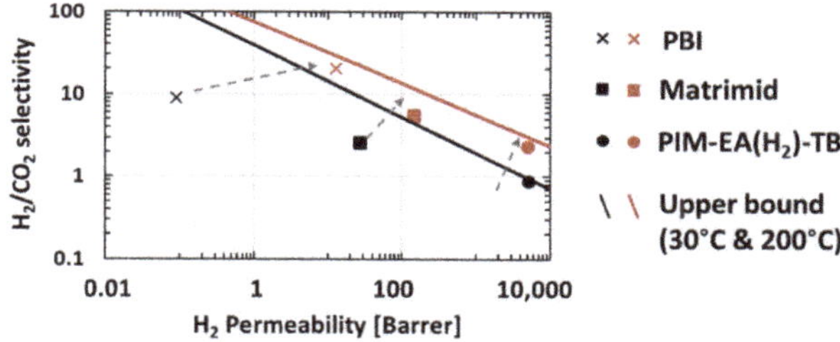

Figure 8. H_2/CO_2 separation performances of PBI (Cross), Matrimid (Square) and PIM-EA(H_2)-TB (circle) at 1 bar/30 °C (Black dot) and at 10 bar/200 °C (Red dot). Upper bound at 200 °C recalculated from [7].

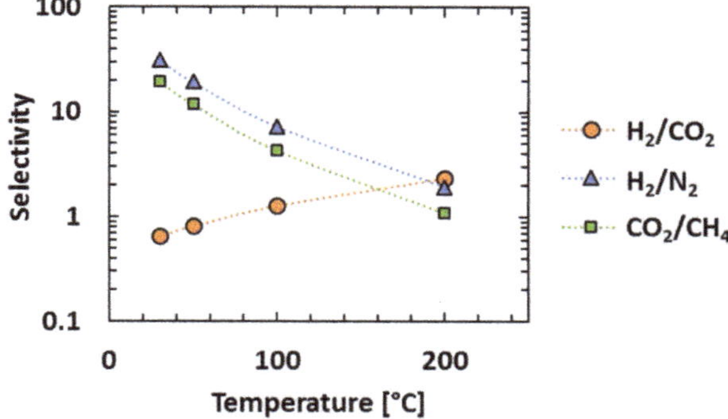

Figure 9. Selectivity of PIM-EA(H_2)-TB as the function of temperature at 10 bar.

4. Conclusions

Transport properties of permeability, diffusivity and solubility of PIM-EA(H_2)-TB have been determined for H_2, N_2, CH_4 and CO_2 over a range of pressures and temperatures. This PIM presents high CO_2 and H_2 permeability coefficients, which allows it to have good ideal selectivity over N_2. PIM-EA(H_2)-TB exhibits the classical behaviour of a glassy polymer, with the decrease of diffusivity coefficient with increasing penetrant molecular size and the increase of sorption coefficient gas with increasing condensability of the permeant. However, no increase in CO_2 permeability due to plasticization is noted over the range of pressure tested. The permeability coefficient of N_2, CH_4 and H_2 increase with increasing temperature while for CO_2 the permeability decreases with increasing temperature, which is classically observed for microporous materials. Therefore, the separation performance of PIM-EA(H_2)-TB for H_2/CO_2 is reversed at high temperature and maintained also at high pressure. This suggests that, after further development to enhance absolute selectivity of H_2 over CO_2, PIMs could become good candidates for membrane materials for use in pre-combustion CO_2 capture. For other gas separations, better performance is obtained at lower temperatures.

Supplementary Materials: The following are available online at http://www.mdpi.com/2077-0375/8/4/132/s1, Table S1: Gas permeability coefficients of N_2, CO_2, H_2 and CH_4 for temperatures between 30 °C and 200 °C and pressure between 1 bar and 20 bar.

Author Contributions: Funding acquisition, N.B.M. and M.-C.F.; Investigation, E.L., R.M.-E. and M.C.; Resources, R.M.-E. and M.C.; Supervision, M.-C.F.; Writing—original draft, E.L.; Writing—review & editing, R.M.-E., M.C., N.B.M. and M.-C.F.

Funding: This work was financially supported by Programme Grant EP/M01486X/1 (SynFabFun) and Grant EP/R000468/1 funded by the Engineering and Physical Sciences Research Council (EPSRC).

Conflicts of Interest: The authors declare no conflict of interest.

References

1. Merkel, T.; Zhou, M.; Baker, R. Carbon dioxide capture with membranes at an IGCC power plant. *J. Membr. Sci.* **2012**, *389*, 441–450. [CrossRef]
2. Franz, J.; Scherer, V. An evaluation of CO_2 and H_2 selective polymeric membranes for CO_2 separation in IGCC processes. *J. Membr. Sci.* **2010**, *359*, 173–183. [CrossRef]
3. Singh, R.; Berchtold, K. H_2 Selective Membranes for Precombustion Carbon Capture. In *Novel Materials for Carbon Dioxide Mitigation Technology*; Elsevier: Amsterdam, The Netherlands, 2015; pp. 177–206.
4. Abanades, J.; Arias, B.; Lyngfelt, A.; Mattisson, T.; Wiley, D.; Li, H.; Hoc, M.; Mangano, E.; Brandani, S. Emerging CO_2 capture systems. *Int. J. Greenh. Gas Control* **2015**, *40*, 126–166. [CrossRef]
5. Li, P.; Chung, T.; Paul, D. Temperature dependence of gas sorption and permeation in PIM-1. *J. Membr. Sci.* **2014**, *450*, 380–388. [CrossRef]
6. Merkel, T.; Gupta, R.; Turk, B.; Freeman, B. Mixed gas permeation of syngas components in poly(dimethylsiloxane) and poly(1-trimethysilyl-1-propyne) at elevated temperatures. *J. Membr. Sci.* **2001**, *191*, 85–94. [CrossRef]
7. Rowe, B.W.; Robeson, L.M.; Freeman, B.D.; Paul, D.R. Influence of temperature on the upper bound: Theoretical considerations and comparison with experimental results. *J. Membr. Sci.* **2010**, *360*, 58–69. [CrossRef]
8. Budd, P.M.; McKeown, N.B.; Ghanem, B.S.; Msayib, K.J.; Fritsch, D.; Starannikova, L.; Belov, N.; Sanfirova, O.; Yampolskii, Y.; Shantarovich, V. Gas permeation parameters and other physicochemical properties of a polymer of intrinsic microporosity: Polybenzodioxane PIM-1. *J. Membr. Sci.* **2008**, *325*, 851–860. [CrossRef]
9. David, O.C.; Gorri, D.; Urtiaga, A.; Ortiz, I. Mixed gas separation study for the hydrogen recovery from $H_2/CO/N_2/CO_2$ post combustion mixtures using a Matrimid membrane. *J. Membr. Sci.* **2011**, *378*, 359–368. [CrossRef]
10. Pesiri, D.R.; Jorgensen, B.; Dye, R.C. Thermal optimization of polybenzimidazole meniscus membranes for the separation of hydrogen, methane, and carbon dioxide. *J. Membr. Sci.* **2003**, *218*, 11–18. [CrossRef]

11. Pinnau, I.; Toy, L.G. Gas and vapor transport properties of amorphous perfluorinated copolymer membranes based on 2,2-bistrifluoromethyl-4,5-difluoro-1,3-dioxole/tetrafluoroethylene. *J. Membr. Sci.* **1996**, *109*, 125–133. [CrossRef]
12. Fuoco, A.; Comesaña-Gándara, B.; Longo, M.; Esposito, E.; Monteleone, M.; Rose, I.; Bezzu, C.; Carta, M.; McKeown, N.; Jansen, J. Temperature Dependence of Gas Permeation and Diffusion in Triptycene-Based Ultrapermeable Polymers of Intrinsic Microporosity. *ACS Appl. Mater. Interfaces* **2018**, *10*, 36475–36482. [CrossRef] [PubMed]
13. Carta, M.; Malpass-Evans, R.; Croad, M.; Rogan, Y.; Jansen, J.C.; Bernardo, P.; Bazzarelli, F.; McKeown, N.B. An Efficient Polymer Molecular Sieve for Membrane Gas Separations. *Science* **2013**, *339*, 303–307. [CrossRef] [PubMed]
14. Hernandez, N.; Iniesta, J.; Leguey, V.M.; Armstrong, R.; Taylor, S.; Madrid, E.; Rong, Y.; Castaing, R.; Malpass-Evans, R.; Carta, M.; et al. Carbonization of polymers of intrinsic porosity to microporous heterocarbon: Capacitive pH measurements. *Appl. Mater. Today* **2017**, *9*, 136–144. [CrossRef]
15. Bernardo, P.; Scorzafave, V.; Clarizia, G.; Tocci, E.; Jansen, J.; Borgogno, A.; Malpass-Evans, R.; McKeown, N.; Carta, M.; Tasselli, F. Thin film composite membranes based on a polymer of intrinsic microporosity derived from Tröger's base: A combined experimental and computational investigation of the role of residual casting solvent. *J. Membr. Sci.* **2019**, *569*, 17–31. [CrossRef]
16. Carta, M.; Croad, M.; Malpass-Evans, R.; Jansen, J.C.; Bernardo, P.; Clarizia, G.; Friess, K.; Lanč, M.; McKeown, N. Triptycene Induced Enhancement of Membrane Gas Selectivity for Microporous Tröger's Base Polymers. *Adv. Mater.* **2014**, *26*, 3526–3531. [CrossRef] [PubMed]
17. Tocci, E.; Lorenzo, L.D.; Bernardo, P.; Clarizia, G.; Bazzarelli, F.; Mckeown, N.B.; Carta, M.; Malpass-Evans, R.; Friess, K.; Pilnáček, K.; et al. Molecular Modeling and Gas Permeation Properties of a Polymer of Intrinsic Microporosity Composed of Ethanoanthracene and Tröger's Base Units. *Macromolecules* **2014**, *47*, 7900–7916. [CrossRef]
18. Lau, C.; Konstas, K.; Doherty, C.; Carta, M.; Lasseuguette, E.; Ferrari, M.; McKeown, N.; Hill, M. Tailoring compatibility to deplasticize ageless, ultrapermeable blends for molecular sieving. In Proceedings of the ICOM 2017, San Francisco, CA, USA, 29 July–4 August 2017.
19. Benito, J.; Sánchez-Laínez, J.; Zornoza, B.; Martín, S.; Carta, M.; Malpass-Evans, R.; Téllez, C.; McKeown, N.B.; Coronas, J.; Gascón, I. Ultrathin Composite Polymeric Membranes for CO_2/N_2 separation with Minimum Thickness and High CO_2 permeance. *ChemSusChem* **2017**, *10*, 4014–4017. [CrossRef]
20. Baker, R. *Membrane and Technology and Applications*; John Wiley & Sons: London, UK, 2004.
21. Robeson, L. The upper bound revisited. *J. Membr. Sci.* **2008**, *320*, 390–400. [CrossRef]
22. Mason, C.R.; Maynard-Atem, L.; Heard, K.W.J.; Satilmis, B.; Budd, P.M.; Friess, K.; Lanč, M.; Bernardo, P.; Clarizia, G.; Jansen, J.C. Polymer of Intrinsic Microporosity Incorporating Thioamide Functionality: Preparation and Gas Transport Properties. *Macromolecules* **2014**, *47*, 1021–1029. [CrossRef]
23. Ghosal, K.; Chern, R.T.; Freeman, B.D.; Daly, W.H.; Negulescu, I.I. Effect of Basic Substituents on Gas Sorption and Permeation in Polysulfone. *Macromolecules* **1996**, *29*, 4360–4369. [CrossRef]
24. Li, P.; Chung, T.; Paul, D. Gas sorption and permeation in PIM-1. *J. Membr. Sci.* **2013**, *432*, 50–57. [CrossRef]
25. Swaidan, R.; Ghanem, B.; Litwiller, E.; Pinnau, I. Physical Aging, Plasticization and Their Effects on Gas Permeation in "Rigid" Polymers of Intrinsic Microporosity. *Macromolecules* **2015**, *48*, 6553–6561. [CrossRef]
26. Thomas, S.; Pinnau, I.; Du, N.; Guiver, M.D. Pure- and mixed-gas permeation properties of a microporous spirobisindane-based ladder polymer (PIM-1). *J. Membr. Sci.* **2009**, *333*, 125–131. [CrossRef]

Article

Poly(1-trimethylsilyl-1-propyne)-Based Hybrid Membranes: Effects of Various Nanofillers and Feed Gas Humidity on CO_2 Permeation

Zhongde Dai, Vilde Løining, Jing Deng, Luca Ansaloni and Liyuan Deng *

Department of Chemical Engineering, Norwegian University of Science and Technology, 7491 Trondheim, Norway; zhongde.dai@ntnu.no (Z.D.); vilde.sl@gmail.com (V.L.); jing.deng@ntnu.no (J.D.); luca.ansaloni@ntnu.no (L.A.)

* Correspondence: liyuan.deng@ntnu.no

Received: 13 August 2018; Accepted: 24 August 2018; Published: 5 September 2018

Abstract: Poly(1-trimethylsilyl-1-propyne) (PTMSP) is a high free volume polymer with exceptionally high gas permeation rate but the serious aging problem and low selectivity have limited its application as CO_2 separation membrane material. Incorporating inorganic nanoparticles in polymeric membranes has been a common approach to improve the separation performance of membranes, which has also been used in PTMSP based membrane but mostly with respect to tackling the aging issues. Aiming at increasing the CO_2 selectivity, in this work, hybrid membranes containing four types of selected nanofillers (from 0 to 3D) were fabricated using PTMSP as the polymer matrix. The effects of the various types of nanofillers on the CO_2 separation performance of the resultant membranes were systematically investigated in humid conditions. The thermal, chemical and morphologic properties of the hybrid membranes were characterized using TGA, FTIR and SEM. The gas permeation properties of the hybrid membranes were evaluated using mixed gas permeation test with the presence of water vapour to simulate the flue gas conditions. Experiments show that the addition of different fillers results in significantly different separation performances; The addition of ZIF-L porous 2D filler improves the CO_2/N_2 selectivity at the expenses of CO_2 permeability, while the addition of TiO_2, ZIF-7 and ZIF-8 increases the CO_2 permeability but the CO_2/N_2 selectivity decreases.

Keywords: CO_2 separation; hybrid membranes; high free volume polymers; PTMSP; ZIF

1. Introduction

Since the late 1960s, membrane-based separation processes have gradually been recognized as feasible alternatives to conventional purification and separation processes [1]. Compared to conventional separation technologies, membrane separation has advantages including small footprint, high energy-efficiency as well as safe, environmentally friendly and easy operation. Membranes have been applied in a wide range of gas separation applications, such as natural sweetening (acid gas removal), biogas upgrading, hydrogen recovery, N_2 production and organic vapour recovery [2]. Nowadays, using membranes for CO_2 capture from flue gas (post-combustion) or syngas (pre-combustion) has attracted much attention [3]. Various types of membrane materials and membrane processes have been developed for CO_2 capture, including polymeric membranes, inorganic membranes and hybrid membranes. Among them, polymeric membranes are the most commonly used and intensively studied owing to its good processability and relatively low cost.

Polymer membranes usually separate gases by preferential permeation under a pressure gradient following the solution-diffusion mechanism. However, fast permeation through a polymeric membrane is usually achieved by compromising selectivity: The permeability and selectivity are subjected to a trade-off as being described by the Robeson upper bound [4]. Developing new membranes with both high permeability and selectivity has thus been a major research goal. Attempts to overcome the performance limitation are generally based on designing and controlling the materials chemistry or nanostructure of the membranes to create more preferential sorption and/or diffusion of the target penetrants. One promising strategy is to synthesize multicomponent hybrid membranes, in which a secondary component (usually inorganic nanoparticles) is introduced and evenly embedded into the polymer matrix. The additives can be porous or non-porous nanoparticles (0 to 3D) [5] and/or liquid plasticizers [6–8]. In the past few years, various nano- or micro-porous particles have been applied in fabricating hybrid membranes, including zeolite, carbon molecular sieve, different metal organic frameworks (MOFs) and covalent organic frameworks (COFs). Non-porous particles include metal oxides, silica, carbon nanotubes and graphenes. Graphene oxide are also used to fabricate hybrid membranes [5,9]. Generally, due to the non-ideal interface between the additives and the polymer phase, adding nanofillers into polymeric matrix of a membrane will result in significant improvement in gas permeability but relatively moderate enhancement in gas selectivity [10].

PTMSP is one of the representative high free volume polymer possessing the highest CO_2 permeability among dense polymeric membranes. The high gas permeability is attributed to the exceptionally high free volume of the polymer [11]. The large and interconnected domains within the polymer matrix allows a rapid diffusion of gas, leading to the high gas permeability across the membrane [11]. The CO_2 permeability is reported to be in the range of 18,000–5200 bar [12–14]. However, the CO_2/N_2 selectivity remains relatively low (<5) [12]. According to process simulation, for membranes with a sufficient selectivity, high gas permeability is a more important property to consider from the economic point of view [15]. The poor selectivity has limited its industrial application. Increasing the selective features of this material is thus of primary importance to boost its practical use in CO_2 separation. Adding inorganic nanofillers into polymeric matrix has been proven to be an effective way of improving the gas separation properties [16,17].

Zeolitic imidazolate frameworks (ZIFs) are a family of the metal organic frameworks (MOFs) usually with zinc or cobalt as metal nodes and imidazolate (or imidazolate derivative) as organic linkers [18]. ZIFs have been extensively used in hybrid membrane fabrications as porous nanoparticles. Among the series of ZIFs, ZIF-8 is one of the most widely studied materials with various applications in gas storage, catalysis and gas separation [19]. Many researchers have reported that adding ZIF-8 into polymeric membranes could not only improve the gas permeability but also enhance selectivity [20–22]. ZIF-7 is also well studied as nanofillers in different polymeric membranes. It is found that ZIF-7 nanoparticles could effectively improve both the CO_2 permeability and CO_2 selectivity over other gases (e.g., N_2 and CH_4) [23–25]. ZIF-L, a two-dimensional zeolitic imidazolate framework, is relatively new, which was firstly developed by Wang et al in 2013 [26,27]. Compared to the conventional ZIF particles, ZIF-L has a distinctive leaf-shaped morphology and cushion-shaped cavity with a dimension of 9.4 Å × 7.0 Å × 5.3 Å between layers, which is well suited to accommodate CO_2 molecules [28]. In addition, the intrinsic high-aspect-ratio of the 2D particles is preferable for improving gas selectivity [29]. It has been used to fabricate hybrid membranes with different polymers for gas separation and pervaporation [30]. However, the comparison of the effect of different ZIFs on the CO_2 transport properties of a given polymeric matrix, especially the high gas permeable matrix, is still not completely disclosed.

Non-porous nanofillers have also been reported to be able to significantly change gas separation performances, typically due to specific affinity with a target molecule (transport via selective surface diffusion phenomena) or interactions with polymeric chain packing, leading to higher free volume within the polymer matrix [31]. Various non-porous additives has been used to fabricate hybrid membranes for gas separation purpose, such as TiO_2, fumed silica and other metal oxides [5]. Due to the characteristic properties such as inexpensive, high hydrophilicity as well as good chemical and thermal stabilities [32,33], TiO_2 has been included as the non-porous nanofiller for comparison with the ZIFs in this study.

Furthermore, as in real gas streams flue gas is always saturated with water vapor and in many cases water vapor is found to strongly influence the properties of the membranes, it is critical to investigate the effects of humidity in the feed gas on the separation performance of the membranes. To the best of our knowledge, the influence of water vapor in separation system on the permeation properties of PTMSP-based hybrid membranes has never be reported.

In the present work, four different nanoparticles were employed as additives to prepare PTMSP-based hybrid membranes, including three porous ZIF nanoparticles (ZIF-8, ZIF-7 and ZIF-L) and one non-porous nanofiller (TiO_2). The resultant hybrid membranes were systematically investigated using various techniques, including TGA, FTIR and SEM and mixed gas permeation tests with controlled humidity (0–100% relative humidity). The separation performance data are analyzed and the influence of the various fillers on the gas transport mechanism through the hybrid membranes is discussed.

2. Experimental

2.1. Materials

PTMSP was supplied by Fluorochem, Morrisville, PA, USA (chemical structure shown in Figure 1). ZIF-8 nanoparticles (trade name of Basolite® Z1200, solid state, Sigma-Aldrich, Steinheim, Germany), chloroform and n-haxane were purchased from Sigma-Aldrich, Steinheim, Germany. ZIF-7 and ZIF-L were synthesized following the procedure described by Zhong et al. [26] and Li et al. [23], respectively. The structure and common preparation route for ZIF-8, ZIF-L and ZIF-7 are given in Figure 2. TiO_2 nanoparticles were kindly supplied by SINTEF Industry, Oslo, Norway. Before use, all the particles were dried in a vacuum oven at 60 °C overnight to ensure the complete removal of moisture. The mixed gases used for permeation testing contains 90 vol% N_2 and 10 vol% CO_2. 99.999 vol% CH_4 was used as sweep gas. All the gases were purchased from AGA, Trondheim, Norway.

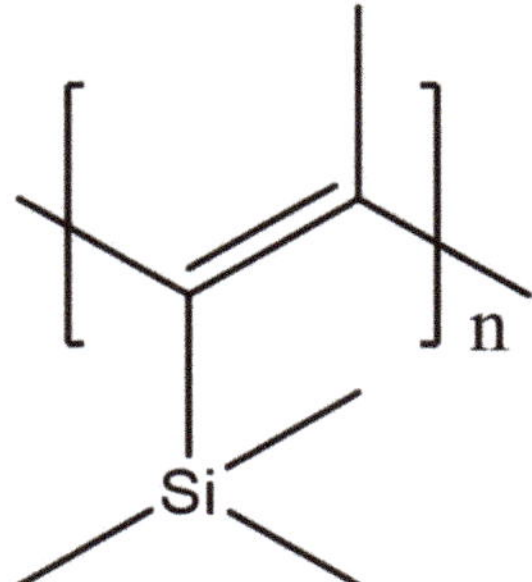

Figure 1. Chemical structure of poly(1-trimethylsilyl-1-propyne) (PTMSP).

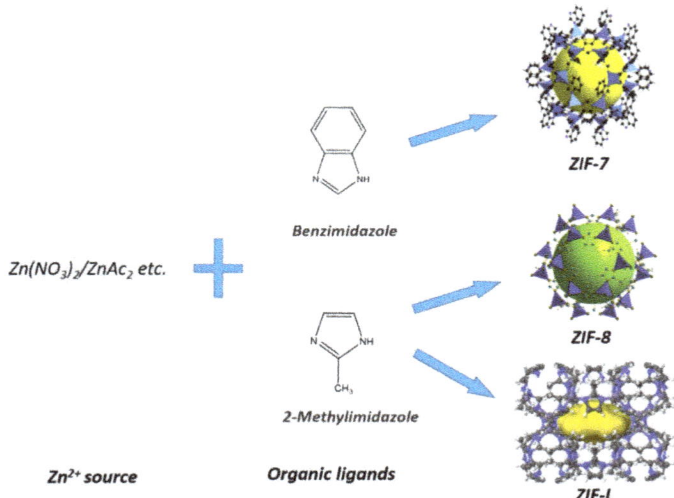

Figure 2. Structure and preparation route of ZIF-7, ZIF-8 and ZIF-L [26].

2.2. Membrane Preparation

Self-standing PTMSP membranes of approximately 50–100 μm were fabricated by casting the polymer/additive mixture on a glass plate using a casting-knife. The polymer/additive blend were prepared as described by Zhang et al. [34]. In brief, solution A (nanoparticles in either chloroform or n-haxane) was sonicated for 4–5 min using a Sonics Vibra-CellTM VC 70T (Leuven, Belgium) and then mixed into solution B (a highly viscous solution of PTMSP in a small amount of solvent). The mixture of A and B was then sonicated again for 4 × 45 min before the solution was cast onto a glass plate. The solution on the glass plate was covered by a glass container to slow down the solvent evaporation.

The content of nanofillers added into the membrane matrix was determined using Equation (1).

$$\Omega_{ad} = \frac{W_{ad}}{W_{ad} + W_{PTMSP}} * 100 \quad (1)$$

where W_{ad} and W_{PTMSP} are the mass of additives and PTMSP polymer, respectively.

2.3. Membrane Characterization

A TG F1 Libra (NETZSCH, Selb, Germany) was used to perform a thermo-gravimetric analysis (TGA) of the hybrid membranes. Around 10 mg of sample were used to perform the test. The samples were analysed in the temperature range from room temperature (RT) to 700 °C at a constant heating rate of 10 °C min^{-1} with N_2 atmosphere.

Fourier Transform Infrared Spectroscopy (FT-IR, NicoletTM iSTM 50, Thermo Fisher, Oslo, Norway) was used to determine the chemical structures of all the components in the hybrid membranes. The wavelengths employed were in the range of 650–4000 cm^{-1} and the given spectra is an average of 16 scans.

Scanning electron microscope (SEM, TM3030Plus, Hitachi, Espoo, Finland) was used to investigate the morphology of the membranes. Both cross section and surface samples were sputter coated (Q150R Rotary-Pumped Sputter Coater/Carbon Coater, Quorum, Laughton, East Sussex, UK) with a thin gold layer to increase the sample conductivity. The cross-section sample was prepared by soaking the membrane sample into liquid N_2.

Gas-separation performance was measured in a humid mixed-gas permeation setup with adjustable RH, as schematically depicted in Figure 3. More details about the testing procedure can be found in our previous reports [35,36]. The RH value is controlled by using four mass flow controllers (El-Flow series, Bronkhorst, Veenendaal, Netherland). A CO_2/N_2 gas mixture (10/90 v/v) constituted the feed gas. CH_4 was used as sweep gas instead of an inert gas (e.g., Helium) since Helium was used as the carrier gas for the gas chromatograph (GC). The feed pressure was set as 2.0 bar while the sweep side pressure was set to 1.05 bar. The composition of permeate stream was monitored by a calibrated gas chromatograph (490 Micro GC, Agilent, Santa Clara, CA, USA). For each permeation test, experiment lasted at least for 6 h until a steady state was reached.

The permeability coefficient (P_i) of the ith penetrant species can be obtained by Equation (2):

$$P_{m,i} = \frac{N_{perm}(1 - y_{H_2O})y_i l}{A\left(\left\langle p_{i,feed}, p_{i,ret}\right\rangle - p_{i,perm}\right)} \quad (2)$$

where N_{perm} is the total permeate flow measured with a bubble flow meter, y_{H_2O} is the molar fraction of water in the permeate flow (calculated according to the RH value and the vapour pressure at the given temperature), y_i is the molar fraction of the species of interest in the permeate and $p_{i,feed}$, $p_{i,ret}$ and $p_{i,perm}$ identify the partial pressures of the ith species in the feed, retentate and permeate, respectively. It is worth mentioning that all the tests were repeated at least two times with the error lower than 5%, as such the error bars are not visible in the figures, thus not presented in the figures.

The separation factor is calculated by Equation (3):

$$\alpha_{i/j} = \frac{y_i/x_i}{y_j/x_j} \quad (3)$$

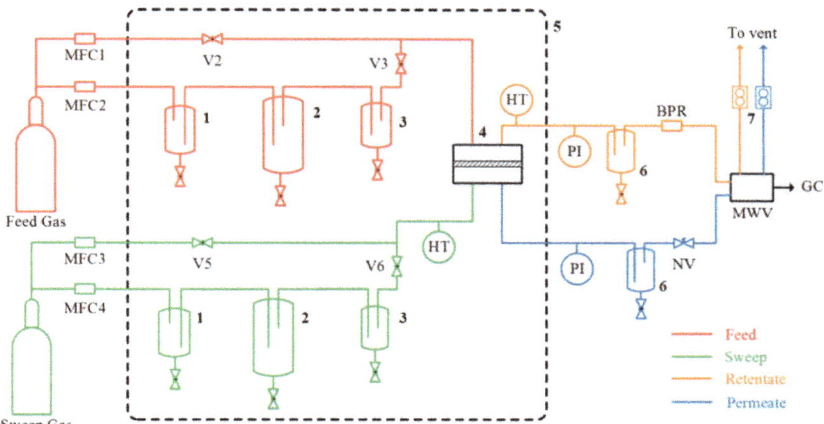

Figure 3. Mixed-gas permeation setup (**1**: MFC-safety trap; **2**: Humidifier; **3**: Droplets trap; **4**: Membrane module; **5**: Heated cabinet; **6**: Water knockout; **7**: Bubble flow meters; **MFC**: Mass flow controller; **NV**: Needle valve; **BPR**: Back-pressure regulator; **PI**: Pressure indicator; **HT**: Humidity and temperature sensor; **MWV**: Multi-way valve; **GC**: Gas chromatograph).

3. Results and Discussion

3.1. Nanofiller Characterization

The morphology of the four nanofillers used in the present study were investigated using SEM, results are shown in Figure 4.

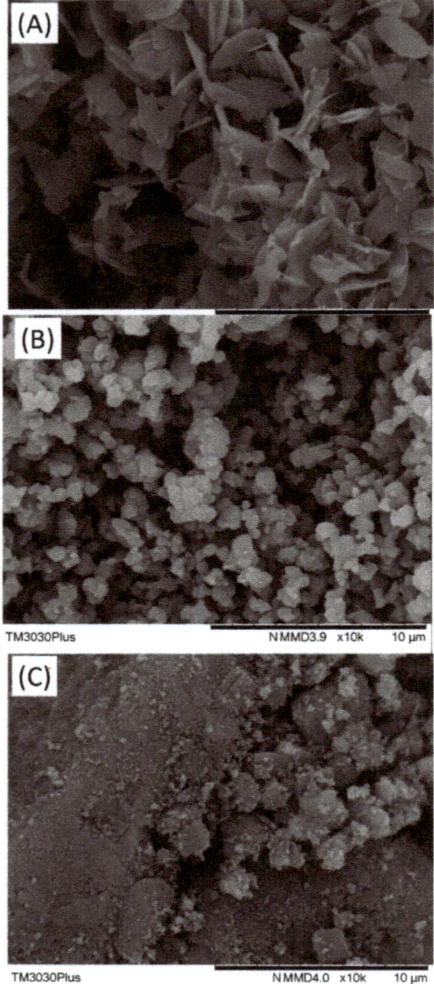

Figure 4. SEM image of the ZIF series used in the present study, (**A**) ZIF-L, (**B**)ZIF-8, and (**C**) ZIF-7.

As it can be seen in Figure 4, ZIF-L presents a leaf-sharp with a width of ~1.8 µm and length of ~4 µm, which is in good agreement with literature results [26]. The SEM image for ZIF-8 particles show obvious aggregation but overall particle size is still in the micrometre range. ZIF-7 was prepared based on the procedure reported in Ref. [23]. The SEM image of ZIF-7 also shows serious aggregation but the individual particles exhibit similar structure as literature [23]. It is worth mentioning that nanoparticles always tend to aggregate in a dry state but in this work the aggregations have been partly broken down to smaller particles via proper dispersion methods (e.g., stirring, ultra-sonication) and using proper solvent.

3.2. Thermal Properties

Thermal stability of membranes is a property that is highly valued in membrane separation. TGA was used to study the influence of the nanofillers on the thermal stability of the hybrid membranes. The TGA results for PTMSP, ZIF-8, ZIF-L, ZIF-7 and TiO_2 and relevant hybrid membranes are shown in Figure 5.

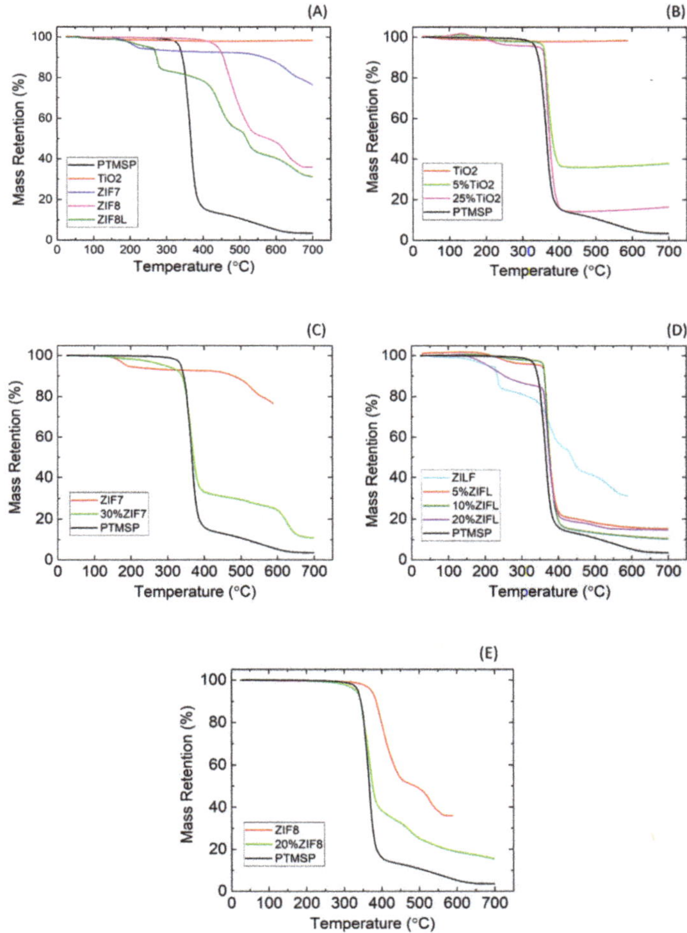

Figure 5. TGA results of neat PTMSP polymer and different nanoparticles (**A**), TGA curves of PTMSP/TiO_2 (**B**), PTMSP/ZIF-7 (**C**), PTMSP/ZIF-L (**D**), and PTMSP/ZIF-8 (**E**).

Among all the tested materials, the TiO_2 nanoparticles present the best thermal stability. The weight-loss of TiO_2 is less than 2 wt % of the original weight at 700 °C, showing that TiO_2 is extremely stable in the given temperature range. A T_{onset} of 439 °C was obtained for the ZIF-8, which is in good agreement with the literature value [21]. At temperatures higher than the T_{onset}, a steep reduction in weight is obtained, which is corresponding to the collapse of the ZIF-8 structure and carbonization under extreme thermal stress. A 66% weight-loss can be observed at 700 °C, which is also consistent with literature data [37]. Considering the ZIF-L, a weight loss of approximately 20% can be found at around 200 °C, which is due to the mass loss arising from the removal of the weakly linked ½ Hmim and guest water molecules [27]. Further increasing the temperature will result in decomposition of ZIF-L into zinc oxide [27]. In the case of ZIF-7 nanoparticles, a slight weight loss was observed at around 177 °C, which was a result of the evaporation of the DMF solvents (b.p. = 153 °C) from the particle cages [38]. Afterwards the sample mass kept stable until around 500 °C, where a second stage of weight loss was observed due to the decomposition of ZIF-7 into zinc oxide [38].

Figure 5B–E presents the TGA curves of the hybrid membranes with the four different types of nanofillers. In the case of TiO_2 and ZIF-8, due to the superior thermal stability of the two nanofillers, adding them into PTMSP resulted in a hybrid membrane with comparable thermal stability to neat PTMSP but with higher residual mass at high temperatures. In the case of the ZIF-7 and ZIF-L, as these two nanofillers has relatively lower $T_{on\text{-}set}$ compared to the PTMSP polymer, adding them into the polymeric phase slightly reduced the overall thermal stability of the hybrid membranes. However, no weight loss can be found below 150 °C, which fulfils the requirements for post-combustion CO_2 capture process.

3.3. Membrane Morphology

The membrane morphology is of critical importance for hybrid membranes. Depending on the affinity between the polymer and the fillers, hybrid membranes may be characterized by different polymer/particle interface and the transport properties of hybrid membranes are strongly dependent on the nanoscale morphology of the membranes [16]. Figure 6 display the cross section of the different membranes fabricated in the present study.

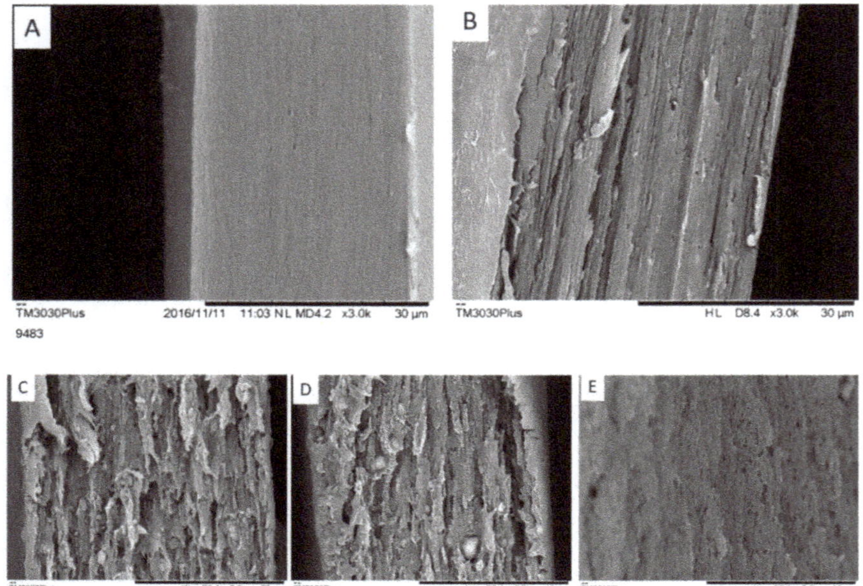

Figure 6. SEM images of membranes. (**A**) PTMSP, (**B**) PTMSP/30 wt % ZIF-7, (**C**) PTMSP/20 wt % ZIF-8, (**D**) PTMSP/20 wt % ZIF-L, and (**E**) PTMSP/20 wt % TiO_2.

Figure 6A shows the SEM results for the neat PTMSP membrane, where a homogeneous morphology can be observed, as expected for a pristine polymeric phase. In the case of the PTMSP/ZIF-7 sample (Figure 6B), no obvious aggregation can be found from the membrane surface. However, from the cross-section image, the ZIF-7 nanoparticles seem to aggregate to one side of the membrane. In the case of the membrane with ZIF-8 and ZIF-L dispersed (Figure 6C,D), aggregation can be found in both membranes. The compatibility between the nanofillers and PTMSP seems not sufficient. In the TiO_2 case, aggregates can be found on the membrane surface but no obvious aggregates are found from the cross-section images. Furthermore, different from other nanofillers, the TiO_2 seems created additional voids or gas diffusion paths in the PTMSP membrane, which is expected to be beneficial for gas permeability but may be to an extent negative to the selectivity.

3.4. FTIR

FTIR was used to study the chemical structure of the nanofiller, PTMSP as well as the hybrid membranes. Results are shown in Figure 7. Detailed peak assignments are listed in Table 1.

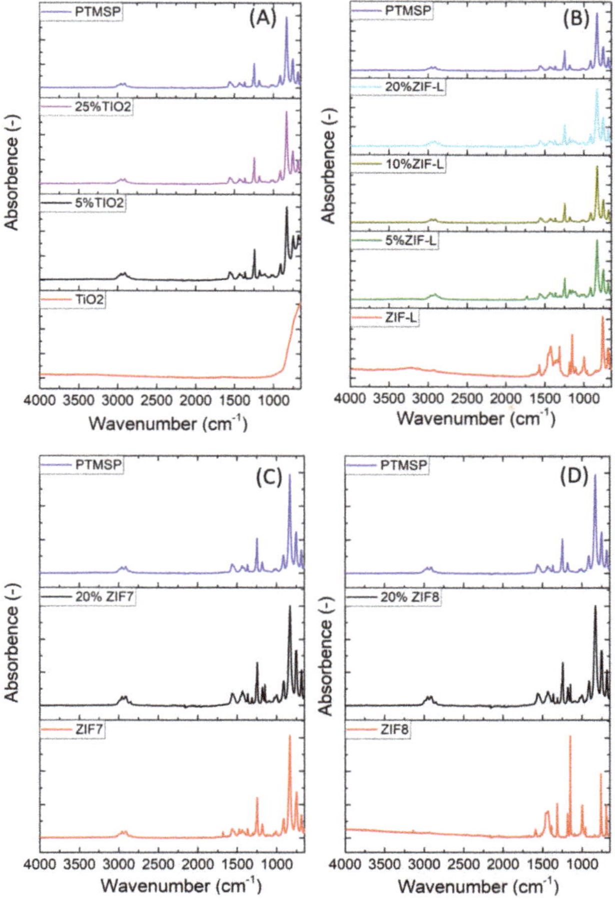

Figure 7. FT-IR spectrum of PTMSP/nanofiller hybrid membranes. (**A**) PTMSP/TiO_2, (**B**) PTMSP/ZIF-L, (**C**) PTMSP/ZIF-7 and (**D**) PTMSP/ZIF-8.

Table 1. Detailed peak assignment of PTMSP and various additives.

	Peak Position (cm^{-1})	Peak Assignment	Ref.
PTMSP	1540	stretching of the double C=C bond	[39]
	1240	deformation of the SiC–H bond	
	820	stretching of the C–Si bond	
ZIF7	1455	C–C stretching	[23]
	777	C–H stretching	
ZIF8/ZIFL	1584	stretching of C–N bond found in the 2-methylimidazole ring	[40]
	1350–1500	ring stretching	
	900–1350	coupled with in-plane ring bending	
	800	out-of-plane bending	
	1146	C–H vibrations	
	1310	C–H vibrations	
TiO_2	768	symmetric stretching vibrations in the Ti–O bond	[41]

Figure 7 depicts the FT-IR of pure PTMSP and different particles. Detailed peak assignments are also listed in Table 1. It is found that the neat PTMSP results are in accordance with the IR spectra given by Khodzhaeva et al. [39]. Three characteristic peaks can be found for PTMSP at 1540, 1240 and 820 cm^{-1}, which are due to stretching of double C=C bond, deformation of Si–C–H bond and stretching of the Si–C bond. In the case of ZIF-7 sample. Two peaks can be observed at 1455 and 777 cm^{-1}, corresponding to C–C bond and C–H bond stretching [23]. Considering ZIF-8 and ZIF-L samples, due to the similar chemical structure, identical peaks were obtained for these two samples. The characteristic peak for ZIF-8/ZIF-L locates at 1584 cm^{-1}, which is due to the stretching of C–N bond found in the 2-methylimidazole ring [40]. The TiO_2-particles have a rather non-distinct IR-spectrum, being approximately a line, until it gradually increases as it comes closer to 1000 cm^{-1}. Murashkevich et al. [41] reported that the very broad peak will reach its maximum around 526 cm^{-1} and that there should be a "shoulder" around 768 cm^{-1} due to symmetric stretching vibrations in the Ti–O bond. From Figure 7 it is observed that the TiO_2 results matches with the literature value quite well.

The FTIR spectrum of the hybrid membrane with the four different nanofillers are also showing in Figure 7. Based on the results, no new peaks or peak position shift can be found, which denoting that no chemical interaction happens between the nanofillers and the polymeric matrix. In the case of TiO_2, due to the fact that the FTIR spectrum of TiO_2 is relatively simpler compared to PTMSP, the PTMSP spectrum is dominating in the hybrid membranes. For the other three additives, the peak intensity of the nanofillers changes in the hybrid membrane, which has good correlation with the content of the fillers in the hybrid membranes.

3.5. Mixed Gas Permeation Results

A CO_2/N_2 gas mixture (10/90 v/v) was used as feed gas to investigate gas separation properties of obtained hybrid membranes. The mixed gas permeation tests of all the studied membranes were carried out at room temperature with controlled RH level.

3.5.1. Comparison of Two Different Solvents

It is well known that the solvents used to fabricate polymeric membranes can greatly affect the membrane properties [42] and that nanoparticles have different dispersion properties in different organic solvents [43]. In the present study, we found that ZIF-7 had relatively better dispersion property in chloroform while ZIF-8, ZIF-L and TiO_2 could be dispersed better in cyclohexane, thus both solvents were employed in fabrication of hybrid membrane.

CO_2/N_2 separation performances of the two neat PTMSP membranes prepared using chloroform and cyclohexane solvents were firstly studied as the blank data for the comparison with the hybrid membranes. The permeation results are shown in Figure 8. The lower boiling point (61 °C) of chloroform led to a higher CO_2 permeability (33,169 bar) compared to the one observed for cyclohexane (b.p. 81 °C) (20,338 bar), even though a lower CO_2/N_2 separation factor was obtained (2.7 for chloroform versus 6.9 for cyclohexane, Figure 8B). The results achieved for both solvents are similar to literature values [44–46]. The different solvent volatility is believed has led to different rearrangement of the polymeric chain in the final matrix: in the case of the less volatile solvent, the longer time needed to obtain the solid film may result in a higher polymer chain packing density.

Figure 8 also presents the effect of the relative humidity level in the feed gas. As can be seen, the CO_2 separation performances appear only to a small extent being affected by an increase of the humidity content in the gaseous stream, as both membranes showing limited change in CO_2 permeability (~10% drop) followed by a slight increase (~10%) in CO_2/N_2 selectivity. Interestingly, literatures on hydrophobic polymers [47,48] showed that water vapour significantly affects the transport of incondensable gases through the polymer matrix, despite of the limited H_2O uptake to the membrane. However, the strong effect of humidity is not observed for PTMSP, most likely due to that the different transport mechanism; in the PTMSP-based membranes, separation is related to the free volume size and distribution within the polymeric matrix. Scholes et al. [49] also reported a limited effect of water vapour on the CO_2 permeability of PTMSP casted using toluene as solvent, which suggests that the negligible effect of water vapour is independent of the type of solvent used during the membrane fabrication procedure.

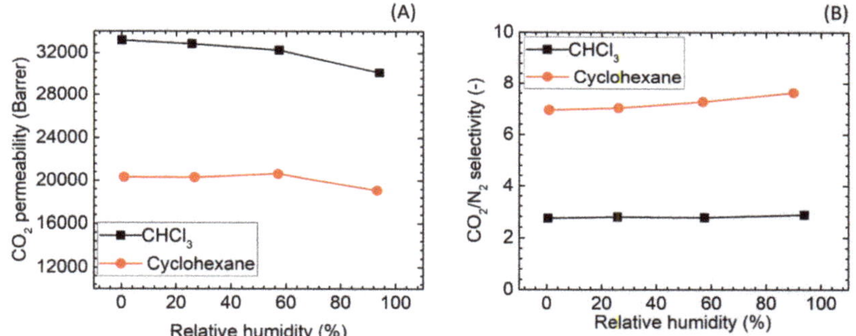

Figure 8. Gas separation performances of PTMSP membrane prepared from two different solvents, (**A**) CO_2 permeability and (**B**) CO_2/N_2 selectivity.

3.5.2. PTMSP/ZIF-8 Hybrid Membranes

In the case of ZIF-8 hybrid membrane, 20 wt % ZIF-8 was added into PTMSP matrix. The amount of the ZIF-8 was determined based on literature values.

As shown in Figure 9, at dry state, adding ZIF-8 into PTMSP results in a sharp increase in CO_2 permeability (27,781 bar, +37%) with a decrease of CO_2/N_2 selectivity (4.5, −36%) compared to the neat PTMSP membrane. The higher permeability and lower selectivity at dry state can be attributed to the micro-voids between the nanofillers and polymer phase, which is observed also from the SEM images (Figure 6C). Similar results have also been achieved for membranes with ZIF-8 embedded in PIM-1 [50], in which a comparable loading of ZIF-8 (28 wt %) showed a trend similar to the one observed in this study for both CO_2 permeability and CO_2/N_2 selectivity (the CO_2 permeability was tuned to values in the same range of PTMSP by ethanol swelling).

From Figure 9B it can be seen that with a small amount of water vapour in the feed gas (RH $\approx$ 25%), the CO_2 permeability dramatically decreased to 17,029 bar, reaching a performance that is 15% lower than that of the neat PTMSP membranes at the same RH condition (20,315 bar). Water adsorption in ZIF-8 was reported to be relatively low compare to other polar molecules [51]. However, the imidazolium linker introduces a –NH functionality on the outer surface of the ZIF cage, causing the H_2O molecules to cover preferentially the crystal structure rather than to be adsorbed in the cage [51]. This effect may reduce the free volume at the polymer/particles interface, negatively affecting the gas transport through the membrane. Experiments show that further increase of the RH value causes a limited drop in CO_2 permeability for the PTMSP/ZIF-8 hybrid membranes. Surprisingly, the humidity level had a very limited effect on the selectivity of the hybrid membrane, suggesting that the gas transport is mainly through the polymeric matrix and at the polymer/particles interface, not through the pores of the fillers, which should give a much higher selectivity.

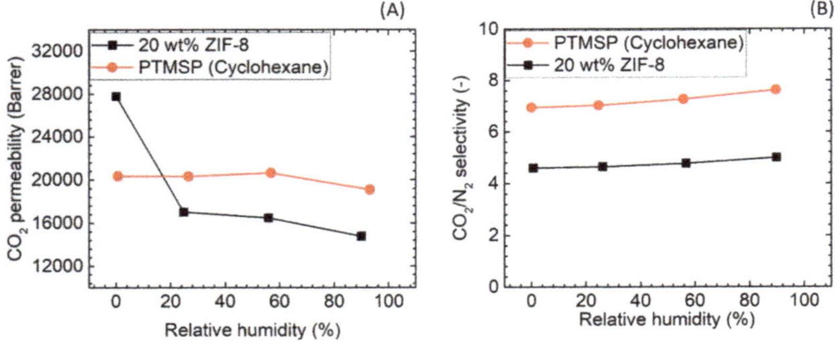

Figure 9. Gas separation performance of PTMSP/ZIF-8 hybrid membranes. (**A**) CO_2 permeability and (**B**) CO_2/N_2 selectivity.

3.5.3. PTMSP/ZIF-L Hybrid Membranes

In order to fully exploit the sieving ability of the 2D morphology of nanofillers, the assessment of the influence of porous 2D nanosheets in membranes has attracted great attention in recent years [52–55]. However, little effort has been dedicated to ZIF-based porous nanosheets, or in particular, not in gas separation membranes. To the best of our knowledge, only one report was found using ZIF-L as nanofiller in hybrid membranes for gas separation [28].

In this work, we investigated the effect of ZIF-L loadings in PTMSP membrane by preparing samples containing three different ZIF-L loadings (5 wt %, 10 wt % and 20 wt %). As can be seen in Figure 10, adding 5 wt % and 10 wt % ZIF-L into PTMSP matrix significantly improves CO_2 permeability and the 5 wt % loading seems the most effective. Surprisingly, a further increase of the ZIF-L content to 20 wt % causes a dramatic decrease in CO_2 permeability (1489 bar), which is one order of magnitude lower than the neat PTMSP value. The opposite trend is observed for the CO_2/N_2 selectivity: at a ZIF-L content of 5 wt % and 10 wt %, the CO_2/N_2 selectivity value is almost unchanged. When 20 wt % ZIF-L is presented in the PTMSP matrix, the CO_2/N_2 selectivity increases from 7 to 13~14. A similar behaviour (initial permeability increase followed by significant decrease) is also reported for other porous 2D nanosheets [53,55]. A graphic representation of the possible gas transport mechanism is proposed and shown in Figure 11. Compared to PTMSP membranes, membranes made of neat ZIF-L particles have relatively lower CO_2 permeability [26]. At lower loadings, increased gas permeability mainly comes from the micro-voids between the ZIF-L and PTMSP. As the addition of ZIF-L introduces more less permeable regions in the PTMSP matrix, which increases the gas diffuse path way in the membrane and part of the gases diffuse through the ZIF-L phase, leading to much lower permeability and relatively higher CO_2/N_2 selectivity. It has been reported that the neat ZIF-8 or ZIF-L membranes normally show relatively low CO_2/N_2 selectivity (in the range of <5) [56,57]. However, according to literature, when the ZIF-8 nanoparticles were added into polymer matrix, gas selectivity of the hybrid membranes are significantly higher [20,21]. Therefore, the intrinsic selectivity of the ZIF-8 material is possibly higher than the selectivity obtained from the ZIF membranes, in which the defects and micro voids between the ZIF-8 crystals may reduce the overall selectivity.

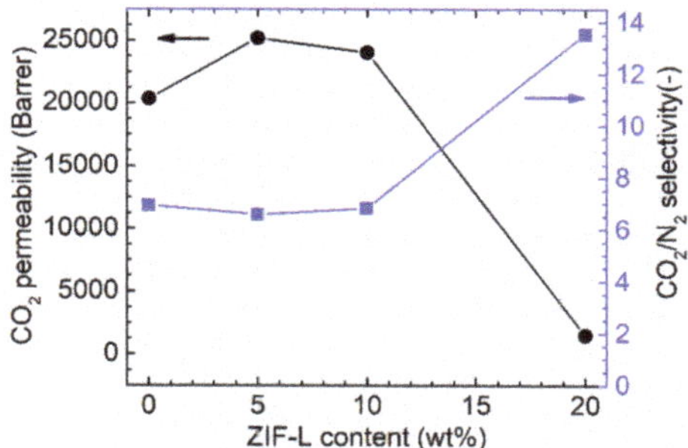

Figure 10. Gas separation performances of PTMSP/ZIF-L hybrid membranes as a function of the ZIF-L loading.

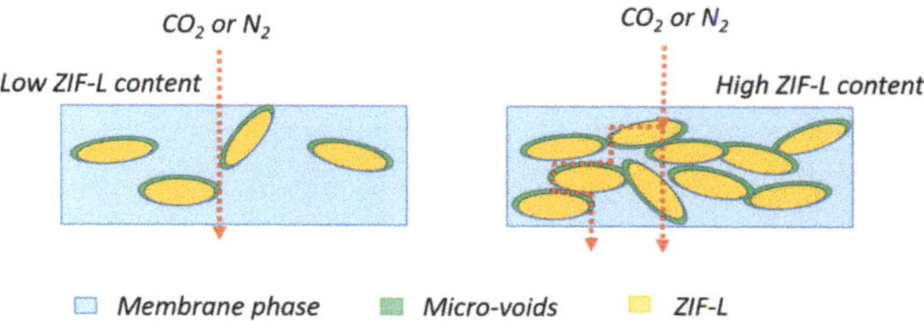

Figure 11. Possible gas transport mechanism in PTMSP/ZIF-L membranes.

The influence of water vapour on the separation performance of the ZIF-L-based hybrid membranes is shown in Figure 12. Increasing the water vapour content in the gaseous streams resulted in a decrease in CO_2 permeability for all the membranes containing ZIF-L fillers. For the 5 and 10 wt % loading the drop is limited to 17–20% of the dry value and the CO_2 transport remains above the value of the neat PTMSP membranes. In the case of 20 wt % ZIF-L loading, a similar decrease (15.7%) was observed. In terms of the neat PTMSP membrane, the CO_2 permeability decreased only 6% with the RH level increasing from 0% RH to 100% RH. Similar to the ZIF-8 case, the presence of water vapour shows a negligible effect on the CO_2/N_2 selectivity, independently from the particles loading within the membrane matrix.

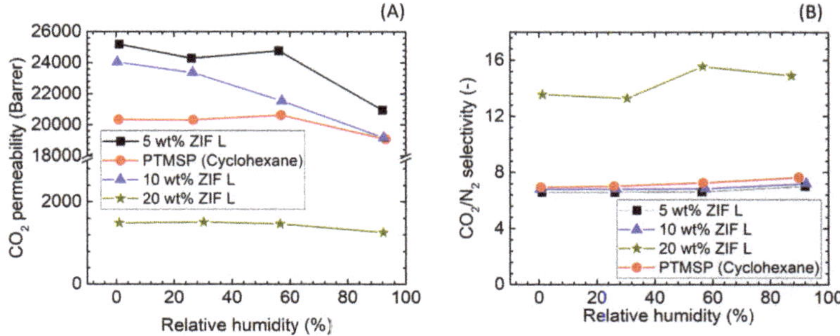

Figure 12. Gas separation performances of PTMSP/ZIF-L hybrid membranes as a function of relative humidity in the gaseous stream. (**A**) CO_2 permeability and (**B**) CO_2/N_2 selectivity.

3.5.4. PTMSP/ZIF-7 Hybrid Membranes

Based on literature, a higher ZIF-7 loading in a hybrid membrane can be beneficial to improving CO_2 selectivity over other gases (e.g., N_2 and/or CH_4). According to Li et al., the highest selectivity was obtained at a ZIF-7 content of 34 wt % [23]. In another report, the highest CO_2/CH_4 selectivity was obtained at a ZIF-7 content of 35 wt % [24]. Therefore, in this study, 30 wt % ZIF-7 nanoparticles were determined as the optimal loading in the PTMSP/ZIF-7 hybrid membrane. The gas permeation results of the resultant hybrid membranes are shown in Figure 13.

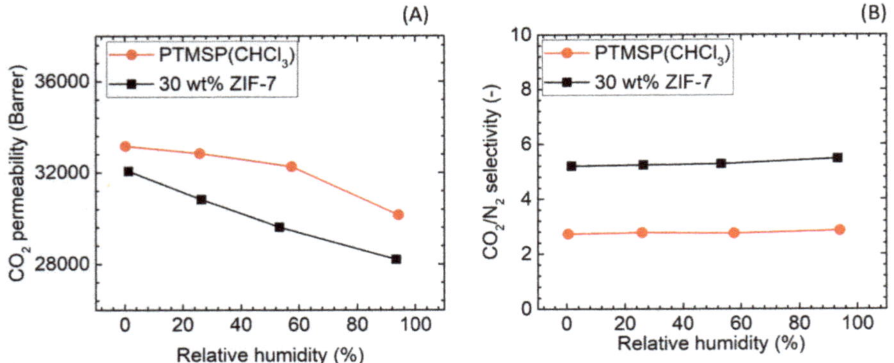

Figure 13. Gas separation performances of PTMSP/ZIF-7 hybrid membranes. (**A**) CO_2 permeability and (**B**) CO_2/N_2 selectivity.

As indicated in Figure 13, even with the ZIF-7 loading of up to 30 wt % in the hybrid membrane, only a minor change in the CO_2 permeability (from 33,169 bar to 32,065 bar) was observed. Similar results have been reported in Pebax/ZIF-7 membranes, in which a ZIF-7 content of 22~34 wt % resulted in a reduced CO_2 permeability and much improved CO_2/N_2 selectivity [23–25]. In another study, ZIF-7 particles were added into the polybenzimidazole (PBI) matrix to improve CO_2/H_2 separation performances. According to their results, adding 50 wt % of ZIF-7 nanoparticles into PBI membranes only increased CO_2 permeability from 0.4 to 1.8 Bar, indicating that the intrinsic ZIF-7 permeability is in the low region [58]. Nevertheless, adding ZIF-7 nanoparticles into Pebax and PBI has been reported to be rather efficient in promoting CO_2 selectivity over other gases [23,25]. In this work, the CO_2/N_2 selectivity was doubled by adding 30 wt % of ZIF-7 into PTMSP matrix (from 2.7 to 5.2). Although the absolute value is still too low, it is a significant improvement compared to the neat PTMSP membranes.

3.5.5. PTMSP/TiO_2 Hybrid Membranes

Compared to the ZIF series, the addition of TiO_2 nanoparticles showed quite different effects. At dry state, adding 5 wt % TiO_2 nanoparticles into PTMSP matrix significantly improved the CO_2 permeability (28,432 bar, 40% increased) but the increase in TiO_2 content (25 wt %) did not show further increase in the permeability (Figure 14). It is frequently reported that the increase of CO_2 permeability in hybrid membranes containing TiO_2 is related mainly to the formation of interfacial voids at the polymer/particles interface and to the disruption of polymer chain packing induced by the nanoparticles rather than to its high CO_2 adsorption capacity [5]. This observation is also confirmed in this study by the reduction of the CO_2/N_2 selectivity by the addition of TiO_2 nanofillers in PTMSP hybrid membranes.

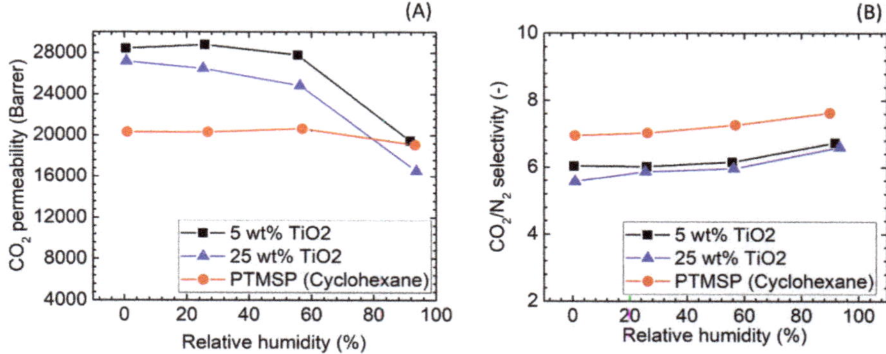

Figure 14. Gas separation performance of PTMSP/TiO_2 hybrid membranes (Cyclohexane used as solvent). (**A**) CO_2 permeability and (**B**) CO_2/N_2 selectivity.

Interestingly, the presence of water vapour in the feed gas reduced the gas permeability to a significant extent. It was found that, when increasing the RH value from 75% to 100%, both membranes with 5 and 25 wt % TiO_2 nanoparticles exhibited a notable decrease (35–40%) in the CO_2 permeability, which may be due to the competitive sorption of water vapour that occupies the additional free volume created by the TiO_2 nanoparticles. Similar to all the cases in this work, the presence of water vapour showed a negligible effect on the selective feature of the hybrid membranes.

The gas separation performances of the hybrid membranes various additives are listed in Table 2.

Table 2. Summary of PTMSP hybrid membrane separation performances.

Solvent	Nanofiller	Nanofiller Content (wt %)	RH (%)	CO_2 Permeability	CO_2/N_2 Selectivity (-)
$CHCl_3$	-	-	0.2	33,169.3	2.7
$CHCl_3$	-	-	94.1	30,152.0	2.9
$CHCl_3$	ZIF-7	30	1.3	32,065.0	5.2
$CHCl_3$	ZIF-7	30	93.4	28,205.3	5.5
Cyclohexane	-	-	0.9	20,338.7	6.9
Cyclohexane	-	-	93.2	19,074.8	7.6
Cyclohexane	TiO_2	5	0.5	28,432.2	6.0
Cyclohexane	TiO_2	5	91.7	19,465.6	6.7
Cyclohexane	TiO_2	25	0.7	27,222.0	5.6
Cyclohexane	TiO_2	25	93.6	16,550.1	6.6
Cyclohexane	ZIF-8	20	0.7	27,781.7	4.6
Cyclohexane	ZIF-8	20	89.9	14,764.1	5.0
Cyclohexane	ZIF-L	5	1.1	25,191.4	6.6
Cyclohexane	ZIF-L	5	92.0	20,949.6	7.0
Cyclohexane	ZIF-L	10	0.5	24,046.1	6.8
Cyclohexane	ZIF-L	10	92.5	19,175.1	7.2
Cyclohexane	ZIF-L	20	1.1	1,489.2	13.5
Cyclohexane	ZIF-L	20	92.3	1,255.1	14.9

As can be seen from Table 2, adding ZIF-8 and TiO_2 into PTMSP improves CO_2 permeability but the improvement is at the expenses of the CO_2/N_2 selectivity. Adding ZIF-7 into PTMSP leads to lower CO_2 permeability and slightly higher CO_2/N_2 selectivity. In the case of ZIF-L, at lower particles content, higher CO_2 permeability and lower CO_2/N_2 selectivity was observed but a further increase of the ZIF-L loading causes a significant improvement of the CO_2/N_2 selective feature but with a dramatic drop in the gas permeability coefficient. In a similar fashion for all the investigated nanoparticles, the presence of water vapour negatively affects the CO_2 permeability of the hybrid membranes, with the ZIF-8 and the TiO_2 based membranes being the most affected. No significant effect is, however, observed for the selective feature of the hybrid membranes.

4. Conclusions

In the present study, a broad range of fillers were used for the fabrication of hybrid membranes for CO_2 capture. Three different ZIF-type porous nanofillers (2D and 3D) and TiO_2 nanoparticles (0D) were utilized to fabricate PTMSP-based hybrid membranes, aiming at the improvement of the CO_2 separation performance. The TGA analysis shows that the hybrid membranes are characterized by high thermal stability compared to the pristine polymer, fulfilling the requirements for most CO_2 separation applications. In addition, the SEM images indicates that the particles are homogeneously distributed within the polymer matrix but the poor affinity with the polymer phase caused the formation of interfacial voids.

The solvent used for the membrane preparation was found very important: the CO_2 permeability decreased by 40% while the selectivity increased by 150% simply by changing the solvent in the membrane casting solution from chloroform to cyclohexane. The loadings of the different nanofillers makes big differences in the separation property enhancement. Adding ZIF-8 and TiO_2 particles into PTMSP matrix has always led to an increment in CO_2 permeability. On the other hand, the addition of ZIF-7 into the hybrid membranes resulted in a relatively lower CO_2 permeability but higher CO_2/N_2 selectivity. The most notable variation was obtained from the addition of ZIF-L, the porous 2D nanosheets. At a low ZIF-L loading, the CO_2 permeability was enhanced but further increase of the particle loading caused in a dramatic decrease in CO_2 permeability but doubled CO_2/N_2 selectivity compared to the neat PTMSP membrane. For all the hybrid membranes, increasing the water vapour content in the gaseous streams has caused a reduction in CO_2 permeability with limited effect on the CO_2/N_2 selectivity.

To sum up, despite the different morphology and properties of the selected nanofillers in this study (porous/non-porous, 0 to 3D nanostructures), adding the nanofillers into PTMSP either leads to higher permeability with lower CO_2/N_2 selectivity, or vice versa.

Author Contributions: L.A. and Z.D. designed the experiment. V.L. conducted gas permeation tests and SEM. J.D. carried out FTIR and TGA. Z.D. analysed the data and wrote the manuscript. L.D. conceived the study and supervised all aspects of the project. All authors worked on manuscript review and editing.

Funding: This research was funded by the European Union's Horizon 2020 Research and Innovation program under Grant Agreement No. 727734 at NTNU.

Conflicts of Interest: The authors declare no conflict of interest.

References

1. Baker, R.W.; Low, B.T. Gas separation membrane materials: A perspective. *Macromolecules* **2014**, *47*, 6999–7013. [CrossRef]
2. Galizia, M.; Chi, W.S.; Smith, Z.P.; Merkel, T.C.; Baker, R.W.; Freeman, B.D. 50th Anniversary Perspective: Polymers and Mixed Matrix Membranes for Gas and Vapor Separation: A Review and Prospective Opportunities. *Macromolecules* **2017**, *50*, 7809–7843. [CrossRef]
3. Luis, P.; Van Gerven, T.; Van der Bruggen, B. Recent developments in membrane-based technologies for CO_2 capture. *Prog. Energy Combust. Sci.* **2012**, *38*, 419–448. [CrossRef]
4. Robeson, L.M. The upper bound revisited. *J. Membr. Sci.* **2008**, *320*, 390–400. [CrossRef]
5. Janakiram, S.; Ahmadi, M.; Dai, Z.; Ansaloni, L.; Deng, L. Performance of Nanocomposite Membranes Containing 0D to 2D Nanofillers for CO_2 Separation: A Review. *Membranes* **2018**, *8*, 24. [CrossRef] [PubMed]
6. Dai, Z.; Ansaloni, L.; Deng, L. Recent advances in multi-layer composite polymeric membranes for CO_2 separation: A review. *Green Energy Environ.* **2016**, *1*, 102–128. [CrossRef]
7. Dai, Z.; Noble, R.D.; Gin, D.L.; Zhang, X.; Deng, L. Combination of ionic liquids with membrane technology: A new approach for CO_2 separation. *J. Membr. Sci.* **2016**, *497*, 1–20. [CrossRef]

8. Deng, J.; Deng, J.; Bai, L.; Zeng, S.; Zhang, X.; Nie, Y.; Deng, L.; Zhang, S. Ether-functionalized ionic liquid based composite membranes for carbon dioxide separation. *RSC Adv.* **2016**, *6*, 45184–45192. [CrossRef]
9. Mahdi, A.S.J.; Dai, Z.; Ansaloni, L.; Deng, L.Y. Performance of Mixed Matrix Membranes Containing Porous 2D and 3D fillers for CO_2 Separation: A Review. *Membranes* **2018**, in press.
10. Rezakazemi, M.; Amooghin, A.E.; Montazer-Rahmati, M.M.; Ismail, A.F.; Matsuura, T. State-of-the-art membrane based CO_2 separation using mixed matrix membranes (MMMs): An overview on current status and future directions. *Prog. Polym. Sci.* **2014**, *39*, 817–861. [CrossRef]
11. Budd, P.M.; Ghanem, B.S.; Makhseed, S.; Mckeown, N.B.; Msayib, K.J.; Tattershall, C.E. Polymers of intrinsic microporosity (PIMs): Robust, solution-processable, organic nanoporous materials. *Chem. Commun.* **2004**, *2*, 230–231. [CrossRef] [PubMed]
12. Bazhenov, S.D.; Borisov, I.L.; Bakhtin, D.S.; Rybakova, A.N.; Khotimskiy, V.S.; Molchanov, S.P.; Volkov, V.V. High-permeance crosslinked PTMSP thin-film composite membranes as supports for CO_2 selective layer formation. *Green Energy Environ.* **2016**, *1*, 235–245. [CrossRef]
13. Chapala, P.P.; Bermeshev, M.V.; Starannikova, L.E.; Belov, N.A.; Ryzhikh, V.E.; Shantarovich, V.P.; Finkelshtein, E.S. A novel, highly gas-permeable polymer representing a new class of silicon-containing polynorbornens as efficient membrane materials. *Macromolecules* **2015**, *48*, 8055–8061. [CrossRef]
14. Srinivasan, R.; Auvil, S.R.; Burban, P.M. Elucidating the mechanism (s) of gas transport in poly [1-(trimethylsilyl)-1-propyne](PTMSP) membranes. *J. Membr. Sci.* **1994**, *86*, 67–86. [CrossRef]
15. Lin, H.; He, Z.; Sun, Z.; Kniep, J.; Ng, A.; Baker, R.W.; Merkel, T.C. CO_2-selective membranes for hydrogen production and CO_2 capture—Part II: Techno-economic analysis. *J. Membr. Sci.* **2015**, *493*, 794–806. [CrossRef]
16. Chung, T.S.; Jiang, L.Y.; Li, Y.; Kulprathipanja, S. Mixed matrix membranes (MMMs) comprising organic polymers with dispersed inorganic fillers for gas separation. *Prog. Polym. Sci.* **2007**, *32*, 483–507. [CrossRef]
17. Goh, P.; Ismail, A.F.; Sanip, S.M.; Ng, B.C.; Aziz, M. Recent advances of inorganic fillers in mixed matrix membrane for gas separation. *Sep. Purif. Technol.* **2011**, *81*, 243–264. [CrossRef]
18. Banerjee, R.; Phan, A.; Wang, B.; Knobler, C.; Furukawa, H.; O'keeffe, M.; Yaghi, O.M. High-throughput synthesis of zeolitic imidazolate frameworks and application to CO_2 capture. *Science* **2008**, *319*, 939–943. [CrossRef] [PubMed]
19. Hwang, S.; Hwang, S.; Chi, W.S.; Lee, S.J.; Im, S.H.; Kim, J.H.; Kim, J. Hollow ZIF-8 nanoparticles improve the permeability of mixed matrix membranes for CO_2/CH_4 gas separation. *J. Membr. Sci.* **2015**, *480*, 11–19. [CrossRef]
20. Guo, A.; Ban, Y.; Yang, K.; Yang, W. Metal-organic framework-based mixed matrix membranes: Synergetic effect of adsorption and diffusion for CO_2/CH_4 separation. *J. Membr. Sci.* **2018**, *562*, 76–84. [CrossRef]
21. Ordoñez, M.J.C.; Balkus, K.J., Jr.; Ferraris, J.P.; Musselman, I.H. Molecular sieving realized with ZIF-8/Matrimid® mixed-matrix membranes. *J. Membr. Sci.* **2010**, *361*, 28–37. [CrossRef]
22. Dai, Y.; Johnson, J.R.; Karvan, O.; Sholl, D.S.; Koros, W.J. Ultem®/ZIF-8 mixed matrix hollow fiber membranes for CO_2/N_2 separations. *J. Membr. Sci.* **2012**, *402*, 76–82. [CrossRef]
23. Li, T.; Pan, Y.; Peinemann, K.V.; Lai, Z. Carbon dioxide selective mixed matrix composite membrane containing ZIF-7 nano-fillers. *J. Membr. Sci.* **2013**, *425–426*, 235–242. [CrossRef]
24. Khoshkharam, A.; Azizi, N.; Behbahani, R.M.; Ghayyem, M.A. Separation of CO_2 from CH_4 using a synthesized Pebax-1657/ZIF-7 mixed matrix membrane. *Pet. Sci. Technol.* **2017**, *35*, 667–673. [CrossRef]
25. Azizi, N.; Hojjati, M.R. Using Pebax-1074/ZIF-7 mixed matrix membranes for separation of CO_2 from CH_4. *Pet. Sci. Technol.* **2018**, *36*, 993–1000. [CrossRef]
26. Zhong, Z.; Yao, J.; Chen, R.; Low, Z.; He, M.; Liu, J.Z.; Wang, H. Oriented two-dimensional zeolitic imidazolate framework-L membranes and their gas permeation properties. *J. Mater. Chem. A* **2015**, *3*, 15715–15722. [CrossRef]
27. Chen, R.; Yao, J.; Gu, Q.; Smeets, S.; Baerlocher, C.; Gu, H.; Wang, H. A two-dimensional zeolitic imidazolate framework with a cushion-shaped cavity for CO_2 adsorption. *Chem. Commun.* **2013**, *49*, 9500–9502. [CrossRef] [PubMed]
28. Kim, S.; Shamsaei, E.; Lin, X.; Hu, Y.; Simon, G.P.; Seong, J.G.; Wang, H. The enhanced hydrogen separation performance of mixed matrix membranes by incorporation of two-dimensional ZIF-L into polyimide containing hydroxyl group. *J. Membr. Sci.* **2018**, *549*, 260–266. [CrossRef]
29. Kim, W.G.; Lee, J.S.; Bucknall, D.G.; Koros, W.J.; Nair, S. Nanoporous layered silicate AMH-3/cellulose acetate nanocomposite membranes for gas separations. *J. Membr. Sci.* **2013**, *441*, 129–136. [CrossRef]

30. Liu, G.; Jiang, Z.Y.; Cao, K.T.; Nair, S.; Cheng, X.X.; Zhao, J.; Gomaa, H.; Wu, H.; Pan, F. Pervaporation performance comparison of hybrid membranes filled with two-dimensional ZIF-L nanosheets and zero-dimensional ZIF-8 nanoparticles. *J. Membr. Sci.* **2017**, *523*, 185–196. [CrossRef]
31. Merkel, T.C.; Freeman, B.D.; Spontak, R.J.; He, Z.; Pinnau, I.; Meakin, P.; Hill, A.J. Ultrapermeable, reverse-selective nanocomposite membranes. *Science* **2002**, *296*, 519–522. [CrossRef] [PubMed]
32. Liang, C.Y.; Uchytil, P.; Petrychkovych, R.; Lai, Y.C.; Friess, K.; Sipek, M.; Reddy, M.M.; Suen, S.Y. A comparison on gas separation between PES (polyethersulfone)/MMT (Na-montmorillonite) and PES/TiO_2 mixed matrix membranes. *Sep. Purif. Technol.* **2012**, *92*, 57–63. [CrossRef]
33. Moghadam, F.; Omidkhah, M.R.; Farahani, E.V.; Pedram, M.Z.; Dorosti, F. The effect of TiO_2 nanoparticles on gas transport properties of Matrimid5218-based mixed matrix membranes. *Sep. Purif. Technol.* **2011**, *77*, 128–136. [CrossRef]
34. Zhang, C.; Dai, Y.; Johnson, J.R.; Karvan, O.; Koros, W.J. High performance ZIF-8/6FDA-DAM mixed matrix membrane for propylene/propane separations. *J. Membr. Sci.* **2012**, *389*, 34–42. [CrossRef]
35. Dai, Z.; Dai, Z.; Ansaloni, L.; Ryan, J.J.; Spontak, R.J.; Deng, L. Nafion/IL hybrid membranes with tuned nanostructure for enhanced CO_2 separation: Effects of ionic liquid and water vapor. *Green Chem.* **2018**, *20*, 1391–1404. [CrossRef]
36. Dai, Z.; Aboukeila, H.; Ansaloni, L.; Deng, J.; Baschetti, M.G.; Deng, L. Nafion/PEG hybrid membrane for CO_2 separation: Effect of PEG on membrane micro-structure and performance. *Sep. Purif. Technol.* **2018**, in press. [CrossRef]
37. James, J.B.; Lin, Y.S. Kinetics of ZIF-8 Thermal Decomposition in Inert, Oxidizing and Reducing Environments. *J. Phys. Chem. C* **2016**, *120*, 14015–14026. [CrossRef]
38. Kang, C.H.; Lin, Y.F.; Huang, Y.S.; Tung, K.L.; Chang, K.S.; Chen, J.S.; Hung, W.S.; Lee, K.R.; Lai, J.Y. Synthesis of ZIF-7/chitosan mixed-matrix membranes with improved separation performance of water/ethanol mixtures. *J. Membr. Sci.* **2013**, *438*, 105–111. [CrossRef]
39. Khodzhaeva, V.; Zaikin, V. Fourier transform infrared spectroscopy study of poly (1-trimethylsilyl-1-propyne) aging. *J. Appl. Polym. Sci.* **2007**, *103*, 2523–2527. [CrossRef]
40. Hu, Y.; Kazemian, H.; Rohani, S.; Huang, Y.; Song, Y. In situ high pressure study of ZIF-8 by FTIR spectroscopy. *Chem. Commun.* **2011**, *47*, 12694–12696. [CrossRef] [PubMed]
41. Murashkevich, A.; Lavitskaya, A.; Barannikova, T.; Zharskii, I. Infrared absorption spectra and structure of TiO_2-SiO_2 composites. *J. Appl. Spectrosc.* **2008**, *75*, 730–734. [CrossRef]
42. Isanejad, M.; Azizi, N.; Mohammadi, T. Pebax membrane for CO_2/CH_4 separation: Effects of various solvents on morphology and performance. *J. Appl. Polym. Sci.* **2017**, *134*, 44531. [CrossRef]
43. Nguyen, V.S.; Rouxel, D.; Vincent, B. Dispersion of nanoparticles: From organic solvents to polymer solutions. *Ultrason. Sonochem.* **2014**, *21*, 149–153. [CrossRef] [PubMed]
44. Lau, C.H.; Konstas, K.; Doherty, C.M.; Kanehashi, S.; Ozcelik, B.; Kentish, S.E.; Hill, A.J.; Hill, M.R. Tailoring physical aging in super glassy polymers with functionalized porous aromatic frameworks for CO_2 capture. *Chem. Mater.* **2015**, *27*, 4756–4762. [CrossRef]
45. Peter, J.; Peinemann, K.V. Multilayer composite membranes for gas separation based on crosslinked PTMSP gutter layer and partially crosslinked Matrimid® 5218 selective layer. *J. Membr. Sci.* **2009**, *340*, 62–72. [CrossRef]
46. Shao, L.; Samseth, J.; Hägg, M.B. Crosslinking and stabilization of nanoparticle filled poly (1-trimethylsilyl-1-propyne) nanocomposite membranes for gas separations. *J. Appl. Polym. Sci.* **2009**, *113*, 3078–3088. [CrossRef]
47. Comesaña-Gandara, B.; Ansaloni, L.; Lee, Y.; Lozano, A.; De Angelis, M. Sorption, diffusion and permeability of humid gases and aging of thermally rearranged (TR) polymer membranes from a novel ortho-hydroxypolyimide. *J. Membr. Sci.* **2017**, *542*, 439–455. [CrossRef]
48. Ansaloni, L.; Minelli, M.; Baschetti, M.G.; Sarti, G. Effect of relative humidity and temperature on gas transport in Matrimid®: Experimental study and modeling. *J. Membr. Sci.* **2014**, *471*, 392–401. [CrossRef]
49. Scholes, C.A.; Jin, J.; Stevens, G.W.; Kentish, S.E. Competitive permeation of gas and water vapour in high free volume polymeric membranes. *J. Polym. Sci. Part. B Polym. Phys.* **2015**, *53*, 719–728. [CrossRef]
50. Bushell, A.F.; Attfield, M.P.; Mason, C.R.; Budd, P.M.; Yampolskii, Y.; Starannikova, L.; Rebrov, A.; Bazzarelli, F.; Bernardo, P.; Carolus Jansen, J.; et al. Gas permeation parameters of mixed matrix membranes based on the polymer of intrinsic microporosity PIM-1 and the zeolitic imidazolate framework ZIF-8. *J. Membr. Sci.* **2013**, *427*, 48–62. [CrossRef]

51. Zhang, K.; Lively, R.P.; Zhang, C.; Koros, W.J.; Chance, R.R. Investigating the Intrinsic Ethanol/Water Separation Capability of ZIF-8: An Adsorption and Diffusion Study. *J. Phys. Chem. C* **2013**, *117*, 7214–7225. [CrossRef]
52. Shete, M.; Kumar, P.; Bachman, J.E.; Ma, X.; Smith, Z.P.; Xu, W.; Mkhoyan, K.A.; Long, J.R.; Tsapatsis, M. On the direct synthesis of Cu(BDC) MOF nanosheets and their performance in mixed matrix membranes. *J. Membr. Sci.* **2018**, *549*, 312–320. [CrossRef]
53. Kang, Z.; Peng, Y.; Hu, Z.; Qian, Y.; Chi, C.; Yeo, L.Y.; Tee, L.; Zhao, D. Mixed matrix membranes composed of two-dimensional metal-organic framework nanosheets for pre-combustion CO_2 capture: A relationship study of filler morphology versus membrane performance. *J. Mater. Chem. A* **2015**, *3*, 20801–20810. [CrossRef]
54. Cheng, Y.; Wang, X.; Jia, C.; Wang, Y.; Zhai, L.; Wang, Q.; Zhao, D. Ultrathin mixed matrix membranes containing two-dimensional metal-organic framework nanosheets for efficient CO_2/CH_4 separation. *J. Membr. Sci.* **2017**, *539*, 213–223. [CrossRef]
55. Rodenas, T.; Luz, I.; Prieto, G.; Seoane, B.; Miro, H.; Corma, A.; Kapteijn, F.; Llabrés i Xamena, F.X.; Gascon, J. Metal—Organic framework nanosheets in polymer composite materials for gas separation. *Nat. Mater.* **2015**, *14*, 48. [CrossRef] [PubMed]
56. Huang, A.; Liu, Q.; Wang, N.; Caro, J. Highly hydrogen permselective ZIF-8 membranes supported on polydopamine functionalized macroporous stainless-steel-nets. *J. Mater. Chem. A* **2014**, *2*, 8246–8251. [CrossRef]
57. Zhang, X.; Liu, Y.; Li, S.; Kong, L.; Liu, H.; Li, Y.; Han, W.; Yeung, K.L.; Zhu, W.; Yang, W. New membrane architecture with high performance: ZIF-8 membrane supported on vertically aligned ZnO nanorods for gas permeation and separation. *Chem. Mater.* **2014**, *26*, 1975–1981. [CrossRef]
58. Yang, T.; Xiao, Y.; Chung, T.S. Poly-/metal-benzimidazole nano-composite membranes for hydrogen purification. *Energy Environ. Sci.* **2011**, *4*, 4171–4180. [CrossRef]

 *membranes*

Article

Study of the Effect of Inorganic Particles on the Gas Transport Properties of Glassy Polyimides for Selective CO_2 and H_2O Separation

Sara Escorihuela [1,2], Lucía Valero [1,2], Alberto Tena [2,*], Sergey Shishatskiy [2,*], Sonia Escolástico [1], Torsten Brinkmann [2] and Jose Manuel Serra [1,*]

1 Instituto de Tecnología Química, UniversitatPolitècnica de València-Consejo Superior de Investigaciones Científicas, Avda. Los Naranjos, s/n 46022 Valencia, Spain; saesro@itq.upv.es (S.E.); valeroromerolucia@gmail.com (L.V.); soesro@upvnet.upv.es (S.E.)

2 Helmholtz-ZentrumGeesthacht, Institute of Polymer Research, Max-Planck-Str.1, 21502 Geesthacht, Germany; torsten.brinkmann@hzg.de

* Correspondence: alberto.tena@hzg.de (A.T.); sergey.shishatskiy@hzg.de (S.S.); jmserra@itq.upv.es (J.M.S.)

Received: 6 November 2018; Accepted: 5 December 2018; Published: 9 December 2018

Abstract: Three polyimides and six inorganic fillers in a form of nanometer-sized particles were studied as thick film solution cast mixed matrix membranes (MMMs) for the transport of CO_2, CH_4, and H_2O. Gas transport properties and electron microscopy images indicate good polymer-filler compatibility for all membranes. The only filler type thatdemonstrated good distribution throughout the membrane thickness at 10 wt.% loading was $BaCe_{0.2}Zr_{0.7}Y_{0.1}O_3$ (BCZY). The influence of this filler on MMM gas transport properties was studied in detail for 6FDA-6FpDA in a filler content range from one to 20 wt.% and for Matrimid® and P84® at 10 wt.% loading. The most promising result was obtained for Matrimid®—10 wt.% BCZY MMM, which showed improvement in CO_2 and H_2O permeabilities accompanied by increased CO_2/CH_4 selectivity and high water selective membrane at elevated temperatures without H_2O/permanent gas selectivity loss.

Keywords: mixed matrix membranes; carbon dioxide; water vapor permeability; polyimides; inorganic fillers; gas separation membranes; water transport

---

1. Introduction

Mixed matrix membranes (MMMs) [1,2] are considered as a very promising route to overcome limitations of the Robeson upper bound of the permeability/selectivity relationship by combining good mechanical, but rather disappointing gas transport properties of polymers, with excellent diffusion and sorption properties of inorganic porous media having very poor mechanical properties, e.g., flexibility [3].

Gas separation membranes have been on the market since 1980 [4] and have proven their reliability [5]. Unfortunately, since the boom of membranes introduction into the market at the end of the 20th century, not too many new membranes reached practical application level for gas separation processes. The reason for it lays in the versatility of existing membrane technology that allows the combination of membrane separation stages with other unit operations and to achieve goals of the separation process [6]. New membranes introduced to the market should exhibit significantly better properties, as compared to those already commercially available. At this point, the combination of properties of polymers and inorganic substances able to selectively transport gas or vapor molecules becomes very appealing [7].

The requirements for polymers to be used in membranes are: Adequate gas transport properties (balance between permeability and selectivity), processability as a thin film, and high reproducible manufacturing from batch to batch. For the formation of MMMs, good adhesion between the polymer matrix and the inorganic filler is essential, especially in case of polymers with high T_g (glass transition temperature) [8], good mechanical properties and stability of properties as a function of time.

The inorganic fillers to be used in MMMs should have (i) particles as small as possible since selective layers of modern polymer based gas separation membranes have a thickness of 100 nm and less; (ii) good affinity to the polymer; and (iii) gas transport properties matching, for the target gas in the separation process, those of the matrix polymer [9]. Several fillers have been studied in order to be combined with a polymer matrix, i.e., zeolites, carbon mesoporous silica, and metal organic frameworks (MOF) [10–15]. Regarding this last particular example, MOFs appear to be especially attractive, as concerns compatibility with polymers [16]. However, rather few reported works on MMFs with MOFs evidence an enhancement of CO_2 permeability and CO_2/CH_4 selectivity, as compared to the pure polymer membrane [17].

Depending on the compatibility of the filler and the polymer matrix, four different cases (Figure 1) and the expected effect on the separation properties can be described [18–22]. Case 1 is an ideal situation, where the filler is perfectly incorporated into the polymer matrix. This situation can result in an improvement of the separation properties of the mixed matrix material. In Case 2, there is a rigidification of the polymer matrix in the area around the filler. This normally results in an increase of the selectivity, due to the increased rigidity, but also in a decrease of the permeability. This is normally confirmed by an increase on the T_g of the MMMs. Case 3 exhibits a creation of an interphase due to the incompatibility between the particle and the polymer matrix. This results not only in an increase of the permeability due to the bigger fractional free volume, but also in a decrease of the selectivity. In Case 4, the polymer matrix penetrates the pore or free volume of the filler. In this case, both permeability and selectivity decrease.

In this work, several MMMs were studied for gas transport properties with focus on CO_2, CH_4 and H_2O as components in various industrially relevant gas mixtures, for example, natural gas, biogas as well as gas streams in chemical and petrochemical industries [23–26]. Biogas and natural gas are considered as the most environmentally friendly resources for large scale electric energy production and as sources with significant methane content, which is involved in several relevant reactions such as combustion, steam reforming, or halogenation. For example, in Germany, natural gas contains about 95% CH_4 and not more than 2% CO_2 concentration in the gas pipelines [27–29]. On the other hand, biogas is a mixture of several gases produced by the anaerobic decomposition of organic matter. Biogas mainly consists of methane (50–70%), carbon dioxide (30–50%), and other compounds including hydrogen sulfide (H_2S), water, and other trace gas compounds [30]. The membranes considered in this study could be used to separate CO_2 from these gases. Additionally, these membranes could also be applied to remove water at low temperature in several combustion processes in order to recover it and on-site reuse for other purposes.

In the current study, three polyimides and six inorganic fillers were used to prepare MMMs as thick films and the corresponding gas transport properties for CO_2, CH_4, and H_2O were systematically studied. Polyimides were selected due to their outstanding gas transport properties (low permeability coefficients, high selectivity, and good thermal stability) for several gas pairs as CO_2/CH_4 [31] or O_2/N_2 [32], and in case of the studied polymers, excellent film forming properties. The six inorganic materials used as nano-sized particles were selected, taking into account the expected good affinity for gas molecules as CO_2 and water vapor [33]. BCZY, LaWO, 8YSZ, and La_2O_3 were selected for the CO_2 and H_2O affinity due to their basicity and/or the important water absorption that present [34–36], whereas zeolites have been previously used as fillers in MMM for high-temperature CO_2 separation [37].

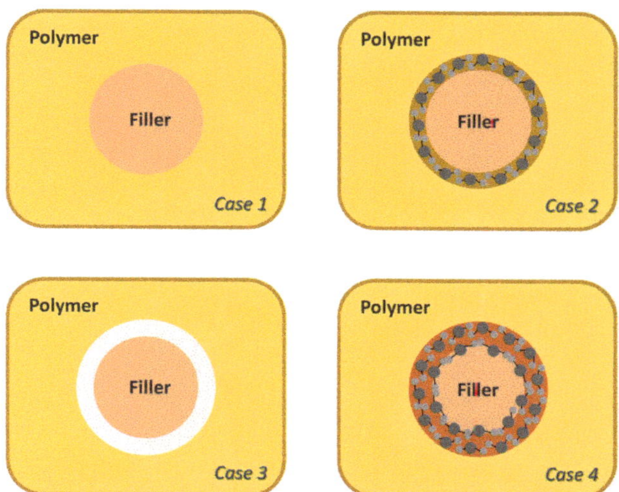

Figure 1. Schematic diagram of various structures for MMMs [38,39].

2. Experimental Section

2.1. Materials

2.1.1. Polymers

Three different polyimides were employed in the present study: 6FDA-6FpDA, Matrimid® 5218, and P84®. The 6FDA-6FpDA polyimide was synthesized following the classical in-situ silylation two steps method [40]. A detailed description of this synthesis can be found in a previous work [41]. Polyimides P84® and Matrimid® 5218 were purchased from HP Polymer GmbH (Lenzing, Austria) and Huntsman Advanced Materials (Salt Lake City, UT, USA), respectively.

2.1.2. Solvents

Tetrahydrofuran (THF), N-methyl-2-pyrrolidon (NMP), dimethylformamide (DMF), dimethylacetamid (DMAc), toluene, chloroform, and isopropanol for analysis grade were purchased from Merck (Darmstadt, Germany) and used as received.

2.1.3. Particles

Six inorganic fillers were employed: 8 mol.% Yttria Stabilized Zirconia (8YSZ), La_2O_3, $La_{5.4}WO_{12}$ (LaWO), $BaCe_{0.2}Zr_{0.7}Y_{0.1}O_3$ (BCZY) and two zeolites (ITQ-2 and Beta). 8YSZ powder was provided by Tosoh Corporation (Tokyo, Japan). La_2O_3 was synthesized by co-precipitation from lanthanum nitrate ($La(NO_3)_3$) and subsequent calcination at 800 °C for five h. LaWO, provided by CerPoTech (Tiller, Norway) in powder form was calcined at 800 °C for six h. BCZY powder, also provided by CerPoTech, was calcined at 950 °C for sixh. Nanocrystalline Beta zeolite (BEA material) and ITQ-2 (delaminated MCM-22 zeolite material) [42] were synthesized by the ITQ (Instituto de Tecnología Química, Valencia, Spain) and are here used after calcination (organics removal) in its acidic form. All the fillers were ball-milled previously for 24 h. Table 1 shows a summary of the fillers properties.

Table 1. Properties of the used particles. 8YSZ [43], La_2O_3 [44], LaWO [45], BCZY [46,47], ITQ-2 [42], and Beta [48].

Particles	8YSZ	La_2O_3	LaWO	BCZY	ITQ-2	Beta
Description	8% mol of Y_2O_3 stabilized ZrO_2 (Tosoh)	Co-precipitation from $La(NO_3)_3$. Calcined 800 °C/5 h	$La_{5.4}WO_{12}$ (CerPoTech). Calcined 950 °C/6 h	$BaCe_{0.2}Zr_{0.7}Y_{0.1}O_3$ (CerPoTech). Calcined 950 °C/6 h	ITQ-2 zeolite. Si/Al = 50	Zeolite nano-crystalline. Si/Al = 12.5
Structure						
Density (g/cm^3)	5.95	6.56	6.58	6.14	-	-
BET area (m^2/g)	6.0	2.9	9.4	31.4	>700	>700
Size (nm)	20–80	60–100	30–120	30–100	Thin sheets (2.5 thick)	10–30
Uses	Solid electrolyte in solid oxide fuel cells (SOFC)	Ferroelectric materials and as feedstock for catalysts	Asymmetric membranes for hydrogen separation	Asymmetric membranes for hydrogen separation	Catalysis	Catalysis

2.2. Membranes Fabrication

2.2.1. Inorganic Particles Dispersion

Different solvents such as THF, NMP, DMF, DMAc, toluene, chloroform, and isopropanol were tested for the dispersion of the particles. The ultrasonic devices used to disperse the fillers were the digital Sonifier®, models 250 and 450 (BRANSON Ultrasonics Corporation, Danbury, CT, USA). The dispersion was carried out with the pulse/pause mode, in particular, pulse off for one second and pulse on for one second for a total duration of 30 min. In addition, the dissolution container was inside an ice bath to avoid the sample heating during the dispersion process. In all the cases, the best dispersions and the most visually stable over time (smallest degree of sedimentation) were obtained by using NMP. The zeolites were fully suspended, whereas in the dispersions obtained with 8YSZ, La_2O_3, LaWO, and BCZY with some sedimentation with time observed.

2.2.2. Membranes Formation

Mix matrix membranes (MMMs) were made with 250 mg of polymer and inorganic fillers and 2.25 g of solvent. First, the inorganic fillers were dried at 120 °C for 24 h before the membrane preparation. Then, the fillers were dispersed in 1 g of NMP, as a solvent, by using an ultrasound device. At the same time, the polymer was dissolved with the rest of the solvent. The particle suspension was finally added to the polymer solution, obtaining a homogeneous solution of polymer and fillers. Subsequently, membranes were prepared by following the solvent evaporation method. The mixed matrix solutions were poured into metal rings placed on a heating plate at 70 °C for 12 h. Then, the membranes were heat treated following the steps: (a) 100 °C under vacuum for 1.5 h, (b) 200 °C under vacuum for 2 h, and (c) cooling to room temperature under vacuum.

2.3. Samples Characterization

Thermogravimetric analysis (TGA) experiments were carried out on the thermal analysis instrument NETZSCH TG 209 F1 Iris (Netzsch GmbH, Selb, Germany) in order to evaluate the thermal stability of the MMM and quantify the percentage of fillers. Disc samples with weights of between five and 15 mg were cut from the pieces obtained as described in Section 2.2. The TGA experiments consisted of two steps: (i) First, the sample was heated from 30 °C to 800 °C at 10 °C/min under an argon flow (dynamic scan); and (ii) once 800 °C was reached, the temperature was maintained for 30 min under synthetic air (static scan), in order to burn out the organic from the samples. The precision for the weight determination is ±0.1 μg. Differential scanning calorimetry (DSC) analysis was used to determine the T_g of polymers. DCS experiments were carried out with a calorimeter DSC 1 (Mettler Toledo, Columbus, OH, USA) at a heating rate of 10 K/min under nitrogen atmosphere to prevent oxidation. The glass transition is determined, with a precision of ±0.2 K, in the second heating cycle to avoid the history effect of sample.

The apparent molecular weight of the polymers and the MMMs was determined by Gel permeation Chromatography (GPC) after calibration with polystyrene standards. GPC measurements were performed at 40 °C having DMAc as eluent on a Waters instrument (Waters GmbH, Eschborn, Germany) equipped with polystyrene gel columns of different pore sizes, using a refractive index (RI) detector.

The XRD analysis were performed by using a D8 DISCOVER X-ray diffractometer (Bruker, Billerica, MA, USA). The range of measured Bragg angles was from 2 to 82°, with an increase of 10°. A 50 kV voltage and 1000 μA current was used.

The morphology of prepared MMMs and particle distribution throughout the membrane cross-section were analyzed using a scanning electron microscope (SEM) "Merlin" (Zeiss, Oberkochen, Germany). Samples for cross-sectional images were prepared by breaking the membrane immersed into liquid nitrogen and subsequent coating with a 4 nm carbon layer.

The permeability, diffusivity and solubility coefficients of CH_4, CO_2 and H_2O vapor in the manufactured membranes were determined by using the well-known constant volume, variable pressure method, i.e., the "time-lag method" [40]. The basic principle is the measurement of the transitory response at the downstream part of a membrane to a pressure step at the upstream part, that is, the time-lag (t_0), which is graphically determined as the intersection of the line drawn through the linear region of the pressure increase curve to intersection with the time axis [49]. The diffusion coefficient (D) is linked to the time-lag (t_0) through the Equation (1).

$$t_0 = \frac{l^2}{6 \cdot D} \tag{1}$$

where l is the thickness of the membrane. The permeability coefficient (P) can be obtained from the range where the permeate pressure increases linearly (Equation (2)).

$$P = D \cdot S = \frac{V_p l (p_{p2} - p_{p1})}{ART\Delta t (p_f - (p_{p2} + p_{p1}))} \tag{2}$$

where V_p is the constant permeate volume, l is the film thickness, A is the effective area of the membrane, R is the gas constant, Δt is the time of the permeate pressure increase from p_{p1} to p_{p2}, and p_f is the feed pressure. Finally, solubility coefficient (S) can be obtained with the permeability and the diffusion coefficient by means of Equation (3).

$$S = \frac{P}{D} \tag{3}$$

The measurements were made at different temperatures and at one bar of feed pressure. For each gas measurement, the facility was evacuated until no desorption from the membrane was observed and the gas to be measured was subsequently refilled. The feed and permeate sides of the membrane are connected to a vacuum pump with valves and additional valves that connect the feed side with several gases.

Experiments on water vapor transport were carried out as follows: The pressure vessel keeping feed pressure of a gas or vapor under study at a constant level during the acquisition of the time-lag curve was filled with water vapor corresponding to the saturation pressure at a given temperature. For the beginning of the experiment, the feed pressure vessel was connected to the previously evacuated measurement cell by opening vacuum valves. It caused a drop of the vapor pressure to approximately

70% of vapor activity at a given temperature. Due to the design of the vacuum system of the "time-lag" facility, it is not possible to carry out vapor measurements at activities higher than 70%. After the time-lag curve was recorded the system was completely evacuated and experiment repeated for 3 times. For the CH_4 and CO_2, gas properties were studied at temperatures up to 80 °C.

3. Results and Discussion

3.1. Thermal Properties

Thermogravimetric analysis experiments (TGA) were performed in order to evaluate the thermal stability of the MMMs and to additionally determine the real amount of particles in the final sample [50]. The parameter used to evaluate the quantity of particles in the membranes is the residual mass (R_M). In Table 2, the TGA results for three different cases studied in this work are represented: 6FDA-6FpDA with 10 wt.% of the six different particles, 6FDA-6FpDA with different percentage of the BCZY particles, and 6FDA-6FpDA, Matrimid® and P84® with 10 wt.% of BCZY. It can be appreciated that the temperature of the maximum weight loss ($T_{max\ loss}$) is a characteristic property of the polymer matrix, i.e., 550 °C for 6FDA-6FpDA, 560 °C for Matrimid®, and finally, 580 °C for P84®. The value of the residual mass R_M, determined after exposure of the sample to synthetic air at 800 °C, indicates the content of inorganic particles within the sample. The low deviation from the theoretical values observed for 10 wt.% 8YSZ, La_2O_3, LaWO, and BCZY in 6FDA-6FpDA indicates a good dispersion and adhesion of these particles in the polymer matrix [37,51]. On the contrary, in the case of the zeolites (ITQ-2 and Beta), an important difference between nominal and experimental content occurs. This difference is attributed to the nanometer size of the particles that may cause a loss of part of the constituting elements of these particles. On the other hand, 6FDA-6FpDA + 20 wt.% BCZY and Matrimid® + 10 wt.% BCZY present an R_M value higher than the initial percentage of particles that indicates an irregular distribution of the polymer chains around the particles. Additionally, an identical temperature of the maximum weight loss $T_{max\ loss}$ for 6FDA-6FpDA with different percentages of the BCZY particles is observed in Figure 2. From these results, it can be concluded that the incorporation of inorganic particles to the polymer matrix does not affect the thermal stability of the polymer.

Table 2. Thermo gravimetric analysis (TGA) and differential scanning calorimetry (DSC) results for the three different cases studied in this work.

Sample Description	$T_{max\ loss}$	Theor. wt.%	R_M wt.%	T_g
6FDA-6FpDA	550 °C	0	0	311.1 °C
+10 wt.% 8YSZ	550 °C	10	9.8	300.2 °C
+10 wt.% La_2O_3	550 °C	10	8.1	302.4 °C
+10 wt.% LaWO	550 °C	10	8.1	290.6 °C
+10 wt.% BCZY	550 °C	10	10.0	311.2 °C
+10 wt.% ITQ-2	550 °C	10	0	294.7 °C
+10 wt.% Beta	550 °C	10	0	294.5 °C
6FDA-6FpDA	550 °C	0	0	311.1 °C
+1 wt.% BCZY	550 °C	1	0	314.2 °C
+5 wt.% BCZY	550 °C	5	4.9	312.5 °C
+10 wt.% BCZY	550 °C	10	10.0	311.2 °C
+15 wt.% BCZY	550 °C	15	16.8	307.9 °C
+20 wt.% BCZY	550 °C	20	27.2	306.3 °C
6FDA-6FpDA	550 °C	0	0	311.1 °C
+10 wt.% BCZY	550 °C	10	10.0	311.2 °C
Matrimid®	560 °C	0	0	320.2 °C
+10 wt.% BCZY	560 °C	10	13.3	315.7 °C
P84®	580 °C	0	0	322.4 °C
+10 wt.% BCZY	580 °C	10	7.2	318.2 °C

The glass transition temperature (T_g) of these three polymers was determined by differential scanning calorimetry (DSC). An evolution of the T_g with the rigidity of the polymer chains following the order of rigidity 6FDA-6FpDA < Matrimid® < P84® was found. For all the 6FDA-6FpDA-based MMMs, the T_g decreases in comparison to the pure polyimide, except for the BCZY membrane that presents a T_g similar to the reference. The observed reduction of the T_g may indicate a plasticization effect introduced by the filler particles.

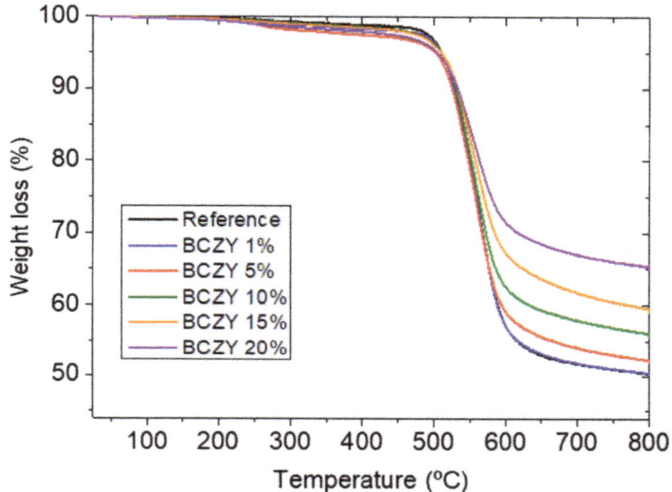

Figure 2. TGA graphs for the results of 6FDA-6FpDA with different percentage of the BCZY particles.

When the behavior of the T_g for different BCZY contents in the polymer matrix was analyzed, only small irregular changes were found, i.e., an increase of T_g for low BCZY percentages between one and 5 wt.% and a similar value for 10 wt.%, which may indicate a small macromolecular chain rigidification [52] (see Table 2). This could have an effect on the gas transport properties, expecting an increase in selectivity but decrease of the permeability coefficients.

3.2. Microstructure Characterization

The X-ray diffraction (XRD) technique is used to evaluate the interaction between the selected particles with the polymer matrix. Figure 2 shows the X-ray spectra of the MMMs made of 6FDA-6FpDA with 10 wt.% of particles and the reference patterns of 8YSZ, $La(OH)_3$, LaWO, and BCZY compounds. Regarding La_2O_3, this inorganic filler is highly hygroscopic, which may affect its crystalline structure. Actually, this was the case in this work, and as revealed in Figure 3, the pattern of La_2O_3 changed to $La(OH)_3$. XRD patterns demonstrate that the particles are well integrated in the polymer matrix, showing the combination of the diffraction pattern of the polymer and particles. Nevertheless, in the case of 8YSZ, La_2O_3, and Beta zeolite, the main peak of the polymer matrix moves to higher values of 2θ, and therefore, the polymer intersegmental distance decreases [53–55]. This may indicate that part of the fraction free volume (FFV) of the polymer can be altered in the vicinity to the embedded particle.

Figure 4 displays the X-ray diffraction pattern of MMMs made of 6FDA-6FpDA, Matrimid® and P84® combined with BCZY 10 wt.% and illustrates how the intersegmental distance between polymer backbones are directly related to the FFV (6FDA-6FpDA > Matrimid® > P84®) [56,57].

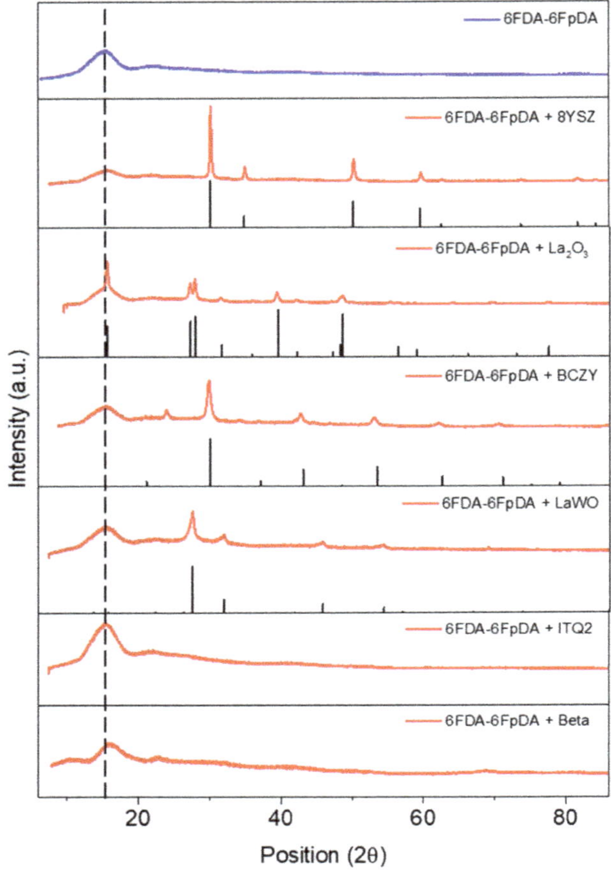

Figure 3. X-ray diffraction patterns for 6FDA-6FpDA with 10 wt.% fillers (red lines) and reference patterns corresponding to 8YSZ, $La(OH)_3$, LaWO, and BCZY crystals (black peaks).

The fillers distribution, sedimentation, and agglomeration in the membranes were analyzed based on scanning electron microscope (SEM) cross-sectional images.

The cross-section of 6FDA-6FpDA MMMs with 10 wt.% fillers (Figure 5) shows that part of the inorganic fillers forms agglomerates. Theoretically, the size of particles (Table 1) should be in the nanometer range, but most of the fillers occur as agglomerates of embedded single nanoparticles. Additionally, sedimentation of particles can also be appreciated, i.e., there is a certain sedimentation for most fillers. These two phenomena may be associated with the dynamics of the membrane formation process. In general, the process for solvent evaporation is slow, and it gives enough time for particles for sedimentation to the bottom, which is something very common in thick film MMMs. This is related to the solvent used, NMP, that exhibits a high boiling point and to the high density of most of the studied filler particles (Table 1). When all the inorganic fillers are compared, it is ascertained that La_2O_3 showed the largest amount of agglomerates, and in general, a poor distribution. In contrast, BCZY is the filler with the best particle distribution throughout the membrane thickness, and the least agglomeration and sedimentation.

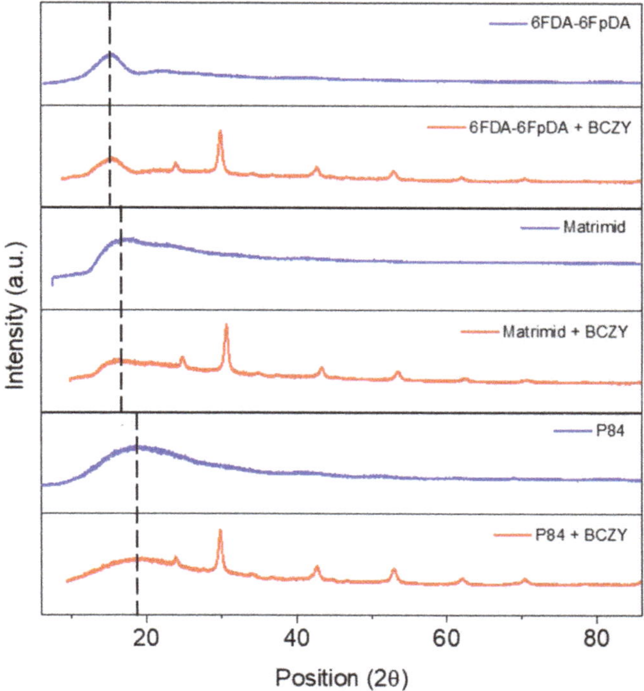

Figure 4. X-ray patterns for different polymer matrix:Pure, blue lines, and with 10 wt.% BCZY, red lines.

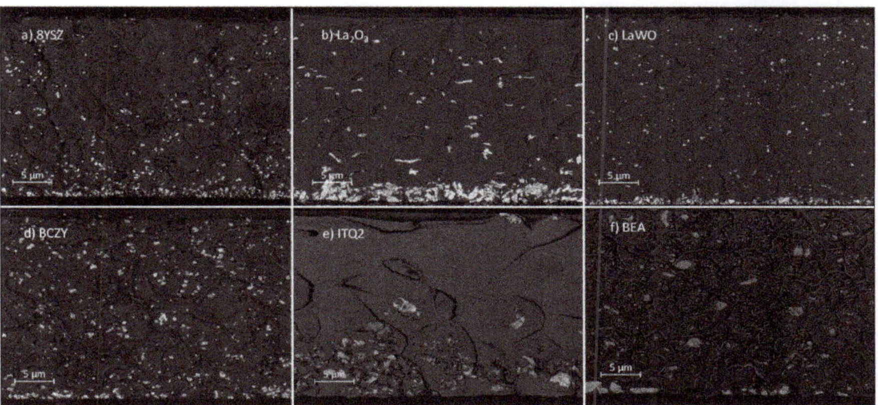

Figure 5. SEM images (fracture cross-sections) for 6FDA-6FpDA MMM and 10 wt.% fillers.

3.3. Gas Transport Properties

Gas transport properties for the three different cases (i.e., particle type, particle content and polymer type) were evaluated. Time lag equipment was used to study, not only the permeability and the selectivity, but also the solubility and the diffusivity coefficients of gases and water vapor in MMMs as a function of temperature.

3.3.1. Influence of the Particle Type

In this section, MMMs composed by 90 wt.% of 6FDA-6FpDA as polymer matrix and 10 wt.% of different fillers are characterized Table 3 shows how the particles influence the polymer matrix reference (6FDA-6FpDA) in terms of permeability and selectivity at 30 °C. Permeability variations are always negative, and on the contrary, the selectivity variations are positive in all the cases. The decrease in permeability with the addition of inorganic fillers can be related to the formation of a densified layer of polymer on the polymer/particle interphase (case 2), although a reduction of the T_g was not observed (Table 2). Polymer densification leads to reduced free volume, and consequently, to higher selectivity for the gas pair with significantly different kinetic diameters of gas molecules. Interestingly, the decrease in CO_2 permeability leads to significant, reverse-proportional increase of the activation energy of CO_2 permeability [58]. This observation allows one to conclude good contact between polymer and inorganic particle, i.e., presence of no gaps in this interface [18,59].

Table 3. CO_2 permeability, CO_2/CH_4 selectivity, percentage variations of permeability and selectivity (6FDA-6FpDA with 10 wt.%fillers at 30 °C). Additionally, the activation energy for CO_2 permeability derived from the data shown in Figure 6.

Membrane Sample Description	CO_2 Permeability (Barrer)	CO_2/CH_4 Selectivity (-)	CO_2 Permeability Variation (%)	CO_2/CH_4 Selectivity Variation (%)	Activation Energy (KJ/mol)
6FDA-6FpDA (Reference)	77.4	48.0	-	-	0.69
+10 wt.% 8YSZ	25.8	53.9	−67	+12	3.73
+10 wt.% La_2O_3	34.1	51.9	−56	+8	2.69
+10 wt.% LaWO	11.9	77.3	−85	+61	5.51
+10 wt.% BCZY	63.8	54.6	−18	+14	1.22
+10 wt.% ITQ-2	28.9	55.1	−63	+15	2.63
+10 wt.% Beta	22.7	64.9	−71	+35	4.98

The permeability of CO_2 and selectivity for the gas pair CO_2/CH_4 was also measured as a function of temperature. CO_2 permeability increases with temperature for all tested MMMs. However, none of them exhibits higher permeability values than the reference membrane, as is observed in Figure 6 (left-hand). Selectivity decreases as a function of temperature, while the effect of the type of filler becomes more visible and relevant at lower temperatures, because the differences between them are more evident. Table 3 displays the activation energy for the different fillers and the reference. The activation energy of the MMMs is higher than the activation energy of the polymeric membrane 6FDA-6FpDA. Hence, the formation of a rigid layer around the particles is confirmed.

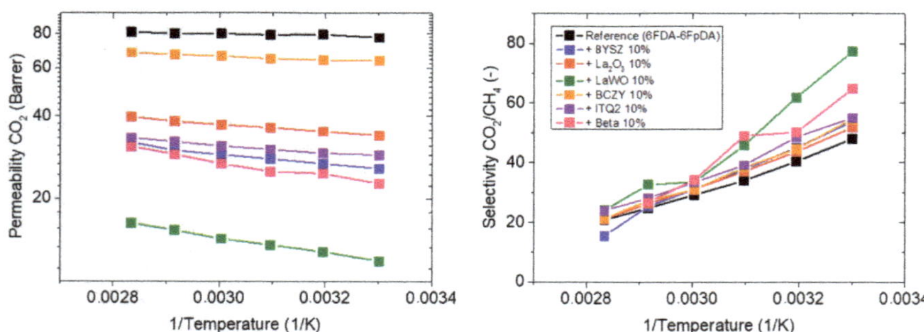

Figure 6. Permeability of CO_2 and CO_2/CH_4 selectivity for the MMM composed by 90 wt.% of 6FDA-6FpDA and 10 wt.% of different fillers as a function of temperature.

Taking into consideration all the MMMs, the sample with BCZY particle exhibits the highest permeability value, as well as a notable increase in CO_2/CH_4 selectivity. All MMMs exhibit worse permeability than the reference, while a visible improvement in selectivity can be ascertained. Gas permeability through a membrane depends on two parameters, diffusivity coefficient and solubility coefficient (Figure 7). The solubility coefficient was not improved by the incorporation of particles. In particular, the membranes containing BCZY and La_2O_3 particles have practically the same CO_2 solubility coefficient as the pure polymer, but the rest of inorganic fillers causes a CO_2 solubility coefficient decrease, and all of them decrease as a function of temperature.

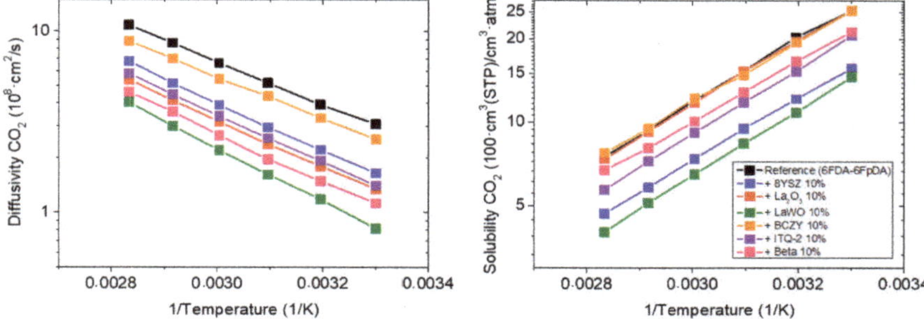

Figure 7. Diffusivity and solubility coefficient of MMMs composed by 90 wt.% of 6FDA-6FpDA and 10 wt.% of different fillers, as a function of temperature.

On the contrary, the diffusivity coefficient, which is mainly determined by the FFV, increases as a function of temperature for all the MMMs. It is assumed that particles are blocking part of the FFV of the polymer matrix, and hence, the final permeability decreases. Observing the evolution of the diffusivity coefficient with temperature, this statement can be confirmed, since the reference diffusivity coefficient is higher than the rest at room temperature, and this difference is bigger at higher temperatures.

As a conclusion, BCZY, among all the studied fillers, was the most promising particle for MMM due to the fact that it is the particle that exhibits the highest permeability value, as well as a notable increase in CO_2/CH_4 selectivity. Therefore, this nano-sized filler was selected for the next step, i.e., evaluation of the influence of the amount of nanoparticles on the separation performance.

3.3.2. Influence of the Particle Content

In order to understand the influence of the inorganic filler concentration on MMM transport properties, membranes were prepared with different BCZY content in weight percent 1%, 5%, 10%, 15% and 20% in the same polymer matrix, i.e., 6FDA-6FpDA. Table 4 summarizes permeability and ideal selectivity data obtained for the aforementioned membranes at 30 °C.

Table 4. CO_2 permeability, CO_2/CH_4 selectivity, percentage variations of permeability, and selectivity (6FDA-6FpDA with different % of BCZY at 30 °C). Additionally, the activation energy for CO_2 permeability derived from permeability vs. temperature data is listed.

Membrane Sample Description	CO_2 Permeability (Barrer)	CO_2/CH_4 Selectivity (-)	CO_2 Permeability Variation (%)	CO_2/CH_4 Selectivity Variation (%)	Activation Energy (KJ/mol)
6FDA-6FpDA (Reference)	77.4	48.0	-	-	0.69
+1 wt.% BCZY	61.4	49.7	−21	+3.6	2.56
+5 wt.% BCZY	45.5	49.6	−41	+3.3	1.89
+10 wt.% BCZY	63.8	54.6	−18	+14	1.22
+15 wt.% BCZY	66.0	47.8	−15	−0.3	0.57
+20 wt.% BCZY	59.7	45.1	−23	−6.0	1.46

As in the previous section, the permeability values for the different filler concentration are lower than the permeability of the reference membrane (6FDA-6FpDA). In contrast, regarding the selectivity variation, the MMM with the highest content of particles has a negative effect on selectivity. This suggests that not only the filler particles are blocking the FFV—decreasing the permeability—but also aggregates influence the selectivity negatively.

Permeability of CO_2 and selectivity of the gas pair CO_2/CH_4 was measured as a function of temperature. Activation energies for the MMMs of 6FDA-6FpDA with different % of BCZY filler are indicated in Table 4. Permeability of CO_2 increases with temperature but decreases with the content of BCZY fillers. All the membranes with BCZY exhibit higher permeability values than with the other fillers studied in the previous section (see Figure 6). This fact indicates that BCZY is the most promising particle. Regarding the selectivity of CO_2/CH_4, it decreases as a function of temperature, and there is not a wide difference between the reference polymer and MMMs. No clear dependence between BCZY content and activation energy can be observed. Therefore, in the next sub-section, the filler BCZY at 10 wt.% is tested for two other polymers that present lower FFV, that is, with lower permeability, in order to check if the addition of particles enables it to improve their separation performance.

3.3.3. Influence of the Polymer Matrix

The influence of different polymeric matrices with different FFV (6FDA-6FpDA, Matrimid® and P84®) on the separation properties is studied, employing the results of the previous studies, that is, the most suitable filler (BCZY) and the most suitable proportion (10 wt.%) [60]. Table 5 shows the polymer matrix influence (6FDA-6FpDA, Matrimid® and P84®) in terms of permeability and selectivity at 30 °C. We hypothesized that, for lower permeable polymers, the particles may positively affect the gas separation properties of the mixed matrix membranes [61–63]. As is shown in the table below, the Matrimid® mixed matrix membrane exhibits higher permeability, as well as selectivity, compared to its reference (case 1 in Figure 1), while the P84® mixed matrix membrane behaves similar to the 6FDA-6FpDA mixed matrix membrane, with a decrease in permeability and an increase in selectivity (case 2 in Figure 1) [59].

Table 5. CO_2 permeability, CO_2/CH_4 selectivity, percentage variations of permeability and selectivity (membranes with different polymeric matrix at 30 °C). Additionally, activation energy is calculated.

Membrane Sample Description	CO_2 Permeability (Barrer)	CO_2/CH_4 Selectivity (-)	CO_2 Permeability Variation (%)	CO_2/CH_4 Selectivity Variation (%)	Activation Energy (KJ/mol)
6FDA-6FpDA (Reference)	77.4	48.0	-	-	0.69
+10 wt.% BCZY	63.8	54.6	−18	+14	1.22
Matrimid®	5.1	40.5	-	-	9.30
+10 wt.% BCZY	6.7	47.5	+31	+17	7.42
P84®	1.4	47.0	-	-	11.64
+10 wt.% BCZY	1.0	64.6	−32	+38	18.33

The permeability of CO_2 and the selectivity of the CO_2/CH_4 gas pair were determined as a function of temperature, see Figure 8. All polymers show good compatibility with BCZY particles, but behave differently, as casted MMMs. 6FDA-6FpDA and P84® show lower permeability coefficients of CO_2 in MMM form while Matrimid® with 10 wt.% of BCZY shows an improvement in both CO_2 permeability and CO_2/CH_4 selectivity. As it can be expected for the CO_2/CH_4 gas pair, the ideal selectivity decreases with temperature for both pure polymer and MMMs.

Figure 9 shows diffusion and solubility coefficients of CO_2 as a function of temperature. It can be ascertained that the permeability of Matrimid® with BCZY particles is higher than the pure polymer Matrimid® because solubility coefficients improve, and diffusivity coefficients remain constant. In general, the activation energy of CO_2 diffusivity (Figure 9) of the polymer membrane is not influenced by the incorporation of BCZY particles suggesting that the diffusion mechanism is apparently not affected.

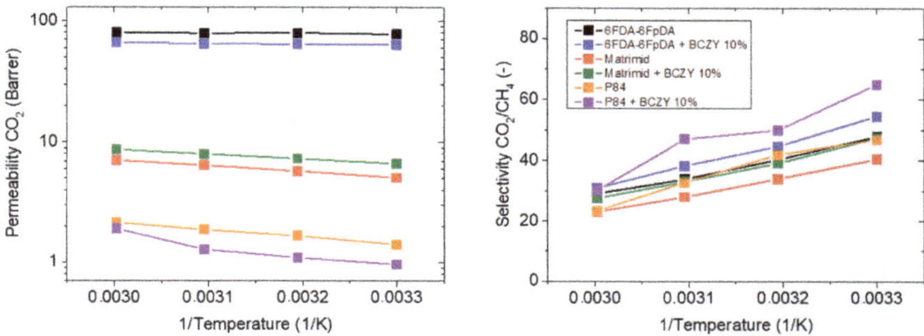

Figure 8. Permeability of CO_2 and CO_2/CH_4 selectivity for the membranes made of 6FDA-6FpDA, Matrimid® and P84® with and without 10 wt.% BCZY as a function of temperature.

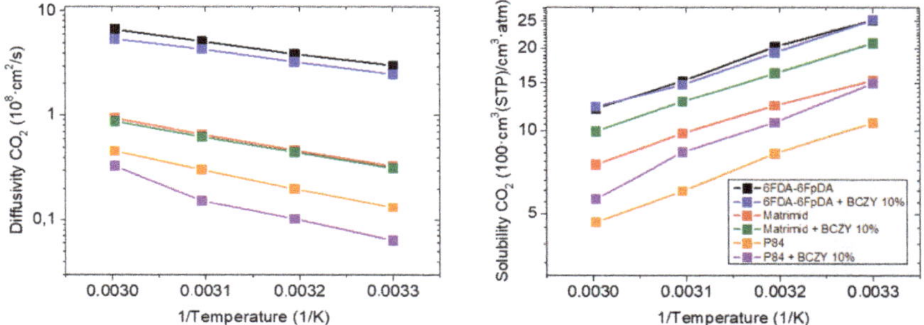

Figure 9. Diffusivity and solubility coefficient of CO_2 of membranes made of 6FDA-6FpDA, Matrimid® and P84® with and without 10 wt.% BCZY as a function of temperature.

3.3.4. Transport of Water Vapor in MMMs

All the MMMs tested until now exhibits high CO_2 solubility coefficients, which may be related to the water vapor transport properties. Additionally, the fillers selected for this work exhibit a high affinity towards water. They can be perspective materials for membranes able to remove water from CO_2-rich gas streams as combustion tail-gas or catalytic converters products. Hence, all prepared membranes were tested for water vapor transport using the "time-lag" setup.

Table 6 summarizes water vapor permeability's and the activation energies for all the MMMs studied in this work. The achieved permeability values are remarkably high, although permeability decreases as a function of temperature. The negative values of the activation energy are related to the exothermic nature of the water solubility and its effect dominates in the overall separation process. In all cases, water permeability decreased in MMMs compared to pure polymers except the case of Matrimid® with 10 wt.% of BCZY. This filler showed the highest permeability coefficient of MMMs when compared to other fillers mixed with 6FDA-6FpDA.Looking at activation energies of water permeability in 10 wt.% MMMs based on three polymers, one may conclude that the "slower" polymers Matrimid® and P84® are more benefitting by the incorporation of BCZY. Namely, P84®-based MMM shows slightly positive E_a (P) for water vaporand Matrimid® MMM exhibits a higher water permeability coefficient than for the pure polymer and a lower corresponding activation energy. These two materials are appealing materials for further application in membrane-assisted water vapor removal from combustion streams or in catalytic reactors in order to shift chemical equilibrium in CO_2-hydrogenation reactions.

Table 6. Steam permeation at 30 °C and activation energies for all the MMMs studied in this work at 30 °C.

	H_2O Permeability (Barrer)	H_2O/CO_2 Selectivity (-)	Activation Energy (KJ/mol)
6FDA-6FpDA	3875	50.06	−3.34
+10 wt.% 8YSZ	1998	77.48	−2.22
+10 wt.% La_2O_3	2381	69.85	−2.79
+10 wt.% LaWO	1287	108.30	−1.35
+10 wt.% BCZY	3319	52.02	−3.31
+10 wt.% ITQ-2	2015	69.70	−1.31
+10 wt.% Beta	1914	84.35	0.69
6FDA-6FpDA	3875	50.06	−3.34
+1 wt.% BCZY	3144	51.20	−2.30
+5 wt.% BCZY	2766	60.83	−2.23
+10 wt.% BCZY	3319	52.02	−3.31
+15 wt.% BCZY	3276	49.67	−2.13
+20 wt.% BCZY	3108	52.03	−2.28
6FDA-6FpDA	3875	50.06	−3.34
+10 wt.% BCZY	3319	52.02	−3.31
Matrimid®	1524	300.60	0.87
+10 wt.% BCZY	1835	276.02	−1.16
P84®	1226	875.71	2.36
+10 wt.% BCZY	821	856.40	1.73

4. Conclusions

A complete study of the influence of inorganic fillers on the gas transport properties of different polyimides, typically applied in gas separation processes, i.e., purification of biogas and natural gas was developed. The strategy of this work was the incorporation of inorganic nanoparticles as MMM fillers. These inorganic materials were selected because the expected good affinity for gas molecules as CO_2 and water vapor. The particles were successfully dispersed and incorporated into the polymer matrix. Consequently, 15 different MMMs were produced as a thick film. Materials were characterized by different techniques including TGA, XRD, DSC, or SEM. Gas transport properties were evaluated for CH_4, CO_2, and H_2O at temperatures from 30 to 80 °C in a time lag equipment.

It was observed that, in general, the inorganic fillers could produce small rigidification in the polymer matrix, although they do not exhibit higher T_g. MMMs studied in this work allowed for improve selectivity, but with a negative impact on permeability. This could be caused by particle aggregation (see Figure 5 for SEM images), blocking part of the FFV but increasing tortuosity of the gases trough the membrane. Regarding the influence of the particle type and/or content, no clear effect of the particles in terms of pore size or particle size was discovered. XRD analysis shows a small decrease in interspacing of the polymer chains with no modification of the particles pattern, meaning that there is not any interaction with polymer.

Taking all the inorganic particles into consideration, BCZY shows the best improvement of selectivity with a small decrease in permeability. In addition, it also exhibits the best distribution, and consequently, it was selected for the rest of the experiments (different percentages and different polymer matrix). For polymer matrixes with lower FFV, such as Matrimid® and P84®, there is a result in improvement of the properties by adding particles, possibly due to the creation of interface between particle and polymer chain. Therefore, MMMs with particles can be used to create interface and we will increase separation properties of slow polymers. Regarding temperature dependence, some changes were observed on activation energy of the process, the incorporation of inorganic fillers does not significantly affect the permeation mechanism determined by the polymer transport properties.

Finally, water permeability was first reported for several polyimides and MMMs of inorganic particles with polyimides, reaching relatively high values. However, the effect of the filler incorporation on the water permeation was not relevant for the polymers except for Matrimid.

Author Contributions: S.E. (Sara Escorihuela), L.V., A.T. and S.E. (Sonia Escolástico) performed the experiments and the data analysis; A.T., S.S. and J.S. Conceived and design the experiments; T.B. and J.M.S. supervised the study and provided scientific discussions. All the authors contribute to the writing paper.

Funding: This work was financially supported by the Spanish Government (SEV-2016-0683, SVP-2014-068356, Project ENE2014-57651-R and IJCI-2016-28330 grants) and GeneralitatValenciana (PROMETEO/2018/006 grant) and Helmholtz-Zentrum Geesthacht (HZG) through the technology transfer project program and by the Helmholtz Association of German Research Centers through the Helmholtz Portfolio MEMBRAIN.

Acknowledgments: Authors are grateful to Susana Valencia (Instituto de Tecnología Química (UPV-CSIC)) for zeolite synthesis; Clarissa Abetz and Anke-Lisa Hoeme (Helmholtz-Zentrum Geesthacht) for excellent SEM images; Silvio Neumann (Helmholtz-Zentrum Geesthacht) for the TGA and DSC measurements and Giovanni Capurso (Helmholtz-Zentrum Geesthacht) for the XRD measurements.

Conflicts of Interest: The authors declare no conflict of interest.

References

1. Santi Kulprathipanja, R.W.N.; Norman, N. Li Separation of Fluids by Means of Mixed Matrix Membranes. U.S. Patent 4,740,219, 26 April 1988.
2. Kulprathipanja, S. Mixed matrix membrane development. *Ann. N. Y. Acad. Sci.* **2003**, *984*, 361–369. [CrossRef] [PubMed]
3. Robeson, L.M. The upper bound revisited. *J. Membr. Sci.* **2008**, *320*, 390–400. [CrossRef]
4. Paul, D.R.; Âmpol'skij, U.P. *Polymeric Gas Separation Membranes*; CRC: Boca Raton, FL, USA, 1994.
5. Baker, R.W. Research needs in the membrane separation industry: Looking back, looking forward. *J. Membr. Sci.* **2010**, *362*, 134–136. [CrossRef]
6. Stünkel, S.; Drescher, A.; Wind, J.; Brinkmann, T.; Repke, J.U.; Wozny, G. Carbon dioxide capture for the oxidative coupling of methane process—A case study in mini-plant scale. *Chem. Eng. Res. Des.* **2011**, *89*, 1261–1270. [CrossRef]
7. Cheng, Y.; Wang, Z.; Zhao, D. Mixed matrix membranes for natural gas upgrading: Current status and opportunities. *Ind. Eng. Chem. Res.* **2018**, *57*, 4139–4169. [CrossRef]
8. Koros, W.; Zhang, C. Materials for next-generation molecularly selective synthetic membranes. *Nat. Mater.* **2017**, *16*, 289–297. [CrossRef]
9. Li, Y.; He, G.; Wang, S.; Yu, S.; Pan, F.; Wu, H.; Jiang, Z. Recent advances in the fabrication of advanced composite membranes. *J. Mater. Chem. A* **2013**, *1*, 10058–10077. [CrossRef]
10. Liu, Y.; Liu, G.; Zhang, C.; Qiu, W.; Yi, S.; Chernikova, V.; Chen, Z.; Belmabkhout, Y.; Shekhah, O.; Eddaoudi, M.; et al. Enhanced CO_2/CH_4 separation performance of a mixed matrix membrane based on tailored mof-polymer formulations. *Adv. Sci.* **2018**, *5*, 1800982. [CrossRef]
11. Bae, T.-H.; Liu, J.; Lee, J.S.; Koros, W.J.; Jones, C.W.; Nair, S. Facile high-yield solvothermal deposition of inorganic nanostructures on zeolite crystals for mixed matrix membrane fabrication. *J. Am. Chem. Soc.* **2009**, *131*, 14662–14663. [CrossRef]
12. Zornoza, B.; Téllez, C.; Coronas, J. Mixed matrix membranes comprising glassy polymers and dispersed mesoporous silica spheres for gas separation. *J. Membr. Sci.* **2011**, *368*, 100–109. [CrossRef]
13. Anson, M.; Marchese, J.; Garis, E.; Ochoa, N.; Pagliero, C. Abs copolymer-activated carbon mixed matrix membranes for CO_2/CH_4 separation. *J. Membr. Sci.* **2004**, *243*, 19–28. [CrossRef]
14. Kim, S.; Chen, L.; Johnson, J.K.; Marand, E. Polysulfone and functionalized carbon nanotube mixed matrix membranes for gas separation: Theory and experiment. *J. Membr. Sci.* **2007**, *294*, 147–158. [CrossRef]
15. Adams, R.; Carson, C.; Ward, J.; Tannenbaum, R.; Koros, W. Metal organic framework mixed matrix membranes for gas separations. *Microporous Mesoporous Mater.* **2010**, *131*, 13–20. [CrossRef]
16. McKeown, N.B. A perfect match. *Nat. Mater.* **2018**, *17*, 216–217. [CrossRef] [PubMed]
17. Dechnik, J.; Sumby, C.J.; Janiak, C. Enhancing mixed-matrix membrane performance with metal–organic framework additives. *Cryst. Growth Des.* **2017**, *17*, 4467–4488. [CrossRef]
18. Bastani, D.; Esmaeili, N.; Asadollahi, M. Polymeric mixed matrix membranes containing zeolites as a filler for gas separation applications: A review. *J. Ind. Eng. Chem.* **2013**, *19*, 375–393. [CrossRef]
19. Carreon, M.A. *Membranes for Gas Separations*; Colorado School of Mines: Golden, CO, USA, 2018.

20. Liu, C.; Greer, D.W.; O'Leary, B.W. Advanced materials and membranes for gas separations: The uop approach. In *Nanotechnology: Delivering on the Promise Volume 2*; American Chemical Society: Washington, DC, USA, 2016; Volume 1224, pp. 119–135.
21. Dechnik, J.; Gascon, J.; Doonan, C.J.; Janiak, C.; Sumby, C.J. Mixed-matrix membranes. *Angew. Chem. Int. Ed.* **2017**, *56*, 9292–9310. [CrossRef] [PubMed]
22. Yang, Y.; Chuah, C.Y.; Nie, L.; Bae, T.-H. Enhancing the mechanical strength and CO_2/CH_4 separation performance of polymeric membranes by incorporating amine-appended porous polymers. *J. Membr. Sci.* **2019**, *569*, 149–156. [CrossRef]
23. Mikkelsen, M.; Jørgensen, M.; Krebs, F.C. The teraton challenge. A review of fixation and transformation of carbon dioxide. *Energy Environ. Sci.* **2010**, *3*, 43–81. [CrossRef]
24. Miltner, M.; Makaruk, A.; Harasek, M. Review on available biogas upgrading technologies and innovations towards advanced solutions. *J. Clean. Prod.* **2017**, *161*, 1329–1337. [CrossRef]
25. Ullah Khan, I.; Hafiz Dzarfan Othman, M.; Hashim, H.; Matsuura, T.; Ismail, A.F.; Rezaei-DashtArzhandi, M.; Wan Azelee, I. Biogas as a renewable energy fuel—A review of biogas upgrading, utilisation and storage. *Energy Convers. Manag.* **2017**, *150*, 277–294. [CrossRef]
26. Montañez-Hernández, L.; Hernandez De Lira, I.; Rafael-Galindo, G.; de Lourdes Froto Madariaga, M.; Balagurusamy, N. Sustainable production of biogas from renewable sources: Global overview, scale up opportunities and potential market trends. *Sustain. Biotechnol. Enzymatic Resour. Renew. Energy* **2018**, 325–354. [CrossRef]
27. Baker, R.W.; Lokhandwala, K. Natural gas processing with membranes: An overview. *Ind. Eng. Chem. Res.* **2008**, *47*, 2109–2121. [CrossRef]
28. Zhang, Y.; Sunarso, J.; Liu, S.; Wang, R. Current status and development of membranes for co2/ch4 separation: A review. *Int. J. Greenh. Gas Control* **2013**, *12*, 84–107. [CrossRef]
29. Rezakazemi, M.; Ebadi Amooghin, A.; Montazer-Rahmati, M.M.; Ismail, A.F.; Matsuura, T. State-of-the-art membrane based CO_2 separation using mixed matrix membranes (mmms): An overview on current status and future directions. *Prog. Polym. Sci.* **2014**, *39*, 817–861. [CrossRef]
30. Angelidaki, I.; Treu, L.; Tsapekos, P.; Luo, G.; Campanaro, S.; Wenzel, H.; Kougias, P.G. Biogas upgrading and utilization: Current status and perspectives. *Biotechnol. Adv.* **2018**, *36*, 452–466. [CrossRef] [PubMed]
31. Yong-Woo, J.; Dong-Hoon, L. Gas membranes for CO_2/CH_4 (biogas) separation: A review. *Environ. Eng. Sci.* **2015**, *32*, 71–85.
32. Murali, R.S.; Sankarshana, T.; Sridhar, S. Air separation by polymer-based membrane technology. *Sep. Purif. Rev.* **2013**, *42*, 130–186. [CrossRef]
33. Kanehashi, S.; Chen, G.Q.; Ciddor, L.; Chaffee, A.; Kentish, S.E. The impact of water vapor on co2 separation performance of mixed matrix membranes. *J. Membr. Sci.* **2015**, *492*, 471–477. [CrossRef]
34. Kreuer, K.D. Proton-conducting oxides. *Annu. Rev. Mater. Res.* **2003**, *33*, 333–359. [CrossRef]
35. Haugsrud, R. Defects and transport properties in Ln_6WO_{12} (Ln = La, Nd, Gd, Er). *Solid State Ion.* **2007**, *178*, 555–560. [CrossRef]
36. Kim, S.; Anselmi-Tamburini, U.; Park, H.J.; Martin, M.; Munir, Z.A. Unprecedented room-temperature electrical power generation using nanoscale fluorite-structured oxide electrolytes. *Adv. Mater.* **2008**, *20*, 556–559. [CrossRef]
37. Fernández-Barquín, A.; Casado Coterillo, C.; Palomino, M.; Valencia, S.; Irabien, A. Lta/poly(1-trimethylsilyl-1-propyne) mixedmatrix membranes for high-temperature CO_2/N_2 separatio. *Chem. Eng. Technol.* **2015**, *38*, 658–666. [CrossRef]
38. Ansaloni, L.; Deng, L. 7–advances in polymer-inorganic hybrids as membrane materials. In *Recent Developments in Polymer Macro, Micro and Nano Blends*; Visakh, P.M., Markovic, G., Pasquini, D., Eds.; Woodhead Publishing: Cambridge, UK, 2017; pp. 163–206.
39. Maghami, M.; Abdelrasoul, A. *Zeolite Mixed Matrix Membranes (Zeolite-Mmms) for Sustainable Engineering*; West Virginia University: Morgantown, VA, USA, 2018.
40. Tena, A.; Shishatskiy, S.; Meis, D.; Wind, J.; Filiz, V.; Abetz, V. Influence of the composition and imidization route on the chain packing and gas separation properties of fluorinated copolyimides. *Macromolecules* **2017**, *50*, 5839–5849. [CrossRef]

41. Escorihuela, S.; Tena, A.; Shishatskiy, S.; Escolástico, S.; Brinkmann, T.; Serra, J.; Abetz, V. Gas separation properties of polyimide thin films on ceramic supports for high temperature applications. *Membranes* **2018**, *8*, 16. [CrossRef] [PubMed]
42. Corma, A.; Fornés, V.; Guil, J.M.; Pergher, S.; Maesen, T.L.M.; Buglass, J.G. Preparation, characterisation and catalytic activity of itq-2, a delaminated zeolite. *Microporous Mesoporous Mater.* **2000**, *38*, 301–309. [CrossRef]
43. Itoh, T.; Mori, M.; Inukai, M.; Nitani, H.; Yamamoto, T.; Miyanaga, T.; Igawa, N.; Kitamura, N.; Ishida, N.; Idemoto, Y. Effect of annealing on crystal and local structures of doped zirconia using experimental and computational methods. *J. Phys. Chem. C* **2015**, *119*, 8447–8458. [CrossRef]
44. Vigneron, F.; Sougi, M.; Meriel, P.; Herr, A.; Meyer, A. Etude par diffraction de neutrons des structures magnétiques de tbbe 13 à basse température. *J. Phys.* **1980**, *41*, 123–133. [CrossRef]
45. Scherb, T.; Kimber, S.A.J.; Stephan, C.; Henry, P.F.; Schumacher, G.; Escolastico, S.; Serra, J.M.; Seeger, J.; Just, J.; Hill, A.H.; et al. Nanoscale order in the frustrated mixed conductor $La_{5.6}WO_{12-\delta}$. *J. Appl. Crystallogr.* **2016**, *49*, 997–1008. [CrossRef]
46. Han, D.; Kishida, K.; Shinoda, K.; Inui, H.; Uda, T. A comprehensive understanding of structure and site occupancy of y in y-doped $BaZrO_3$. *J. Mater. Chem. A* **2013**, *1*, 3027–3033. [CrossRef]
47. Morejudo, S.H.; Zanón, R.; Escolástico, S.; Yuste-Tirados, I.; Malerød-Fjeld, H.; Vestre, P.K.; Coors, W.G.; Martínez, A.; Norby, T.; Serra, J.M.; et al. Direct conversion of methane to aromatics in a catalytic co-ionic membrane reactor. *Science* **2016**, *353*, 563–566. [CrossRef] [PubMed]
48. IZA Structure Comission. Available online: http://www.iza-structure.org/ (accessed on 20 June 2018).
49. Lilleparg, J.; Georgopanos, P.; Emmler, T.; Shishatskiy, S. Effect of the reactive amino and glycidyl ether terminated polyethylene oxide additives on the gas transport properties of pebax® bulk and thin film composite membranes. *RSC Adv.* **2016**, *6*, 11763–11772. [CrossRef]
50. Zhang, C.; Dai, Y.; Johnson, J.R.; Karvan, O.; Koros, W.J. High performance zif-8/6fda-dam mixed matrix membrane for propylene/propane separations. *J. Membr. Sci.* **2012**, *389*, 34–42. [CrossRef]
51. Fernández-Barquín, A.; Casado Coterillo, C.; Palomino, M.; Valencia, S.; Irabien, A. Permselectivity improvement in membranes for CO_2/N_2 separation. *Sep. Purif. Technol.* **2016**, *157*, 102–111. [CrossRef]
52. Sabetghadam, A.; Seoane, B.; Keskin, D.; Duim, N.; Rodenas, T.; Shahid, S.; Sorribas, S.; Guillouzer, C.L.; Clet, G.; Tellez, C.; et al. Metal organic framework crystals in mixed-matrix membranes: Impact of the filler morphology on the gas separation performance. *Adv. Funct. Mater.* **2016**, *26*, 3154–3163. [CrossRef] [PubMed]
53. Khayet, M.; García-Payo, M.C. X-ray diffraction study of polyethersulfone polymer, flat-sheet and hollow fibers prepared from the same under different gas-gaps. *Desalination* **2009**, *245*, 494–500. [CrossRef]
54. Recio, R.; Palacio, L.; Prádanos, P.; Hernández, A.; Lozano, Á.E.; Marcos, Á.; de la Campa, J.G.; de Abajo, J. Gas separation of 6fda–6fpda membranes: Effect of the solvent on polymer surfaces and permselectivity. *J. Membr. Sci.* **2007**, *293*, 22–28. [CrossRef]
55. Tylkowski, N.A.G.B.a.B. *Chemical Synergies: From the Lab to in Silico Modelling*; De Gruyter: Berlin, Germany, 2018.
56. Calle, M.; Lozano, A.E.; de Abajo, J.; de la Campa, J.G.; Álvarez, C. Design of gas separation membranes derived of rigid aromatic polyimides. 1. Polymers from diamines containing di-tert-butyl side groups. *J. Membr. Sci.* **2010**, *365*, 145–153. [CrossRef]
57. Liu, Y.; Huang, J.; Tan, J.; Zeng, Y.; Ding, Q.; Zhang, H.; Liu, Y.; Xiang, X. Barrier and thermal properties of polyimide derived from a diamine monomer containing a rigid planar moiety. *Polym. Int.* **2017**, *66*, 1214–1222. [CrossRef]
58. Yampolskii, Y.; Shishatskii, S.; Alentiev, A.; Loza, K. Correlations with and prediction of activation energies of gas permeation and diffusion in glassy polymers. *J. Membr. Sci.* **1998**, *148*, 59–69. [CrossRef]
59. Jamil, A.; Ching, O.P.; Shariff, A.B.M. Current status and future prospect of polymer-layered silicate mixed-matrix membranes for CO_2/CH_4 separation. *Chem. Eng. Technol.* **2016**, *39*, 1393–1405. [CrossRef]
60. Bae, T.-H.; Long, J.R. CO_2/N_2 separations with mixed-matrix membranes containing mg2(dobdc) nanocrystals. *Energy Environ. Sci.* **2013**, *6*, 3565–3569. [CrossRef]
61. Castarlenas, S.; Téllez, C.; Coronas, J. Gas separation with mixed matrix membranes obtained from mof uio-66-graphite oxide hybrids. *J. Membr. Sci.* **2017**, *526*, 205–211. [CrossRef]

62. Galve, A.; Sieffert, D.; Vispe, E.; Téllez, C.; Coronas, J.; Staudt, C. Copolyimide mixed matrix membranes with oriented microporous titanosilicate jdf-l1 sheet particles. *J. Membr. Sci.* **2011**, *370*, 131–140. [CrossRef]
63. Vinoba, M.; Bhagiyalakshmi, M.; Alqaheem, Y.; Alomair, A.A.; Pérez, A.; Rana, M.S. Recent progress of fillers in mixed matrix membranes for CO_2 separation: A review. *Sep. Purif. Technol.* **2017**, *188*, 431–450. [CrossRef]

Review

Perfluoropolymer/Molecular Sieve Mixed-Matrix Membranes

Gianni Golemme [1,*] and Anna Santaniello [2]

1 Department of Environmental and Chemical Engineering, University of Calabria, Via P. Bucci 45 A, 87036 Rende, Italy

2 Physics Department, University of Calabria, Via P. Bucci 22 C, 87036 Rende, Italy; anna.santaniello@fis.unical.it

* Correspondence: giovanni.golemme@unical.it; Tel.: +39-0984-496702

Received: 30 December 2018; Accepted: 21 January 2019; Published: 23 January 2019

Abstract: Despite the outstanding chemical, thermal and transport properties of amorphous and glassy perfluorinated polymers, only few works exist on the preparation and transport properties of perfluoropolymer/molecular sieves mixed-matrix membranes (MMMs), probably because of their poor compatibility. In this review, the compatibilization of ceramic molecular sieves with perfluorinated matrices is considered first, examining the effect of the surface treatment on the gas transport properties of the filler. Then the preparation of the defect-free hybrid membranes and their gas separation capabilities are described. Finally, recent modelling of the gas transport properties of the perfluoropolymer MMMs is reviewed. The systematic use of molecular sieves of different size and shape, either permeable or impermeable, and the calculation of the bulk transport properties of the molecular sieves—i.e., the unrestricted diffusion and permeability—allow to understand the nature of the physical phenomena at work in the MMMs, that is the larger the perfluoropolymer fractional free volume at the interface, and restricted diffusion at the molecular sieves. This knowledge led to the formulation of a new four-phase approach for the modelling of gas transport. The four-phase approach was implemented in the frame of the Maxwell model and also for the finite element simulation. The four-phase approach is a convenient representation of the transport in MMMs when more than one single interfacial effect is present.

Keywords: glassy amorphous perfluoropolymers; mixed matrix membranes; zeolitic molecular sieves; gas separation; interfacial compatibilization; fractional free volume; restricted diffusion; barriers to mass transport; four phases Maxwell model; finite element modelling of transport

---

1. Introduction

Fluoropolymers display better chemical and solvent resistance with respect to similar hydrocarbon polymers, therefore they are widely used to prepare membranes for battery separators, microfiltration and ultrafiltration, membrane contactors and chemical reactors [1–5]. The best thermal, chemical and solvent resistances are obtained in perfluorinated polymers (PFPs), where the strength of C–C and C–F bonds and the shielding of the carbon skeleton by fluorine atoms prevent chemical attacks [6–8].

Perfluoropolyacids (e.g., Nafion®) are semi-crystalline and hydrophilic PFPs. They are used as proton exchange membranes in fuel cells and for electrolysis, such as in the chlor-alkali process. Perfluoropolyacid-based mixed-matrix membranes (MMMs), i.e., mixtures of a filler in a polymeric matrix [9], were reviewed elsewhere [7]. In this work, instead, the few pioneering examples of MMMs made of molecular sieves embedded in amorphous and glassy PFP matrices will be considered.

Poly(tetrafluoroethylene) [10] is insoluble because of its semi-crystalline nature, but the presence of bulky or atactic substituents on the carbon chain affords amorphous, glassy and hydrophobic polymers which can be dissolved in fluorinated solvents, lending themselves to the preparation of membranes via solvent evaporation. Hydrophobic PFPs are apolar thanks to the symmetry around the carbon atoms, and the strong attraction of F atoms towards all electrons in the molecule minimizes the interactions with light and the polarizability of the material. As a consequence, hydrophobic and amorphous PFPs display very low refractive indexes [11], the only intermolecular interactions are very weak dispersion forces, and their cohesive energy density and solubility parameters are smaller than those of the corresponding hydrocarbon polymers [12,13]. These features correspond to peculiar macroscopic properties of amorphous and hydrophobic PFPs, i.e., no solubility in common solvents, low specific surface energy, hydrophobic and organophobic behavior, poor adhesion of almost any substance, including glues and adhesives. The main difficulty to overcome to prepare hydrophobic PFP-based MMMs is the homogeneous mixing of finely dispersed nano-fillers in the polymer, and a good adhesion at the interface.

The first commercial soluble and glassy PFPs appeared about 30 years ago (Figure 1): co-polymers of perfluoro-1,3-dioxoles (Teflon® AF, grades 1600 and 2400, from DuPont; Hyflon® AD, grades 40, 60 and 80, from Ausimont—now Solvay Advanced Polymers) [14], and poly(perfluoro-4-vinyloxy-1-butene), also known with the commercial name Cytop® (Asahi Glass) [11].

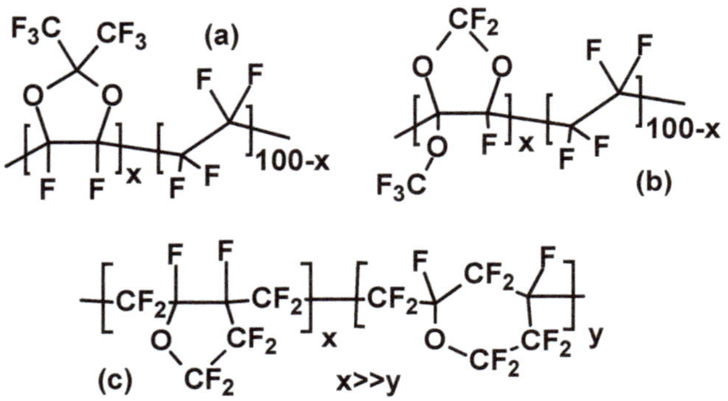

Figure 1. Repeat units of amorphous perfluorinated polymers: (**a**) Teflon® AF, grade 2400: x = 87, grade 1600: x = 65; (**b**) Hyflon® AD, grade 80X: x = 85, grade 60X: x = 60; (**c**) Cytop®. *x* and *y* represent the molar composition and do not imply a sequence length.

Theoretical studies indicate very high potential energy barriers for the rotation around the bonds linking two dioxolane rings in Teflon AF (Figure 1a) and Hyflon AD (Figure 1b). The polymer chains have limited or no rotational freedom, and the stiffness of the backbone increases with the bulkiness (Teflon AF > Hyflon AD, see Figure 1) and the percent of dioxole monomer [15]. In turn, stiffness of the backbone and amount of bulky groups correlate with the glass transition temperature (T_g), the gas permeability, and the fractional free volume (FFV), i.e., the fraction of the volume which is not occupied by the electronic clouds of the polymer: Teflon AF2400 (T_g = 240 °C) > Teflon AF1600 (T_g = 160 °C) > Hyflon AD 80 X (T_g = 135 °C) > Hyflon AD 60X (T_g = 110 °C) > Cytop (T_g = 108 °C) [6,16,17]. Teflon® AF 2400, the most permeable polymer of this group, has a free volume of more than 30% [6,18,19], and it is one of the most permeable polymers known today [9].

Gas transport through dense membranes is governed by Equation (1) [20,21]:

$$J_i = P_i(p_{il} - p_{i0})/l \tag{1}$$

where J_i is the volume (molar) flux of component i (e.g., in units $cm^3(STP)/(cm^2 \cdot s)$), l the membrane thickness, p_{i0} the partial pressure of component i on the feed side, p_{il} the partial pressure of component i on the permeate side, and the permeability P_i represents the ability of molecules i to permeate through the membrane material.

The mass transport through dense, non-porous membranes obeys a solution–diffusion mechanism [20,21]. In its simplest form, the permeability P_i of the substance i is a function of its solubility in the membrane (S_i, a thermodynamic quantity) and of its diffusivity in the membrane (D_i, a kinetic quantity), as defined in Equation (2):

$$P_i = S_i \cdot D_i \quad (2)$$

Selectivity measures a membrane's ability to separate the components of a mixture. Ideal selectivity (α_{ij}) is the ratio of permeabilities of pure gases i and j, and is defined by Equation (3):

$$\alpha_{ij} = P_i/P_j \quad (3)$$

The combination of Equation (2) and Equation (3) yields Equation (4):

$$\alpha_{ij} = (S_i/S_j) \cdot (D_i/D_j) = \alpha_S \cdot \alpha_D \quad (4)$$

which highlights the fact that the overall permeability selectivity α_{ij} is the product of the solubility selectivity α_S and of the diffusion selectivity α_D.

The size of a certain molecule (Table 1) governs its mobility in a polymer: the larger the size, the smaller D. Instead, the higher its condensability (e.g., as measured by boiling point or critical temperature) the larger its solubility in the polymer. In rubbers, the solubility selectivity dominates over diffusion, because the liquid-like motions of the polymer chains determine a high penetrant mobility and have a leveling effect on the different sizes of the penetrants. The frozen macromolecular motions of glassy polymers instead limit diffusion, and the mobility selectivity dominates over solubility. Therefore, glassy polymers are usually size selective and poorly permeable. However, when the glassy polymer backbone is extremely stiff, and an effective polymer packing cannot be reached, very high FFVs and large permeabilities are observed, such as in certain polyacetylenes [22,23], in the polymers of intrinsic micro-porosity (PIMs) [24], and in glassy amorphous PFPs, as described above.

Table 1. Kinetic diameter (d_k) [25] and critical temperature (T_c) [26] of six gases.

Gas	He	H_2	CO_2	O_2	N_2	CH_4
d_k/Å	2.6	2.89	3.3	3.46	3.64	3.80
T_c/K	5.2	33.18	304.20	154.58	126.19	190.6

Glassy amorphous and hydrophobic PFP membranes perform best in those gas separations (He/H_2, H_2/hydrocarbons, He/N_2, He/CO_2, He/CH_4, N_2/CH_4) where the diffusion selectivity in favour of the smaller molecule is offset, in part, by the solubility selectivity in favour of the tardier, more condensable species [9]. In such cases the modest affinity of the penetrants to PFPs further reduces the unfavorable contribution of solubility to the overall selectivity, when compared to hydrocarbon-based polymers [6]. Another interesting province of PFPs is natural gas sweetening, where CO_2 is at the same time smaller and more soluble than CH_4. The outstanding resistance to solvents (aliphatic and aromatic hydrocarbons, water, amines) and to plasticization of Hyflon AD60X is used in commercial high flux membranes for the upgrading of natural gas containing large fractions of CO_2, that would spoil the separation properties of cellulose acetate or polyimide membranes [5].

In the last few years, other amorphous and glassy PFPs have been synthesized and studied for gas separation: atactic poly(hexafluoropropene) [27], new perfluoro-1,3-dioxoles [28], homo- and co-polymers of perfluoro(2-methylene-1,3-dioxolane)s (Figure 2) [29–31]. The latter class of PFPs looks extremely interesting for the preparation of gas separation membranes, with impressive separation factors, beyond the 2008 Robeson upper bound for the H_2/CH_4, He/CH_4, CO_2/CH_4, and N_2/CH_4 separations [30,31].

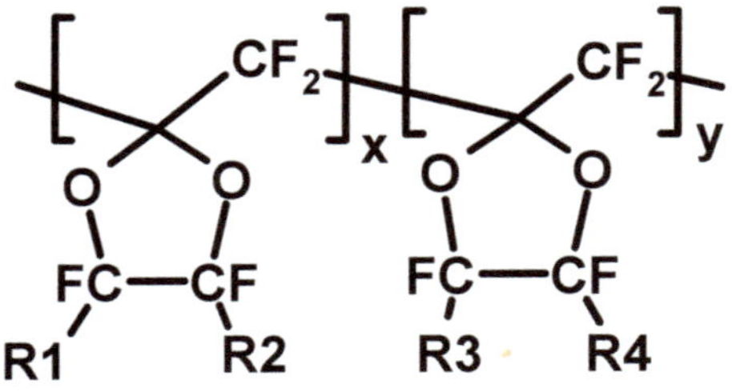

Figure 2. Repeat units of homo- and co-polymers of perfluoro(2-methylene-1,3-dioxolane)s. R1–R4 = F, CF_3.

In gas separation, the outstanding properties of perfluoropolymers are potentially useful to overcome the limited resistance of hydrocarbon-based polymers, as in the purification of natural gas [32], but unfortunately their intrinsic permselectivity often represents a limit for the potential applications of gas separation membranes [9]. The trade-off between gas permeability and selectivity of polymers can be overcome by using MMMs, i.e., membranes in which small particles of a filler are dispersed in a continuous polymeric phase. Mixed-matrix membranes combine the advantages of both components, since they couple the ease of production and the good mechanical properties of polymeric membranes with the outstanding transport properties of suitable fillers [33].

Fumed silica, titania, and magnesium oxide are among the non-porous fillers used to prepare MMMs. Porous fillers offer an additional degree of freedom for the improvement of the separation capabilities of the polymeric matrices. The porous fillers used for MMMs have been recently reviewed [34]: zeolites, metal-organic frameworks (MOFs), carbon nanotubes, activated carbon, microporous organic polymers, mesoporous oxides, porous bi-dimensional graphene, and graphene oxides, and more. Molecular sieves are crystalline materials with uniform, well-defined micropores (<2 nm), which are able to sort molecules according to their size; they include zeolites, made of SiO_4 and AlO_4 tetrahedra, zeotypes containing TO_4 tetrahedra other than SiO_4 and AlO_4 (e.g., PO_4, GeO_4, TiO_4), and analogous non-zeolitic materials containing elements with different coordination number, such as TiO_6 octahedra in ETS-4 and ETS-10 [34]. In the following, the preparation and the transport properties of MMMs made of molecular sieves dispersed in glassy and hydrophobic PFP is reviewed. First, the strategy for the compatibilization of molecular sieves with the perfluoropolymers is considered. Then the preparation and the separation properties of several MMMs are described. Finally, the macroscopic modelling of the gas transport properties of PFP MMMs is analysed, together with the microscopic features of the membranes. The new four phase approach developed to account for the physical phenomena observed in the PFP MMMs can be used for other MMMs with complex interfacial effects.

2. Surface Modification of Molecular Sieves and Surface Permeability

A good adhesion of the glassy PFP to the surface of the molecular sieve is a key factor to avoid the formation of defective gaps at the interface, which would spoil the separation performances of the MMMs, acting as non-selective by-passes for the transport of gases [35]. Unlike rubbers, the achievement of a good adhesion is complicated by the mismatch between the vitrified glassy polymer, which keeps shrinking during the removal of the residual solvent, and the rigid zeolitic molecular sieve. It has to be considered that hydrophilic hydroxyls covering the outer surface of the molecular sieves [36] strongly repel hydrophobic PFPs and perfluorinated solvents.

The several strategies used to disperse molecular sieves in glassy, hydrocarbon-based polymers —polysulfones, polyimides, polybenzimidazoles, etc.—rely on the formation of the strongest possible interactions between the two phases. Covalent bonds form when the dangling 3-aminopropyl groups of zeolites treated with 3-aminopropylalkoxysilanes open and react with the imide rings of polyimides. The covalent bonds strongly facilitate the adhesion of the polymer and the production of defect-free polymer-filler interfaces [37]. In other cases, hydrogen bonds or polar interactions are intense enough to guarantee an intimate contact between the two phases. The strongly polar surface of zeolite 4A is able to adhere to the hydrophobic but highly polarizable poly(1-trimethylsilyl-1-propyne) [38]. Unfortunately, hydrophobic PFPs are chemically inert, apolar, and not polarizable, therefore the strategy was chosen to make the zeolitic molecular sieves fluorophilic, via a post-synthetic grafting of perfluorinated alkyl tails.

The reaction of a moderate excess of 1H,1H,2H,2H-perfluorodecyltrichlorosilane with the external −OH groups of the molecular sieves (e.g., silicalite-1, SAPO-34) made the water contact angle increase from 89° to 150° [39]. The control on the reaction parameters (e.g., dry conditions, low temperature, hydrophobic solvents, short reaction times) produced a continuous mono-layer of vertically aligned perfluorinated moieties tightly packed on the outer surface of the molecular sieves, as indicated by the results of X-ray photoelectron spectroscopy [40]. The modified sieve crystals exposed the terminal $-CF_3$ groups of the grafts, and their surface properties became comparable to those of Teflon particles, as indicated by the formation of homogeneous suspensions in perfluorinated solvents. The practically unchanged BET surface area (e.g., 611 vs. 616 $m^2 \cdot g^{-1}$ before the grafting of SAPO-34 [41]) indicated that the external modification did not plug the pores of the particles.

The effect of the external grafting on the transport of penetrants in and out of silicalite-1 crystals was tested with H_2 at cryogenic temperatures [36], and with *n*-heptane at room temperature [40]. The first relevant result of the experiments was that the rate of sorption in both cases depended on the modification of the outer surface. This can only happen if the surface permeability is the rate determining step in adsorption, i.e., that barriers for the transport of mass exist on the outer surface of silicalite-1, and that the sorption/desorption rates of species of different size (H_2, *n*-heptane), at different temperatures, was determined by the presence of these barriers. The second surprising result was that the bulky 1H,1H,2H,2H-perfluorodecyl tails on the surface enhanced the sorption/desorption rates of both H_2 and *n*-heptane with respect to the as-made molecular sieve. The explanation of this effect was the reduction of the hydrophilic character of the outer surface, where the terminated crystal framework, rich of silanol ≡Si–OH groups, promotes the adsorption of moisture (Figure 3). This hypothesis was supported by a theoretical study predicting, at room temperature, the'formation of a water layer on the surface of silicalite-1, extending into the silicalite pores for about 1.0–1.5 nm [41].

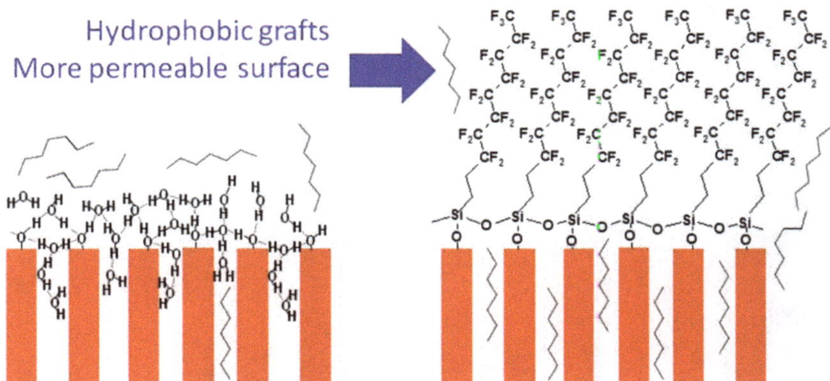

Figure 3. Pictorial representation of the sorption of n-heptane at the surface of silicalite-1 before and after the modification with 1H,1H,2H,2H-perfluorodecyltrichlorosilane. Reproduced from Reference [40] with permission. Copyright of Elsevier B.V.

3. Preparation of Hydrophobic PFP/Molecular Sieve MMMs

As made zeolitic molecular sieves are usually impermeable, because their pores are filled with the templates, the organic substances directing their synthesis. At this stage, their BET area was measured, in order to determine the stoichiometric amount of the modification agent needed [40]. Then the sieves were calcined to free their pores from the organics. If the size of the crystals was small, they were first embedded in a cross-linked network of an organic polymer. The larger average distance among them greatly reduced the sintering into aggregates [42]. Also, as made, impermeable molecular sieves were used to prepare MMMs, with the aim to study the influence of the sole polymer phase on the gas transport properties [43,44]. Next, the molecular sieves were made fluorophilic as described in Section 2 above, and then were used for the preparation of membranes according to Scheme 1. In a typical procedure, the molecular sieves were homogeneously suspended in a polymeric solution by repeated cycles of mechanical stirring and ultrasound treatments (1). The resting suspension was poured into a vessel with a flat, levelled bottom (2), to allow for the evaporation of the solvent at room temperature (3). The residual solvent lingering in the peeled MMM was eliminated in a vacuum oven (4). The similar values in the density of the fillers (1.53–1.87 g·cm^{-3}), of the solvents (about 1.7 g·cm^{-3}) and of the polymers (1.7–2.0 g·cm^{-3}) is probably the reason behind the absence of sedimentation problems during the slow evaporation step (3).

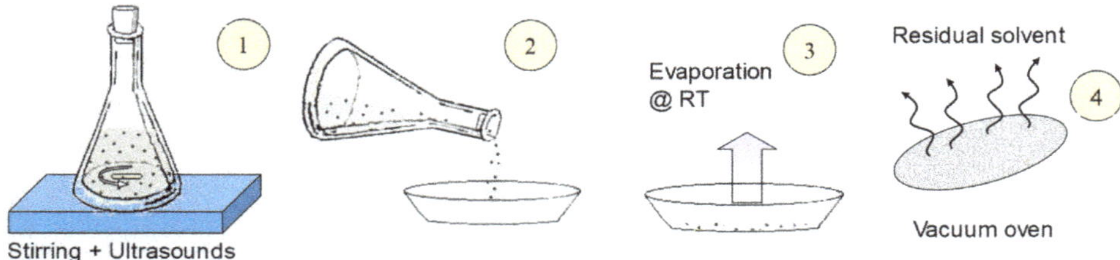

Scheme 1. Sequences in the preparation of self-supported perfluorinated polymer (PFP)/molecular sieve MMMs.

Defect-free MMMs containing various amounts of silicalite-1 crystals of different size (from 0.08 to 1.5 µm) and aspect ratio (the ratio between the largest and the smallest dimension, AR in the following) were produced with Teflon AF1600, Teflon AF2400 and Hyflon AD60X, up to a volume fraction of 45% (Figure 4) [39,45].

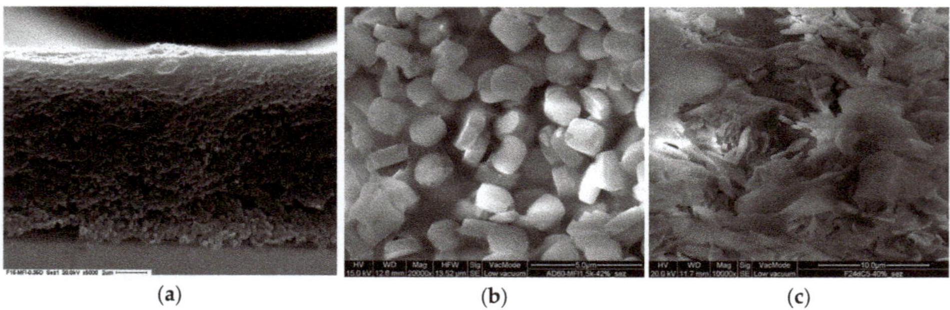

Figure 4. SEM cross-sections of PFP MMMs containing silicalite-1 crystals: (**a**) 30 wt% silicalite-1 (ca. 0.35 µm, AR 2) embedded in a Teflon AF1600 matrix; (**b**) 42 wt% silicalite-1 (1.5 µm, AR 3) embedded in a Hyflon AD60X matrix; (**c**) 40 wt% silicalite-1 (ca. 3 µm, AR 8) embedded in a Teflon AF2400 matrix. Reproduced from Reference [45] with permission. Copyright of John Wiley & Sons, Ltd.

Other MMMs were prepared by embedding 20–35% SAPO-34 crystals of different size (0.2–2 μm) and shape (AR 2–9), both permeable and impermeable, in Hyflon AD60X (Figure 5) [43]. Increasing fractions of molecular sieves made the membranes more rigid, but a good flexibility was still retained at 35 v% loading (Figure 6).

Figure 5. SEM cross-sections of Hyflon AD60X MMMs containing 35 v% SAPO-34 crystals of different size and aspect ratio: (**a**) 1.5 μm, AR 9; (**b**) 2 μm, AR 2; (**c**) 0.2 μm, AR 2. Reproduced from Reference [43] with permission. Copyright of the Korean Society of Industrial and Engineering Chemistry and Elsevier B.V.

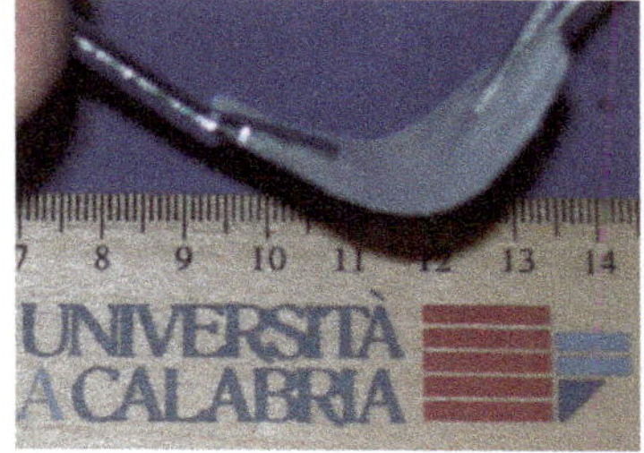

Figure 6. Bending of the Hyflon AD60X MMM containing 35 v% SAPO-34 crystals of 0.2 μm, AR 2. Reproduced from Reference [43] with permission. Copyright of the Korean Society of Industrial and Engineering Chemistry and Elsevier B.V.

4. Transport Properties of Hydrophobic PFP/Molecular Sieve MMMs

4.1. PFP/Silicalite-1 MMMs

Silicalite-1 is a hydrophobic medium pore zeolite (5.4–5.6 Å) with channels delimited by 10-member rings of SiO_4 units [46]. It is the siliceous form of ZSM-5 (structure topology MFI), with a three-dimensional pore network of straight and zigzag channels. It was chosen as the porous filler of PFPs because it can be easily prepared in a wide range of sizes and shapes, it is one of the most studied molecular sieves, with several gas sorption and mobility data available in the literature [47], and also because it only contains a modest amount of extra-framework cations and is hydrophobic [46]. Hydrophobicity represents an advantage for the modelling of gas transport because the large amount of water adsorbed by hydrophilic zeolites is an obstacle for the transport of gas, and in addition the uncertain amount of moisture, which rapidly adsorbs even during short expositions to the atmosphere, turns hydrophilic zeolite pore networks into ill-defined systems.

As a first important result, PFPs MMMs prepared with the fluorophilic silicalite-1 were all defect-free [39,45]: no gap was visible at the polymer-zeolite interface (Figure 4), and gas permeability data excluded the presence of defects. Transport features were dependent on the loading, and the size and the shape of the crystals.

A Hyflon AD60X MMM containing 42 w% silicalite-1 of 1.5 μm size (Figure 4a,b) gave an interesting combination of permeability and N_2/CH_4 ideal selectivity, in excess of the 2008 upper bound [9] (Figure 7). The ideal CO_2/CH_4 selectivity was also improved in a similar way, getting closer to the 2008 upper bound for this gas pair [45]. These results were not expected, because the sizes of CO_2, N_2 and CH_4 (Table 1) are all much smaller than the pore size of the MFI topology.

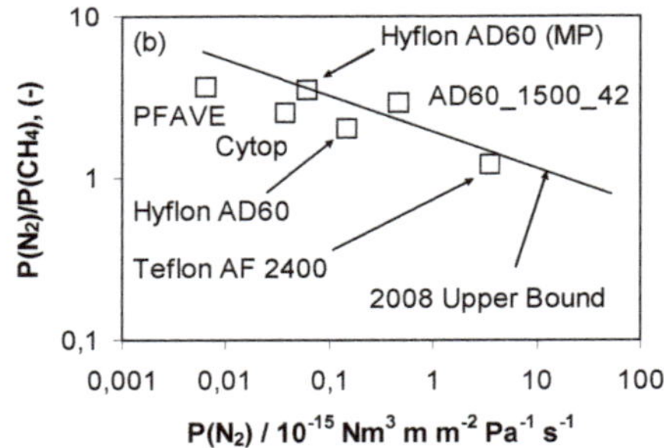

Figure 7. Selectivity vs. permeability plot for the N_2/CH_4 separation at 25 °C of perfluorinated polymers and a Hyflon AD60X MMM containing 42 w% (45 v%) Silicalite-1 (AD60_1500_42). Reproduced from Reference [45] with permission. Copyright of John Wiley & Sons, Ltd.

Teflon AF2400 MMMs were tested for the n-butane/CH_4 separation [44]. Permeability and ideal separation factor of an MMM containing 40 w% silicalite-1 (1.5 μm size and AR 3, AF24_1500_40 in Figure 8) were higher than those of Teflon AF2400 hybrid membranes containing fumed silica [48]. The increase of polymer permeability upon the introduction of an impermeable filler (in that case fumed silica) was not expected. Positron annihilation (PALS) experiments demonstrated that the larger permeability of Teflon AF2400 hybrid membranes stems from free volume elements of larger size, probably caused by a looser packing of the polymer at the interface with silica [48]. Larger free-volume elements had been observed already for hybrid membranes containing fumed SiO_2 of other glassy, high FFV polymers, such as poly(1-trimethylsilyl-1-propyne) (PTMSP) [49] and poly(4-methyl-2-pentyne) (PMP) [50], characterized by rigid backbones like Teflon AF2400. The decrease of the polymer density in Teflon AF1600 and Teflon AF2400 MMMs containing fumed silica was confirmed by detailed studies using the non-equilibrium lattice fluid (NELF) model, based on a non-equilibrium thermodynamics approach to glassy polymers [51,52]. The larger butane permeability measured when a porous filler is present with respect to the case in which fumed silica is incorporated indicated the active role of silicalite-1 in the transport of butane; also the larger C_4H_{10}/CH_4 selectivity indicated the active participation of silicalite-1, in which the strongly adsorbed butane excludes methane [53].

When the AF24_1500_40 membrane was fed with equimolar butane-methane mixtures, a reversal of the selectivity from methane to butane on increasing the butane partial pressure (from 50 to 106 kPa at 25 °C) was observed. It is also interesting to note the advantages of a larger aspect ratio of the silicalite-1 crystals on the transport properties of the MMMs. The AF24_dC5_40 membrane, containing same loading (40 w%) of platy silicalite-1 crystals (3 μm, AR 8, Figure 4c), in all conditions afforded better butane/CH_4 selectivity in the mixture experiments than the AF24_1500_40 membrane, with the same permeability [45].

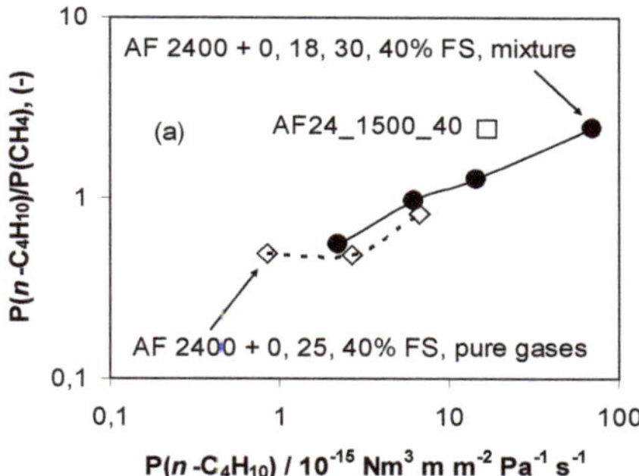

Figure 8. Selectivity vs. permeability plot of Teflon AF 2400 based hybrid membranes for the n-C_4H_{10}/CH_4 pair at 25 °C: pure gas data of fumed silica filled AF2400 (◇: 0, 25 and 40%, Δp = 0.33–0.44 bar) from Reference [48]; pure gas data of membrane AF24_1500_40 (□, Δp = 1.00 bar); mixed gas data (2% n-butane feed, Δp = 11.2 bar) of fumed silica filled AF2400 (•: 0, 18, 30 and 40%), from Reference [48]; for all measurements the downstream pressure was atmospheric. Reproduced from Reference [45] with permission. Copyright of John Wiley & Sons, Ltd.

Teflon AF1600 MMMs containing tiny silicalite-1 crystal (80 nm) were more permeable (He, H_2, CO_2, O_2, N_2 and CH_4) than the polymer, and less selective. This behaviour might be the effect of a larger permeability of silicalite-1, of the increase of the FFV around the crystals, similarly to the Teflon AF2400/fumed silica MMMs, or else of the voids in the crystal aggregates contained in the MMMs.

Larger 350 nm silicalite-1 crystals, instead, distributed homogeneously in the Teflon AF1600 matrix (Figure 4a). A generalized decrease of gas permeability and a steady improvement of the sieving properties of the polymer were observed, as evidenced by the larger gas/CH_4 ideal selectivity [45]. This effect was not expected, because silicalite-1 (pore size 5.4–5.6 Å) does not sieve small gas molecules (Table 1).

If the fillers are perfectly dispersed and do not interact with each other or with the matrix, the permeability of MMM is described by the Maxwell model [54] (Equation (5)):

$$P^{MMM} = P^{M}\frac{P^{F} + 2P^{M} - 2\phi(P^{M} - P^{F})}{P^{F} + 2P^{M} + \phi(P^{M} - P^{F})} \tag{5}$$

In Equation (5), P^{MMM} is the gas permeability of the MMM, P^{F} is the gas permeability of the dispersed phase (silicalite-1), P^{M} is the gas permeability of the continuous phase (the polymer), and ϕ is the volume fraction of the dispersed phase. Since the experimental permeability of the polymer and of the MMMs were available, the gas permeability of silicalite-1 could be easily worked out. The different behaviour of the MMMs with silicalite-1 crystals of different size indicated that probably the filler does influence the physical state of the polymer. Therefore, it was decided to derive a theoretical permeability of the MMM from the gas permeability of filler and polymer, and to compare this value to the experimental results. The polymer permeability was readily measured, whereas the silicalite-1 permeability cannot be measured directly.

4.1.1. Derivation of the Permeability of Silicalite-1

Care should be taken in the choice of the right gas permeability of a zeolitic molecular sieve from literature values. It has to be noted that in the literature a large scatter of the permeability values of different supported zeolitic membranes made of the same zeolite is found [47]. This happens because the gas permeability of zeolite membranes is affected by two major sources of error: (a) the presence of a variable amount of non-zeolitic pores, usually inter-crystalline space at the boundaries of the crystal domains [55], which act as non-selective defects increasing the membrane permeability, and (b) the presence of barriers to the transport of gas on the surface or in the interior of the crystals—crystal defects, interfaces in intergrowth crystals, presence of amorphous matter, moisture and/or coke [36,41,47,56–71]—which instead decrease gas permeability.

In practice, a more reliable method is to obtain the intrinsic gas permeability of a molecular sieve by combining solubility coefficients with the unrestricted intra-crystalline diffusion coefficients in Equation (2). Reproducible solubility of several gases in a variety of zeolites is available from experimental measurements; instead, the intra-crystalline gas diffusion coefficients are not, in general, available. The gas diffusion coefficients in molecular sieves, as obtained experimentally with different techniques, sometimes differ by three orders of magnitude or more. It is found that this scatter is the effect of surface and intra-crystalline barriers. The macroscopic techniques (e.g., sorption kinetics, chromatographic pulse methods) based on gas uptake usually measure a slower gas mobility than microscopic methods (such as pulsed field gradient nuclear magnetic resonance (PFG-NMR) and neutron scattering techniques), which are able to probe unrestricted diffusion inside the crystals [47]. Unfortunately, PFG-NMR and neutron scattering can only be used with some gases, and they are not readily available for investigation. Molecular dynamics (MD) simulations [47,72–76] represent a convenient way to evaluate the mobility of penetrants inside micro- and meso-porous media (the Maxwell–Stefan diffusion coefficient), in equilibrium conditions (self-diffusion), in the presence of concentration gradients, and with other co-adsorbed species.

The diffusion coefficients of CO_2 and CH_4 in silicalite-1 obtained with a transient pulse-response technique [77] were combined with the experimental solubility of the same gases [78] in Equation (2) in order to obtain the permeability coefficients in the molecular sieve [45].

4.1.2. Modelling of the Transport Properties of Teflon AF1600/MFI MMMs

It turned out that silicalite-1 is definitely more permeable to CO_2 and CH_4 than Teflon AF 1600. As a consequence, the Maxwell Equation (5) predicted a larger permeability of the silicalite-1 containing MMMs with respect to the polymer. It was therefore surprising to discover that the experimental permeability of the MMM containing 30 w% of 350 nm silicalite crystals was smaller than the one of Teflon AF 1600 for both CO_2 and CH_4. The authors concluded that the silicalite permeability was strongly reduced by some barrier to the transport of gas at the surface or in the interior of the crystals [45]. In that work the average diffusion coefficients of gases through the Teflon AF 1600 MMMs was measured by means of the Daynes–Barrer time-lag method [79]. In this method the transient permeation behaviour before the attainment of the steady state is used to obtain average diffusion coefficients by means of Equation (6) below:

$$D = l^2/6\theta \tag{6}$$

where l is the thickness of the membrane and θ (the time-lag) is the intercept of the pressure vs. time asymptotic line at long times with the time axis (Figure 9).

Silicalite-1 partially immobilizes the penetrant gases and due to its large adsorptive capacity, it acts like a sink for CO_2 and CH_4. Therefore, additional time is required to accumulate the excess penetrant before steady state can be reached [80–84]. This was probably the origin of the anomalous increase of the CO_2 permeability observed after the apparent attainment of the steady state in a Teflon AF1600/silicalite-1 MMM [39]. The simplified treatment of the transient permeation experiments via Equation (6) underestimated the real, steady-state diffusion. Nevertheless, the time-lag diffusion coefficients of the two gases worked out with Equation (6) were higher in the MMM than through the Teflon AF 1600 polymer. Combined with the reduced permeability of the MMM, this circumstance indicated that the polymer phase in the MMM offered faster diffusion paths to gases, i.e., the polymer in the defect-free MMM was endowed with a larger free volume.

These earlier studies on PFP MMMs demonstrated that porous ceramic fillers, after the grafting of the perfluorinated tails, can be homogeneously dispersed in perfluorinated polymers. The presence of the filler disrupts the effective molecular packing of the polymer chains in the bulk, creating an interface with larger FFV. The increase of the ideal gas selectivity proved that such an interface is characterized by a good adhesion between the two phases. The macroscopic modelling of gas transport demonstrated also the presence of barriers for the transport of mass at the silicalite-1 crystals.

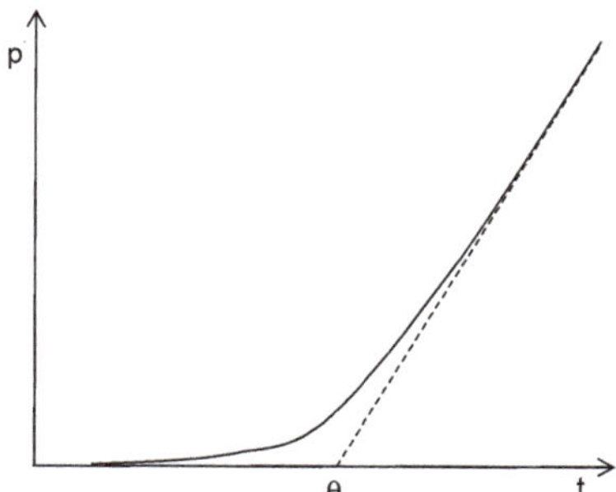

Figure 9. Increase of gas pressure (p) in the downstream receiving volume during a transient permeation experiment. θ is the so-called time-lag.

4.2. *Hyflon AD60X/SAPO-34 MMMs*

SAPO-34 is a family of silicon aluminium phosphate molecular sieves with the same pore topology of chabazite (CHA), a natural zeolite. In SAPO-34, large cavities are connected by narrow ports (3.8 Å) delimited by eight member rings of TO_4 units [85]. Due to its small pore size, comparable to the diameter of methane (Table 1), SAPO-34 is very selective for the CO_2/CH_4 separation [86–88], and also displays interesting N_2/CH_4 selectivity [89].

On the basis of the experience with the PFP/silicalite-1 hybrids, new experiments were planned to make clear the importance for MMMs of interfacial polymer free volume and restricted diffusion in molecular sieves. Several Hyflon AD60X MMMs were prepared with different loadings (20–35 v%) of both permeable and impermeable SAPO-34 of different size and aspect ratios, with the aim to study in detail the effect of these variables on the final transport properties, and to isolate the behaviour of the polymer in the MMMs from the properties of the filler.

The MMMs prepared with the fluorophilic SAPO-34 were all defect-free [43]. No gap was visible at the polymer-zeolite interface (Figure 5), and gas permeability data (He, H_2, N_2, CH_4, CO_2) excluded the presence of defects. Transport features were dependent on the loading, the size and the shape of the crystals, as seen in the diagram of the Robeson's plot for the CO_2/CH_4 gas pair (Figure 10).

In all cases the presence of SAPO-34 enhanced the selectivity of Hyflon AD60X for the CO_2/CH_4 separation. Thanks to the combination of the improved selectivity and of the plasticization resistance of the polymer, the MMMs may represent a viable option for the purification of natural gas containing large amounts of CO_2. SAPO-34 afforded large selectivity gains with respect to the polymer, especially with 35% flat crystals of the largest aspect ratio (+81%). Inspection of Figure 5a revealed that the flat SAPO-34 crystals in this membrane were preferentially oriented with their largest dimensions about parallel to the membrane faces. The largest selectivity was accompanied by a decrease of permeability with respect to Hyflon AD 60X. Low aspect ratio SAPO-34 crystals at the same loading, resulted instead in an increase of both permeability and selectivity, with better results for the small (200 nm) SAPO-34. Since the Maxwell model—Equation (5)—predicts the same behaviour for the permeability of MMMs with the same loading of filler, irrespective of size and shape [54], it was evident that non-ideal effects were at work in Hyflon AD60X/SAPO-34 MMMs.

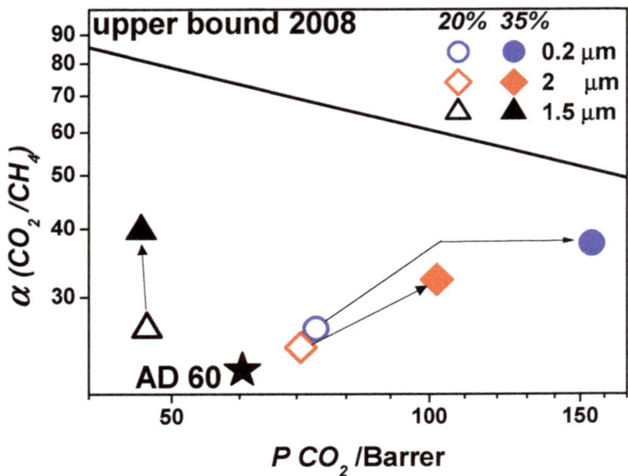

Figure 10. CO_2/CH_4 Robeson plot of Hyflon AD60X (AD60, black star) and of its MMMs containing calcined and permeable SAPO-34: tiles (1.5 μm, AR = 9, triangles), large (2 μm, AR = 2, diamonds), small (0.2 μm, AR = 2, circles); empty symbols 20 v% loading, full symbols 35 v% loading. 1 Barrer is equivalent to 10^{-10} cm^3 (STP) cm $cmHg^{-1} \cdot s^{-1} \cdot cm^{-2}$. Reproduced from Reference [43] with permission. Copyright of the Korean Society of Industrial and Engineering Chemistry and Elsevier B.V.

4.2.1. Derivation of the Gas Permeability of SAPO-34

As in the case of the PFP membranes containing silicalite-1, the approach for the understanding of the underlying physical phenomena was the derivation of the unrestricted, intra-crystalline gas permeability of SAPO-34 (bulk values in the following), and its use in the macroscopic modelling of the transport phenomena in the MMMs [43].

The bulk permeability of He, H_2, N_2, CH_4 and CO_2 in SAPO-34 was assumed to be equal to the one of an all-silica zeolite with the same CHA topology. Bulk permeability was calculated as the product of solubility and diffusion (Equation (2)) [43] on the basis of in silico simulation results [90] of the pressure-dependent gas loading, and of the loading-dependent gas diffusivity in a SiO_2 molecular-sieve with the CHA structure, in the fugacity interval 100–600 kPa.

4.2.2. Modelling of the Transport Properties of Hyflon AD60X/SAPO-34 MMMs

The macroscopic model calculation of the gas transport in the MMMs varied according to the aspect ratio of the SAPO-34 crystals [43]. The resistance model of Cussler [91], derived for regular arrays of parallel ribbons (flakes) (Equation (7)), in which *AR* is the aspect ratio of the filler,

$$P^{MMM} = \frac{P^M}{1 - \phi + \{1/[(1/\phi)(P^F/P^M) + 4(1-\phi)/(AR^2\phi^2)]\}} \tag{7}$$

approximates more closely the morphology of the MMMs containing oriented crystals of larger aspect ratio and is currently used to fit the experimental data in such cases [92,93].

The effective medium theory approach of Maxwell [54] was used for the MMMs containing crystals of smaller aspect ratio. Both the Cussler and the Maxwell models ignore interfacial effects, since they are based exclusively on the gas permeability of the unperturbed polymer, of the molecular sieve, on the aspect ratio of the filler (Cussler model only), and on their volume fractions.

The gas permeability of the MMMs containing flat crystals demonstrated the presence of barriers to the transport of mass at SAPO-34 [43]. The gas permeability of SAPO-34 is compared in Figure 11a with the experimental values of Hyflon AD60X (AD60) and of the MMMs with permeable SAPO-34 crystals of AR 9 (tiles). The unrestricted intra-crystalline permeability of the SAPO-34 crystals was much larger than that of the polymer, but the permeability of the MMMs was lower, with larger volume fractions (0.35 vs. 0.20) reducing the permeability even further. The experimental permeability did not obey the model prediction (Figure 11b), but it was larger than the prediction for an impermeable filler, indicating that SAPO-34 is permeable, and that its permeability is restricted. The presence of barriers for the transport of mass at zeolitic molecular sieves is the norm [47], although this is not acknowledged in most of the literature of MMMs. Diffusion experiments with infrared micro-imaging demonstrated the existence of surface barriers on SAPO-34 crystals, as well as the variability of surface barriers at different crystals from the same batch [64].

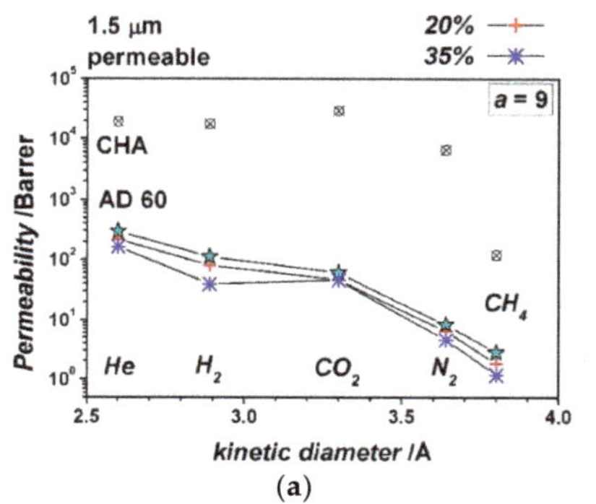

(a)

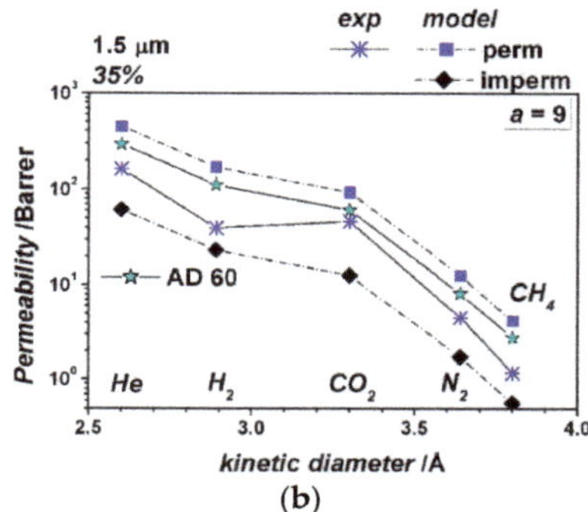

(b)

Figure 11. Experimental (continuous lines) and calculated (dash-dotted lines) gas permeability of Hyflon AD60X (AD60, stars) MMMs: 0%, 20% and 35% *v/v* of SAPO-34 tiles (1.5 μm, aspect ratio AR = 9). (**a**) MMMs and Hyflon AD60X experimental results; SAPO-34 (CHA, cross-hatched circles) permeability calculated after MD simulations [90] for the siliceous CHA structure. (**b**) Comparison of experimental (permeable tiles) and analytical results (Cussler model, both permeable and impermeable tiles), 35 v% loading. 1 Barrer is equivalent to 10^{-10} cm^3 (STP) cm cmHg^{-1}·s^{-1}·cm^{-2}. Reproduced from Reference [43] with permission. Copyright of the Korean Society of Industrial and Engineering Chemistry and Elsevier B.V.

When SAPO-34 is impermeable, gas only permeates through the polymer phase, therefore the behaviour of MMMs prepared with same volume fractions of impermeable SAPO-34 gave valuable information on the state of Hyflon AD60X only in the MMMs [43]. Gas permeability for MMMs containing small (0.2 μm) impermeable SAPO-34 of aspect ratio 2 (Figure 12a) was larger than predicted by the Maxwell model, demonstrating the enlargement of the polymer free-volume at the interface with the filler. Small (0.2 μm) and large (2 μm) SAPO-34 of same aspect ratio were characterized by outer surface area of 39 and 5.3 m^2·cm^{-3}, respectively, therefore the volume of interfacial polymer in the MMMs containing small and large fillers, being roughly proportional to the outer surface area, was several times larger for the former. This circumstance was used to explain why the gas permeability of MMMs containing large impermeable SAPO-34 (Figure 12b) was close to the values predicted by the Maxwell model [43].

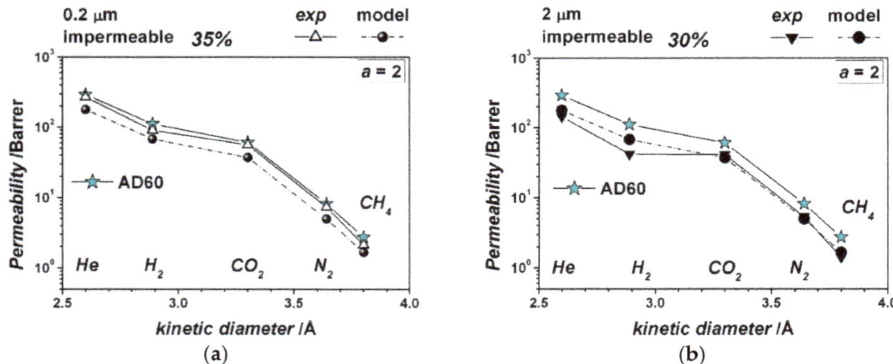

Figure 12. Experimental (continuous lines) and calculated (Maxwell model, dash-dotted lines) gas permeability of Hyflon AD60X (AD60, stars) MMMs with impermeable SAPO-34, aspect ratio AR = 2: (**a**) MMM containing small SAPO-34 (0.2 μm), 35% *v/v*; (**b**) MMM containing large SAPO-34 (2 μm), 30% *v/v*. 1 Barrer is equivalent to 10^{-10} cm^3 (STP) cm $cmHg^{-1}{\cdot}s^{-1}{\cdot}cm^{-2}$. Reproduced from Reference [43] with permission. Copyright of the Korean Society of Industrial and Engineering Chemistry and Elsevier B.V.

Two distinct non-ideal effects, namely the increase of polymeric FFV at the interface with the filler and the barriers at SAPO-34, combined in the Hyflon AD60X/SAPO-34 MMMs [43]. The experimental gas permeability of MMMs containing small (0.2 μm) permeable SAPO-34 of aspect ratio 2 (Figure 13a) was comparable to the prediction of the Maxwell model, because the larger interfacial polymer free-volume compensated the effect of the barriers at SAPO-34. Instead, the experimental gas permeability of MMMs containing large impermeable SAPO-34 (Figure 13b) was lower than the values predicted by the Maxwell model, because in this case the quantity of the interfacial polymer of higher free volume was minimal [43].

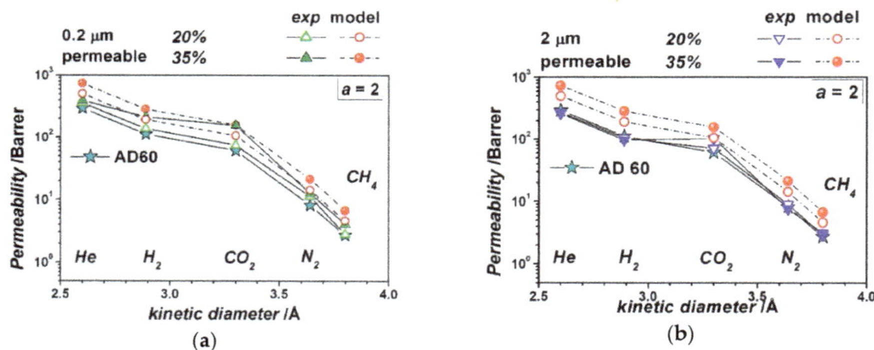

Figure 13. Experimental (continuous lines) and calculated (Maxwell model, dash-dotted lines) gas permeability of Hyflon AD60X (AD60, stars) MMMs with permeable SAPO-34, aspect ratio AR = 2: (**a**) MMMs containing small SAPO-34 (0.2 μm), 20 and 35% *v/v*; (**b**) MMMs containing large SAPO-34 (2 μm), 20 and 35% *v/v*. 1 Barrer is equivalent to 10^{-10} cm^3 (STP) cm $cmHg^{-1}{\cdot}s^{-1}{\cdot}cm^{-2}$. Reproduced from Reference [43] with permission. Copyright of the Korean Society of Industrial and Engineering Chemistry and Elsevier B.V.

The Maxwell (Equation (5)) and the Cussler (Equation (7)) models were used to calculate the expected ideal CO_2/CH_4 selectivity for Hyflon AD60X MMMs containing 35% *v/v* of SAPO-34 with different aspect ratios [43]. If mass transport in SAPO-34 were not restricted by the presence of barriers (filled squares in Figure 14), then the gain in selectivity with respect to the polymer would be irrelevant. If instead the restricted diffusion of mass reduced the effective average permeability of SAPO-34 by two orders of magnitude (empty squares in Figure 14), then the selectivity raised asymptotically to about 45 for AR ~50. This finding demonstrated that very selective fillers do not improve the MMM selectivity if they are too permeable compared to Hyflon AD60X, as predicted in the literature [92,93]. Another interesting information was that most of the selectivity gain could be obtained already with aspect ratio 9 (Figure 14).

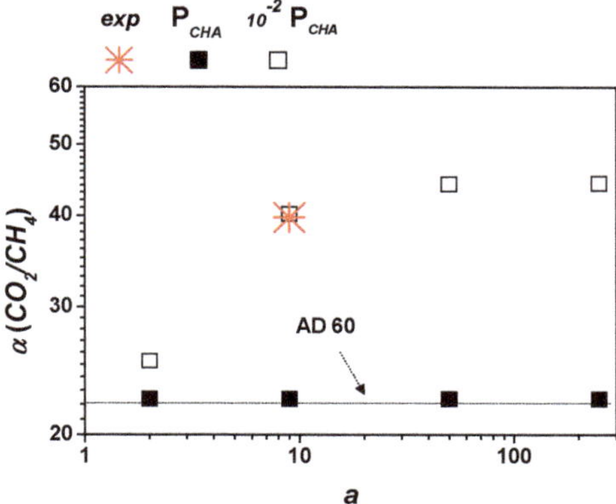

Figure 14. Experimental (asterisk) and calculated (squares, Cussler model) CO_2/CH_4 ideal selectivity of Hyflon AD60X MMMs containing 35% *v/v* of SAPO-34 of different aspect ratios (*a* on the abscissa). The constant selectivity of the polymer [43] is reported as a full line for comparison; full squares refer to the ideal selectivity of MMMs calculated after MD simulations for the siliceous CHA structure [90]; open squares refer to the ideal selectivity of MMMs calculated after MD simulations for the siliceous CHA structure [90], when the derived, unrestricted gas permeabilities used for the model calculation are divided by 100. Reproduced from Reference [43] with permission. Copyright of the Korean Society of the Industrial and Engineering Chemistry and Elsevier B.V.

The modelling of gas transport in Hyflon AD60X/SAPO-34 MMMs with the effective medium approach (Cussler and Maxwell models) made it clear that, besides the unperturbed (bulk) polymer and filler, a region of interfacial polymer and some barrier at SAPO-34 must be considered. Starting from this evidence, Di Maio et al. [44] introduced a new four phase approach of MMMs which explicitly takes into account, besides the matrix and filler bulk phases, the interfacial polymer phase ("matrix*" in Figure 15) and the barriers at SAPO-34. For the sake of simplicity, the latter were concentrated into a skin of reduced permeability encapsulating the core with unrestricted mobility ("filler*" and "filler", respectively, in Figure 15).

The new four-phase approach was implemented for He, H_2, N_2 and CO_2 permeation in Hyflon AD60X/SAPO-34 MMMs in two ways: by means of the effective medium analytical formula (4M Maxwell model), and through the finite element solution (FEM) of the transport differential equations [44].

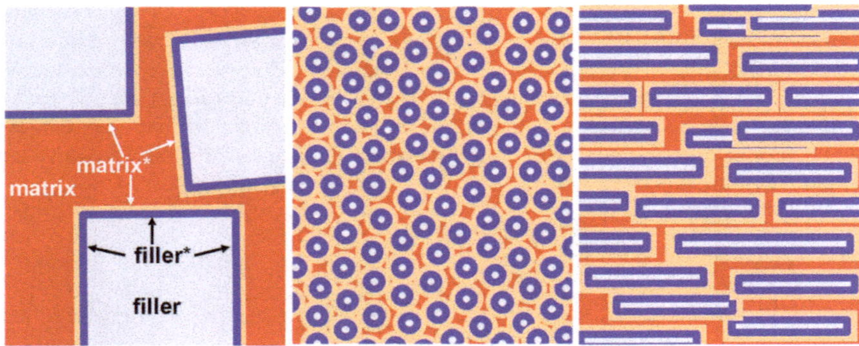

Figure 15. Schematic representation of three Hyflon AD60X MMMs containing SAPO-34 crystals of different size and shape (large, small, and tiles in the text) in the four resistance models (see text). "filler*" (dark-blue) and "matrix*" (light-orange) denote, in all of the three illustrations, the modified diffusion layers of SAPO-34 and Hyflon AD60X, respectively.

The 4M Maxwell model was inspired by a previous three-phase Maxwell model introduced by Koros and co-workers [94,95], and the procedure for the workout of the MMM permeability consisted in three steps. First the SAPO-34 crystals were treated as pseudo-particles made of a very permeable core with a poorly permeable skin ("filler" and "filler*", pseudo-particle 1 in Figure 16a). The permeability of pseudo-particle 1 was calculated by means of the Maxwell Equation (5). Second, a new pseudo-particle made of the whole filler (pseudo-particle 1) surrounded by the modified polymer (pseudo-particle 1 plus "matrix*", pseudo-particle 2 in Figure 16b) was considered, and its permeability was calculated by using again the Maxwell Equation (5). Finally, the MMM permeability was calculated by combining the permeability of the pseudo-particle 2 with the permeability of the bulk polymer (pseudo-particle 2 plus "matrix", Figure 16c). The unknowns regarding the polymer (i.e., the permeability and the thickness of the interfacial polymer, "matrix*" in Figure 16) were removed in part by applying the Maxwell Equation (5) to the experimental permeability data of MMMs containing same volume fractions of identical, but impermeable, molecular sieve particles [95]. The unrestricted permeability of SAPO-34—i.e., the permeability of the SAPO-34 core, "filler" in Figure 16—was calculated as described in Section 4.2.1. More details on the formulations of the relevant analytical equations can be found in the Supplementary Materials of Reference [44]. It has to be noted that the permeability and the thickness of the interfacial polymer surrounding the particles of filler ("matrix*") are not independent of each other, as well as the thickness and the permeability of the barrier skin at SAPO-34 ("filler*"). From the mathematical point of view, in both cases a larger difference in the permeability from the value of the bulk corresponds to a thinner layer, and vice versa. A thickness of 20 nm for the modified polymer layer ("matrix*" in Figure 15) was adopted in order to avoid the overlapping of vicinal layers in the FEM simulation (vide infra), because this length was just less than half of the average distance among the smallest (200 nm) crystals at maximum (35%) volumetric loading. A 20-nm thickness was also chosen for the barrier layer at SAPO-34 ("filler*" in Figure 15). A recent theoretical work on a comparable zeolite—SAS topology, made of large cavities connected by 8-member rings of TO_4 units—found a thickness of surface barriers of the same order of magnitude, i.e., 3, 7 and 14 nm for H_2, CO_2 and CH_4, respectively [96]. For simplicity, gas independent filler* thickness was considered in the 4M model calculation.

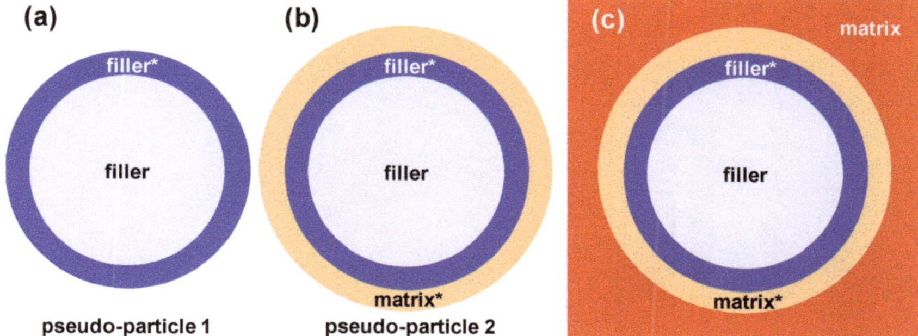

Figure 16. The three frames of the 4M Maxwell model procedure: (**a**) pseudo-particle 1 composed of bulk filler and modified filler; (**b**) pseudo-particle 2, composed of pseudo-particle 1 and modified matrix; (**c**) the whole mixed-matrix membrane, composed of pseudo-particle 2 and bulk matrix.

Numerical simulation of the gas permeation through composite membranes was carried out by solving the finite-element approximation of the partial differential equations governing the relevant solution/diffusion processes [44], using the commercial software COMSOL Multiphysics [97]. In order to simplify the computational model, a body-centred cubic arrangement of oriented filler particles was built to generate symmetrical properties of the system (Figure 17).

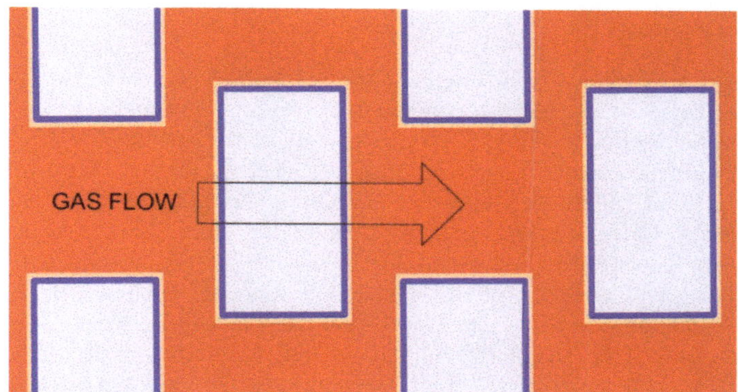

Figure 17. Bi-dimensional scheme of the body-centred cubic arrangement of oriented SAPO-34 particles built by the application software for the representation of the four-resistance model.

Pure gas diffused continuously through the distinct matrix and filler domains. Diffusion was expressed in terms of C_i, the concentration of species i. The following equations:

$$D_{ij}\left(\frac{\partial^2 C_i}{\partial x^2}+\frac{\partial^2 C_i}{\partial y^2}+\frac{\partial^2 C_i}{\partial z^2}\right)=0 \tag{8}$$

in which the diffusivity coefficient D_{ij} of species i in material j is assumed uniform and isotropic, were solved. A concentration jump at the matrix/filler interface formed due to the different gas solubility. At the interface boundaries both uniform mass flux and the following equilibrium condition were applied:

$$\frac{C_i{}^I}{S_i{}^I}=\frac{C_i{}^{II}}{S_i{}^{II}} \tag{9}$$

where S_i is the solubility of species i and the superscripts denote the two solid phases, I and II, in contact.

Computational grids (Figure 18) were built by using the application software.

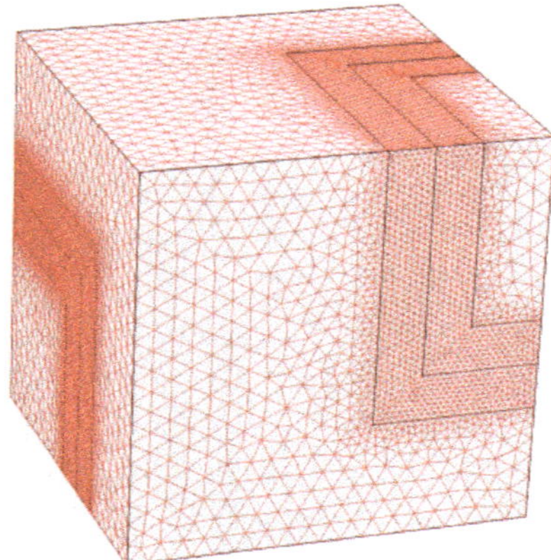

Figure 18. Geometry and discretization grid of the computational cell for a system containing 0.2 μm SAPO-34, AR = 2, volume fraction 0.20. Inlet is from the left front face and outlet from the right rear face. On the other four faces symmetry conditions apply. Reproduced from Reference [44] with permission. Copyright of Elsevier B.V.

The inputs of the FEM simulations were: (i) SAPO-34 bulk diffusivity and solubility data from simulations (see Section 4.2.1.); (ii) the experimental polymer permeability; (iii) the polymer gas diffusivity from the literature [98]; (iv) the permeability of the ten different MMMs containing permeable and impermeable molecular sieves. By imposing 20 nm as the thickness of each modified phases ("matrix*" and "filler*" in Figure 15), the permeability ratios between the pristine and the modified phases were optimized for each gas by a least square iterative minimization of the discrepancies between calculated and experimental permeability. The properties of the modified polymer layer ("matrix*") were obtained in a first optimization step using the data subset including only MMMs incorporating impermeable fillers. Then the results of the former step were used to focus on the properties of the modified SAPO-34 layer ("filler*"). Additional details on the numerical procedure can be found in Reference [44]. The FEM simulations indicated that the permeability in the high FFV polymeric interface ("matrix*") increased of 2.6 $\pm$ 0.3 times, whereas instead the permeability of the barrier skin of SAPO-34 ("filler*") decreased by a factor ranging from 3.5×10^{-5} to 1.1×10^{-4} (Maxwell 4M), or from 3.8×10^{-5} to 4.3×10^{-4} (FEM) [44].

By the above assumptions, the results of both FEM simulations and macroscopic 4M Maxwell model compared well with the experimental data (Table 2 and Figure 19).

Table 2. Features of the Hyflon AD60X MMMs containing permeable SAPO-34. Modified from Reference [44] with permission. Copyright of Elsevier B.V.

Experiment No.	5	6	7	8	9	10
Crystal longest size, μm	0.2	2	0.2	2	1.5	1.5
Particle Aspect Ratio, -	2	2	2	2	9	9
Particle volume fraction, -	0.20	0.20	0.35	0.35	0.20	0.35

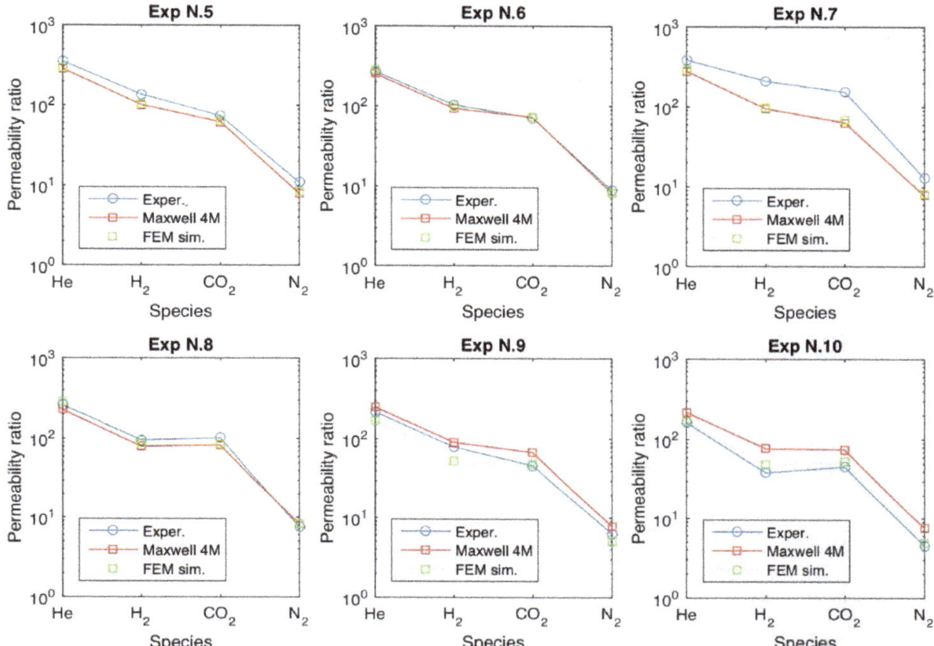

Figure 19. Comparison between the experimental permeability of Hyflon AD60X/SAPO-34 MMMs and their predictions based on the 4-phase approach according to the Maxwell model and the FEM simulations for permeable particles, cases No. 5–10 in Table 2. Reproduced from Reference [44] with permission. Copyright of Elsevier B.V.

Not surprisingly, the best fit was found with the MMMs containing the large fillers (2 µm, AR 2, exp. 6, 8) in which the relative importance of the interfacial effects was minimal (Table 2 and Figure 19). In the case of small SAPO-34, both methods underestimated of the same quantities the permeability of the MMMs (Figure 19, 0.2 µm, AR 2, exp. 5,7), probably due to the formation of percolation paths connecting the modified polymer phase ("matrix*" in Figure 15) in real membranes. The FEM simulation was able to reproduce in a better way the data of the MMMs containing oriented flat crystals (Figure 19, tiles, 1.5 µm, AR 9, exp. 9, 10), because the large aspect ratio and the orientation of the filler were implicit in the model. Instead, the macroscopic 4M Maxwell model, based on spherical particles, showed a worse fit, overestimating the real permeability. Notice that the predictions of the 4M Maxwell model are satisfactory even at relatively high-volume fractions (35%) of the filler.

The overall agreement of the experiments with both the macroscopic 4M Maxwell modelling and the FEM simulations is excellent if we consider the simplifying assumptions adopted, and namely: (i) the constant, average values of solubility and diffusion coefficients for polymer and filler bulk; (ii) the uniform constant features of the surface barrier for all of the three different samples of SAPO-34.

The major advantages of the FEM simulation consist in the quantification of the contributions of the different phases to the overall gas transport and in the possibility to visualize the concentration profiles in a unit cell [44] as well as the streamlines of the gas (Figure 20). The FEM simulation clearly indicated the lesser contribution of SAPO-34 to the overall transport in the MMMs containing small crystals of modest aspect ratio, whereas the predominant role of high free volume polymer led to high permeability. Instead SAPO-34 contribution to transport was maximal when oriented crystals of large aspect ratio were present, consistent with the lesser permeability and the larger selectivity of the MMMs.

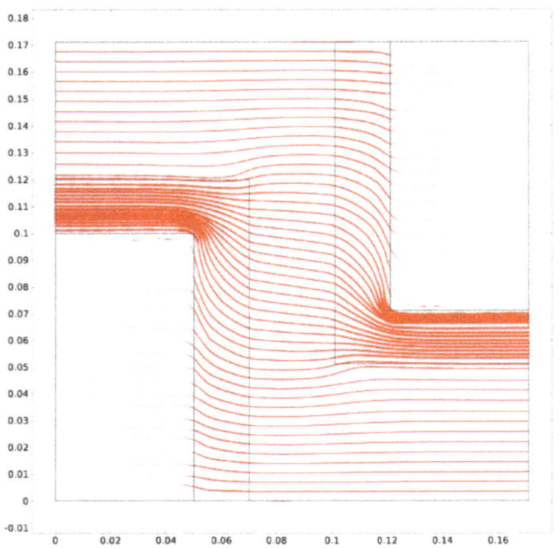

Figure 20. Experiment No. 5 (SAPO 0.2 μm size, volume fraction 20%, AR = 2, see also Table 2 and Figure 18): streamlines of the CO_2 flow in a horizontal plane at half of the cell height. The streamlines are initiated at 50 points uniformly distributed on an imaginary vertical line located halfway along the transport direction and then reconstructed back and forth; their line thickness is proportional to the flow magnitude and axis units are in μm. Reproduced from Reference [44] with permission. Copyright of Elsevier B.V.

4.3. Remarks and Perspectives on the Modelling of Transport in MMMs

The lesson learned from the study of the PFP/molecular sieve systems is that the macroscopic modelling of the transport properties of MMMs requires the understanding of the underlying physical phenomena. If we had only prepared MMMs with the high aspect ratio SAPO-34 (triangles in Figure 10) we might have argued about the prevalence either of a rigidified polymer layer around the filler, or of a reduced permeability region within the sieve surface [35]: the possibility of a high FFV polymer shell surrounding the filler would have not been considered.

The physical state of polymers at MMM interfaces should be investigated in depth to probe the effect of fillers on free volume, adsorption properties, swelling ability, ageing rate. Positron annihilation lifetime spectroscopy (PALS) [99], ^{129}Xe NMR [17], and the NELF model [52] applied to MMMs containing impermeable fillers look like valuable tools for this purpose.

It has to be noted that the new four-phase approach is not restricted to the modelling of PFP MMMs, since it can be applied whenever two non-ideal phenomena are present, including those cases in which a rigidified polymer layer is observed. Besides the consistent interpretation of the physical phenomena, an adequate modelling of the transport properties of real systems allows predictive capabilities to guide the preparation of better MMMs.

Crystalline porous materials, in real life, contain defects and are far from ideal. The modelling of mass transport in MMMs needs theoretical and experimental studies of diffusion in the fillers, both ideal and defective. The nature and the effects of the barriers to mass transport need to be studied in depth. In MMMs containing extremely thin, exfoliated layers, either zeolitic molecular sieves, porous graphenes, MOFs or others, the transport properties of the bulk materials may be altered to a large extent, and interfacial effects become of primary importance.

5. Summary and Conclusions

Grafting perfluorinated tails on the outer surface of zeolitic molecular sieves improved the surface permeability and allowed the preparation of defect-free MMMs with hydrophobic and amorphous PFPs.

Silicalite-1/Hyflon AD60X MMMs displayed some improvement of the CO_2/CH_4 and N_2/CH_4 ideal selectivity with respect to the polymer. The improvements were not expected, because the pores of silicalite-1 are not able to discriminate the above gases on the basis of size. Methane is more permeable than *n*-butane in Teflon AF2400 membranes, but the presence of silicalite-1 in Teflon AF2400 MMMs was able to revert the selectivity in favour of *n*-butane, probably because of the combined effects of a larger interfacial FFV and of the butane-selective molecular sieve. Other things being equal, large aspect ratio silicalite-1 increased the *n*-butane/CH_4 selectivity even further.

The gas transport properties of Teflon AF1600 MMMs containing silicalite-1 crystals were strongly dependent on the size of the filler. Small crystals added permeability (He, H_2, CO_2, O_2, N_2 and CH_4) to the polymer with less selectivity, whereas larger crystals induced a reduction of permeability and an increase of the gas/CH_4 selectivity. In order to clarify this puzzling situation, a change of paradigm was introduced: instead of calculating the permeability of the filler from the performance of the MMMs through the Maxwell model (Equation (5)), the permeability of the filler was obtained from the gas solubility and diffusion, using this value to predict the performance of an ideal MMM through Equation (5). The comparison of the predicted vs. the experimental permeability (CO_2 and CH_4) and the evaluation of the experimental diffusion coefficients (time lag method) indicated the presence of two different non-ideal effects in the Teflon AF1600/silicalite-1 MMMs: a polymer allowing larger gas mobility, i.e., a larger polymer FFV, and a much less permeable filler, i.e., the presence of barriers to the transport of mass at silicalite-1.

A new set of Hyflon AD60X MMMs were prepared with three different SAPO-34 crystals, of different size and aspect ratio, both permeable and impermeable. All of the MMMs where characterized by improvements in the ideal CO_2/CH_4 selectivity, but the permeability was strongly influenced by the size and the shape of SAPO-34. The large aspect ratio crystals preferentially oriented parallel to the plane of the membrane and induced a reduction of the polymer permeability. The filler with modest aspect ratio, instead, determined an increase of the permeability, with the smaller SAPO-34 crystals resulting in the larger gain. The analysis of the transport properties of the MMMs subset containing impermeable SAPO-34 clarified that Hyflon FFV had increased. As in the case of silicalite-1, the permeability of SAPO-34 was obtained from independent solubility and diffusivity values and used to predict the performance of the MMMs through Equation (5). The predicted MMMs permeability was higher than the experimental results, evidencing the presence of barriers to the transport of mass at SAPO-34. On the basis of the experimental evidences, the transport properties of the MMMs were modelled through a new four phase approach, which explicitly considered the pristine two phases, i.e., the unperturbed, bulk polymer and SAPO-34, and two interfacial phases, namely the perturbed, high FFV polymer, and a superficial skin of the crystals, concentrating all of the additional resistance to the transport of mass (the barrier). The four-phase macroscopic modelling, implemented both via an extension of the Maxwell analytical method (4M Maxwell model) and by means of a FEM numerical simulation, nicely reproduced the experimental results. The new four-phase approach lends itself to interpret the transport data of those MMMs evidencing two simultaneous non-ideal behaviours.

The above pioneering works clearly show: the feasibility of FFP/molecular sieve MMMs and their interesting properties for gas separation of technological relevance; the complexity and the competitive character of the processes at work in MMMs; the adequacy of four-phase models for the description of permeation in MMMs.

Funding: This work was supported by the MIUR project ComESto, ARS01_01259.

Conflicts of Interest: The authors declare no conflict of interest.

List of Symbols and Acronyms

4M	four-phase Maxwell model
α	permeability selectivity, -
α_S	solubility selectivity, -
α_D	diffusion selectivity, -
ϕ	volume fraction, -
θ	time-lag, s
AR	aspect ratio, -
Barrer	permeability unit, 10^{-10} cm^3 (STP) cm $cmHg^{-1}\cdot s^{-1}\cdot cm^{-2}$
BET	Brunauer, Emmett and Teller method for the determination of the surface area
C	concentration, mol m^{-3} or cm^3(STP) cm^{-3}
CHA	framework type code of zeolite chabasite
Cytop®	poly(perfluoro-4-vinyloxy-1-butene)
D	diffusivity, m^2/s
d_k	kinetic diameter, Å
FEM	finite element numerical modelling
FFV	fractional free volume, -
Hyflon® AD	co-polymers of tetrafluoroethylene and perfluoro-4-methoxy-1,3-dioxole
J	flux, $mol/(m^2\cdot s)$ or $cm^3(STP)/(cm^3\cdot s)$
l	membrane thickness, μm
MD	molecular dynamics
MFI	framework type code of zeolite ZSM-5
MMM	mixed matrix membrane
MOF	metal organic framework
Nafion®	copolymers of tetrafluoroethylene and sulfonic acid terminated perfluorovinyl ethers
NELF	non-equilibrium lattice fluid
NMR	nuclear magnetic resonance
p	pressure, Pa
P	permeability, Barrer or $Nm^3\cdot m\cdot m^{-2}\cdot Pa^{-1}\cdot s^{-1}$
P^F	permeability of the dispersed phase, Barrer or $Nm^3\cdot m\cdot m^{-2}\cdot Pa^{-1}\cdot s^{-1}$
P^M	permeability of the continuous phase, Barrer or $Nm^3\cdot m\cdot m^{-2}\cdot Pa^{-1}\cdot s^{-1}$
P^{MMM}	permeability of the MMM, Barrer or $Nm^3\cdot m\cdot m^{-2}\cdot Pa^{-1}\cdot s^{-1}$
PALS	positron annihilation lifetime spectroscopy
PFG-NMR	pulsed field gradient NMR
S	solubility, $mol\cdot m^{-3}\cdot Pa^{-1}$ or $cm^3(STP)\cdot cm^{-3}\cdot Pa^{-1}$
T_c	critical temperature, K
T_g	glass transition temperature, K
PFP	perfluorinated polymer
PIM	polymer of intrinsic microporosity
SAPO-34	silicon aluminium phosphate of chasbasite (CHA) structure topology.
SEM	scanning electron microscopy
Silicalite-1	pure SiO_2 zeolite of ZSM-5 (MFI) structure topology
Teflon®	polytetrafluoroethylene
Teflon® AF	co-polymer of tetrafluoroethylene and perfluoro-2,2-dimethyl-1,3-dioxole
STP	standard temperature and pressure

References

1. Cui, Z.; Drioli, E.; Lee, Y.-M. Recent Progress in Fluoropolymer Membranes. *Progr. Polym. Sci.* **2014**, *39*, 164–198. [CrossRef]
2. Marzouk, S.A.M.; Al-Marzouqi, M.H.; Teramoto, M.; Abdullatif, N.; Ismail, Z.M. Simultaneous Removal of CO_2 and H_2S from Pressurized CO_2-H_2S-CH_4 Gas Mixture Using Hollow Fiber Membrane Contactors. *Sep. Purif. Technol.* **2012**, *86*, 88–97. [CrossRef]

3. Feng, S.; Zhong, Z.; Wang, Y.; Xing, W.; Drioli, E. Progress and Perspectives in PTFE Membranes: Preparation, Modification and Application. *J. Membr. Sci.* **2018**, *549*, 332–349. [CrossRef]
4. Buonomenna, M.G.; Golemme, G.; De Santo, M.P.; Drioli, E. Direct Oxidation of Cyclohexene with Inert Polymeric Membrane Reactor. *Org. Process Res. Dev.* **2010**, *14*, 252–258. [CrossRef]
5. Scholes, C.; Stevens, G.; Kentish, S. Membrane Gas Separation Applications in Natural Gas Processing. *Fuel* **2012**, *96*, 15–28. [CrossRef]
6. Merkel, T.C.; Pinnau, I.; Prabhakar, R.; Freeman, B.D. Gas and Vapor Transport Properties of Perfluoropolymers. In *Materials Science of Membranes for Gas and Vapor Separation;* Yampolskii, Y.P., Pinnau, I., Freeman, B.D., Eds.; John Wiley & Sons: Chichester, UK, 2006; Chapter 9; pp. 251–270.
7. Golemme, G. Perfluoropolymer Membranes for Separations and Electrochemical Processes. In *Advanced Materials for Membrane Preparation;* Buonomenna, M.G., Golemme, G., Eds.; Bentham Science: Sharjah, UAE, 2012; Chapter 9; pp. 106–124.
8. Arcella, V.; Merlo, L.; Pieri, R.; Toniolo, P.; Triulzi, F.; Apostolo, M. Fluoropolymers for Sustainable Energy. In *Handbook of Fluoropolymer Science and Technology;* Smith, D.W., Jr., Iacono, S.T., Iyer, S.S., Eds.; John Wiley & Sons: Hoboken, NJ, USA, 2014; Chapter 17; pp. 393–412.
9. Robeson, L.M. The Upper Bound Revisited. *J. Membr. Sci.* **2008**, *320*, 390–400. [CrossRef]
10. Plunkett, R.J. Tetrafluoroethylene Polymers. U.S. Patent 2230654, 1941.
11. Nakamura, M.; Sugiyama, N.; Etoh, Y.; Aosaki, K.; Endo, J. Development of Perfluoro Transparent Resins Obtained by Radical Cyclopolymerization for Leading Edge Electronic and Optical Applications. *Nippon Kagaku Kaishi* **2001**, *12*, 659–668. [CrossRef]
12. Van Krevelen, D.V.; Nijenhuis, K. *Properties of Polymers*, 4th ed.; Elsevier: Amsterdam, The Netherlands, 2009.
13. Hansen, C.M. *Hansen Solubility Parameters: A User's Handbook;* CRC Press: Boca Raton, FL, USA, 2007.
14. Resnick, P.R.; Buck, W.H. Teflon® AF Amorphous Fluoropolymers. In *Modern Fluoropolymers. High Performance Polymers for Diverse Applications;* Scheirs, J., Ed.; Wiley: Chichester, UK, 1997; p. 397.
15. Tokarev, A.V.; Bondarenko, G.N.; Yampol'skii, Y.P. Chain Structure and Stiffness of Teflon AF Glassy Amorphous Fluoropolymers. *Polym. Sci. (Vysokomol. Soedin.) Ser. A* **2007**, *49*, 909–920. [CrossRef]
16. Alentiev, A.Y.; Yampolskii, Y.P.; Shantarovich, V.P.; Nemser, S.M.; Platé, N.A. High Transport Parameters and Free Volume of Perfluorodioxole Copolymers. *J. Membr. Sci.* **1997**, *126*, 123–132. [CrossRef]
17. Golemme, G.; Nagy, J.B.; Fonseca, A.; Algieri, C.; Yampolskii, Y.P. ^{129}Xe NMR Study of Free Volume in Amorphous Perfluorinated Polymers: Comparsion with Other Methods. *Polymer* **2003**, *44*, 5039–5045. [CrossRef]
18. Bondar, V.I.; Freeman, B.D.; Yampol'skii, Y.P. Sorption of Gases and Vapors in an Amorphous Glassy Perfluorodioxole Copolymer. *Macromolecules* **1999**, *32*, 6163–6171. [CrossRef]
19. Dlubek, G.; Pionteck, J.; Rätzke, K.; Kruse, J.; Faupel, F. Temperature Dependence of the Free Volume in Amorphous Teflon AF1600 and AF2400: A Pressure-Volume-Temperature and Positron Lifetime Study. *Macromolecules* **2008**, *41*, 6125–6133. [CrossRef]
20. Wijmans, J.G.; Baker, R.W. The Solution-Diffusion Model: A Review. *J. Membr. Sci.* **1995**, *107*, 1–21. [CrossRef]
21. Wijmans, J.G.; Baker, R.W. The Solution-Diffusion Model: A Unified Approach to Membrane Permeation. In *Materials Science of Membranes for Gas and Vapor Separation;* Yampolskii, Y.P., Pinnau, I., Freeman, B.D., Eds.; John Wiley & Sons: Chichester, UK, 2006; Chapter 5; pp. 159–189.
22. Nagai, K.; Masuda, T.; Nakagawa, T.; Freeman, B.D.; Pinnau, I. Poly[1-(trimethylsilyl)-1-propyne) and Related Polymers: Synthesis, Properties and other Functions. *Prog. Polym. Sci.* **2011**, *26*, 721–798. [CrossRef]
23. Morisato, A.; Pinnau, I. Synthesis and Gas Permeation Properties of Poly(4-methyl-2-pentyne). *J. Membr. Sci.* **1996**, *121*, 243–250. [CrossRef]
24. McKeown, N.B.; Budd, P.M.; Msayib, K.J.; Ghanem, B.S.; Kingston, H.J.; Tattershall, C.E.; Makhseed, S.; Reynolds, K.J.; Fritsch, D. Polymers of Intrinsic Microporosity (PIMs): Bridging the Void between Microporous and Polymeric Materials. *Chem. Eur. J.* **2005**, *11*, 2610–2620. [CrossRef] [PubMed]
25. Breck, D.W. *Zeolite Molecular Sieves: Structure, Chemistry, and Use;* J. Wiley and Sons: New York, NY, USA, 1974.
26. National Institute of Standards and Technology. Available online: https://webbook.nist.gov/chemistry/name-ser/ (accessed on 15 October 2018).

27. Belov, N.A.; Alentiev, A.Y.; Ronova, I.A.; Sinitsyna, O.V.; Nikolaev, A.Y.; Zharov, A.A. Microstructure Relaxation Process of Polyhexafluoropropylene after Swelling in Supercritical Carbon Dioxide. *J. Appl. Polym. Sci.* **2016**, *133*, 43105. [CrossRef]
28. Belov, N.; Nikiforov, R.; Polunin, E.; Pogodina, Y.; Zavarzin, I.; Shantarovich, V.; Yampolskii, Y. Gas Permeation, Diffusion, Sorption and Free Volume of Poly(2-trifluoromethyl-2-pentafluoroethyl-1,3-perfluorodioxole). *J. Membr. Sci.* **2018**, *565*, 112–118. [CrossRef]
29. Okamoto, Y.; Du, Q.; Koike, K.; Mikeš, F.; Merkel, T.C.; He, Z.; Zhang, H.; Koike, Y. New Amorphous Perfluoro Polymers: Perfluorodioxolane Polymers for Use as Plastic Optical Fibers and Gas Separation Membranes. *Polym. Adv. Technol.* **2016**, *27*, 33–41. [CrossRef]
30. Fang, M.; He, Z.; Merkel, T.C.; Okamoto, Y. High-performance Perfluorodioxolane Copolymer Membranes for Gas Separation with Tailored Selectivity Enhancement. *J. Mater. Chem. A* **2018**, *6*, 652–658. [CrossRef]
31. Okamoto, Y.; Zhang, H.; Mikeš, F.; Koike, Y.; He, Z.; Merkel, T.C. New Perfluoro-dioxolane-based Membranes for Gas Separations. *J. Membr. Sci.* **2014**, *471*, 412–419. [CrossRef]
32. Baker, R.W.; Lokhandwala, K. Natural Gas Processing with Membranes: An Overview. *Ind. Eng. Chem. Res.* **2008**, *47*, 2109–2121. [CrossRef]
33. Buonomenna, M.G.; Yave, W.; Golemme, G. Some Approaches for High Performance Polymer Based Membranes for Gas Separation: Block Copolymers, Carbon Molecular Sieves and Mixed Matrix Membranes. *RSC Adv.* **2012**, *2*, 10745–10773. [CrossRef]
34. Chuah, C.Y.; Goh, K.; Yang, Y.; Gong, H.; Li, W.; Karahan, H.E.; Guiver, M.D.; Wang, R.; Bae, T.-H. Harnessing Filler Materials for Enhancing Biogas Separation Membranes. *Chem. Rev.* **2018**, *118*, 8655–8769. [CrossRef] [PubMed]
35. Moore, T.T.; Koros, W.J. Non-Ideal Effects in Organic–Inorganic Materials for Gas Separation Membranes. *J. Mol. Struct.* **2005**, *739*, 87–98. [CrossRef]
36. Kalantzopoulos, G.; Policicchio, A.; Maccallini, E.; Krkljus, I.; Ciuchi, F.; Hirscher, M.; Agostino, R.G.; Golemme, G. Resistance to the Transport of H_2 through the External Surface of As-made and Modified Silicalite-1 (MFI). *Microporous Mesoporous Mater.* **2016**, *220*, 290–297. [CrossRef]
37. Vankelecom, I.F.J.; Van den Broeck, S.; Merckx, E.; Geerts, H.; Grobet, P.; Uytterhoeven, J.B. Silylation to Improve Incorporation of Zeolites in Polyimide Films. *J. Phys. Chem.* **1996**, *100*, 3753–3758. [CrossRef]
38. Fernández-Barquín, A.; Casado-Coterillo, C.; Palomino, M.; Valencia, S.; Irabien, A. LTA/Poly(1-trimethylsilyl-1-propyne) Mixed-Matrix Membranes for High-Temperature CO_2/N_2 Separation. *Chem. Eng. Technol.* **2015**, *38*, 658–666. [CrossRef]
39. Golemme, G.; Bruno, A.; Manes, R.; Muoio, D. Preparation and Properties of Superglassy Polymers—Zeolite Mixed Matrix Membranes. *Desalination* **2006**, *200*, 440–442. [CrossRef]
40. Golemme, G.; Policicchio, A.; Sardella, E.; De Luca, G.; Russo, B.; Liguori, P.F.; Melicchio, A.; Agostino, R.G. Surface Modification of Molecular Sieve Fillers for Mixed Matrix Membranes. *Colloid Surf. A Physicochem. Eng. Asp.* **2018**, *538*, 333–342. [CrossRef]
41. Smirnov, K.S. A Molecular Dynamics Study of the Interaction of Water with the External Surface of Silicalite-1. *Phys. Chem. Chem. Phys.* **2017**, *19*, 2950–2960. [CrossRef]
42. Wang, H.T.; Holmberg, B.T.; Yan, Y.S. Homogeneous Polymer-Zeolite Nanocomposite Membranes by Incorporating Dispersible Template-removed Zeolite Nanocrystals. *J. Mater. Chem.* **2002**, *12*, 3640–3643. [CrossRef]
43. Santaniello, A.; Golemme, G. Interfacial Control in Perfluoropolymer Mixed Matrix Membranes for Natural Gas Sweetening. *J. Ind. Eng. Chem.* **2018**, *60*, 169–176. [CrossRef]
44. Di Maio, F.P.; Santaniello, A.; Di Renzo, A.; Golemme, G. Description of Gas Transport in Perfluoropolymer/SAPO-34 Mixed Matrix Membranes Using Four-Resistance Model. *Sep. Purif. Technol.* **2017**, *185*, 160–174. [CrossRef]
45. Golemme, G.; Jansen, J.C.; Muoio, D.; Bruno, A.; Manes, R.; Buonomenna, M.G.; Choi, J.; Tsapatsis, M. Glassy Perfluoropolymer—Zeolite Hybrid Membranes for Gas Separations. In *Membrane Gas Separation*; Yampolskii, Y.P., Freeman, B.D., Eds.; John Wiley & Sons: Chichester, UK, 2010; Chapter 6; pp. 113–124. [CrossRef]
46. Flanigen, E.M.; Bennett, J.M.; Grose, R.W.; Cohen, J.P.; Patton, R.L.; Kirchner, R.M.; Smith, J.V. Silicalite, a New Hydrophobic Crystalline Silica Molecular Sieve. *Nature* **1978**, *271*, 512–516. [CrossRef]

47. Kärger, J.; Ruthven, D.M.; Theodorou, D.N. *Diffusion in Nanoporous Materials*; Wiley VCH: Weinheim, Germany, 2012; ISBN 9783527310241.
48. Merkel, T.C.; He, Z.; Pinnau, I.; Freeman, B.D.; Meakin, P.; Hill, A.J. Sorption and Transport in Poly(2,2-bis(trifluoromethyl)-4,5-difluoro-1,3-dioxole-*co*-tetrafluoroethylene) Containing Nanoscale Fumed Silica. *Macromolecules* **2003**, *36*, 8406–8414. [CrossRef]
49. Merkel, T.C.; He, Z.; Pinnau, I.; Freeman, B.D.; Meakin, P.; Hill, A.J. Effect of Nanoparticles on Gas Sorption and Transport in Poly(1-trimethylsilyl-1-propyne). *Macromolecules* **2003**, *36*, 6844–6855. [CrossRef]
50. Merkel, T.C.; Freeman, B.D.; Spontak, R.J.; He, Z.; Pinnau, I.; Meakin, P.; Hill, A.J. Ultrapermeable, Reverse-Selective Nanocomposite Membranes. *Science* **2002**, *296*, 519–522. [CrossRef]
51. De Angelis, M.G.; Sarti, G.C. Solubility and Diffusivity of Gases in Mixed Matrix Membranes containing Fumed Silica: Correlations and Predictions Based on the NELF Model. *Ind. Eng. Chem. Res.* **2008**, *47*, 5214–5226. [CrossRef]
52. Ferrari, M.C.; Galizia, M.; De Angelis, M.G.; Sarti, G.C. Gas and Vapor Transport in Mixed Matrix Membranes Based on Amorphous Teflon AF1600 and AF2400 and Fumed Silica. *Ind. Eng. Chem. Res.* **2010**, *49*, 11920–11935. [CrossRef]
53. Krishna, R. Multi-scale Modeling Strategy for Separation of Alkane Mixtures using Zeolites. *Comput. Aided Chem. Eng.* **2002**, *10*, 109–114.
54. Maxwell, J.C. *A Treatise on Electricity and Magnetism*; Dover Publication: New York, NY, USA, 1954.
55. Yu, M.; Noble, R.D.; Falconer, J.L. Zeolite Membranes: Microstructure Characterization and Permeation Mechanisms. *Acc. Chem. Res.* **2011**, *44*, 1196–1206. [CrossRef] [PubMed]
56. Ahunbay, M.G.; Elliott, J.R., Jr.; Talu, O. The Diffusion Process of Methane through a Silicalite Single Crystal Membrane. *J. Phys. Chem. B* **2002**, *106*, 5163–5168. [CrossRef]
57. Ahunbay, M.G.; Elliott, J.R., Jr.; Talu, O. Surface Resistance to Permeation through the Silicalite Single Crystal Membrane: Variation with Permeant. *J. Phys. Chem. B* **2004**, *108*, 7801–7808. [CrossRef]
58. Newsome, D.A.; Sholl, D.S. Atomically detailed simulations of surface resistances to transport of CH_4, CF_4, and C_2H_6 through silicalite membranes. *Microporous Mesoporous Mater.* **2008**, *107*, 286–295. [CrossRef]
59. Combariza, A.F.; Sastre, G. Influence of Zeolite Surface in the Sorption of Methane from Molecular Dynamics. *J. Phys. Chem. C* **2011**, *115*, 13751–13758. [CrossRef]
60. Stavitski, E.; Drury, M.R.; de Winter, D.A.M.; Kox, M.H.F.; Weckhuysen, B.M. Intergrowth Structure of Zeolite Crystals and Pore Orientation of Individual Subunits Revealed by Electron Backscatter Diffraction/Focused Ion Beam Experiments. *Angew. Chem. Int. Ed.* **2008**, *47*, 5637–5640. [CrossRef] [PubMed]
61. Chmelik, C.; Kortunov, P.; Vasenkov, S.; Kärger, J. Imaging of Transient Guest Profiles in Nanoporous Host Materials: A New Experimental Technique to Study Intra-Crystalline Diffusion. *Adsorption* **2005**, *11*, 455–460. [CrossRef]
62. Karwacki, L.; van der Bij, H.E.; Kornatowski, J.; Cubillas, P.; Drury, M.R.; de Winter, D.A.M.; Anderson, M.W.; Weckhuysen, B.M. Unified Internal Architecture and Surface Barriers for Molecular Diffusion of Microporous Crystalline Aluminophosphates. *Angew. Chem. Int. Ed.* **2010**, *49*, 6790–6794. [CrossRef]
63. Hibbe, F.; Chmelik, C.; Heinke, L.; Pramanik, S.; Li, J.; Ruthven, D.M.; Tzoulaki, D.; Kärger, J. The Nature of Surface Barriers on Nanoporous Solids Explored by Microimaging of Transient Guest Distributions. *J. Am. Chem. Soc.* **2011**, *133*, 2804–2807. [CrossRef] [PubMed]
64. Cousin Saint Remi, J.; Lauerer, A.; Chmelik, C.; Vandendael, I.; Terryn, H.; Baron, G.V.; Denayer, J.F.M.; Kärger, J. The Role of Crystal Diversity in Understanding Mass Transfer in Nanoporous Materials. *Nat. Mater.* **2015**, *15*, 401–406. [CrossRef] [PubMed]
65. Karwacki, L.; Kox, M.H.F.; de Winter, D.A.M.; Drury, M.R.; Meeldijk, J.D.; Stavitski, E.; Schmidt, W.; Mertens, M.; Cubillas, P.; John, N.; et al. Morphology-Dependent Zeolite Intergrowth Structures Leading to Distinct Internal and Outer-Surface Molecular Diffusion Barriers. *Nat. Mater.* **2009**, *8*, 959–965. [CrossRef] [PubMed]
66. Fasano, M.; Humplik, T.; Bevilacqua, A.; Tsapatsis, M.; Chiavazzo, E.; Wang, E.N.; Asinari, P. Interplay between Hydrophilicity and Surface Barriers on Water Transport in Zeolite Membranes. *Nat. Commun.* **2016**, *7*, 12762. [CrossRef] [PubMed]
67. Kortunov, P.; Vasenkov, S.; Chmelik, C.; Kärger, J.; Ruthven, D.M.; Włoch, J. Influence of Defects on the External Crystal Surface on Molecular Uptake into MFI-Type Zeolites. *Chem. Mater.* **2004**, *16*, 3552–3558. [CrossRef]

68. Brabec, L.; Kocirik, M. Silicalite-1 Crystals Etched with Hydrofluoric Acid Dissolved in Water or Acetone. *J. Phys. Chem. C* **2010**, *114*, 13685–13694. [CrossRef]
69. Tzoulaki, D.; Schmidt, W.; Wilczok, U.; Kärger, J. Formation of Surface Barriers on Silicalite-1 Crystal Fragments by Residual Water Vapour as Probed with Isobutane by Interference Microscopy. *Microporous Mesoporous Mater.* **2008**, *110*, 72–76. [CrossRef]
70. Texeira, A.R.; Chang, C.C.; Coogan, T.; Kendall, R.; Fan, W.; Dauenhauer, P.J. Dominance of Surface Barriers in Molecular Transport through Silicalite-1. *J. Phys. Chem. C* **2013**, *117*, 25545–25555. [CrossRef]
71. Chisholm, N.O.; Funke, H.H.; Noble, R.D.; Falconer, J.L. Carbon Dioxide/Alkane Separations in a SSZ-13 Membrane. *J. Membr. Sci.* **2018**, *568*, 17–21. [CrossRef]
72. Krishna, R. Diffusion of Binary Mixtures in Zeolites: Molecular Dynamics Simulations Versus Maxwell–Stefan Theory. *Chem. Phys. Lett.* **2000**, *326*, 477–484. [CrossRef]
73. Chempath, S.; Krishna, R.; Snurr, R.Q. Nonequilibrium Molecular Dynamics Simulations of Diffusion of Binary Mixtures Containing Short N-Alkanes in Faujasite. *J. Phys. Chem. B* **2004**, *108*, 13481–13491. [CrossRef]
74. Krishna, R.; Van Baten, J. Unified Maxwell-Stefan Description of Binary Mixture Diffusion in Micro- and Meso-Porous Materials. *Chem. Eng. Sci.* **2009**, *64*, 3159–3178. [CrossRef]
75. Krishna, R. The Maxwell—Stefan Description of Mixture Diffusion in Nanoporous Crystalline Materials. *Microporous Mesoporous Mater.* **2014**, *185*, 30–50. [CrossRef]
76. Krishna, R. Methodologies for Screening and Selection of Crystalline Microporous Materials in Mixture Separations. *Sep. Purif. Technol.* **2018**, *194*, 281–300. [CrossRef]
77. Nijhuis, T.A.; van den Broeke, L.J.P.; Linders, M.J.G.; van de Graaf, J.M.; Kapteijn, F.; Makkee, M.; Moulijn, J.A. Measurement and Modeling of the Transient Adsorption, Desorption and Diffusion Processes in Microporous Materials. *Chem. Eng. Sci.* **1999**, *54*, 4423–4436. [CrossRef]
78. Sun, M.S.; Shah, D.B.; Xu, H.H.; Talu, O. Adsorption Equilibria of C1 to C4 Alkanes, CO_2, and SF_6 on Silicalite. *J. Phys. Chem. B* **1998**, *102*, 1466–1473. [CrossRef]
79. Rutherford, S.W.; Do, D.D. Review of Time Lag Permeation Technique as a Method for Characterisation of Porous Media and Membranes. *Adsorption* **1997**, *3*, 283–312. [CrossRef]
80. Paul, D.R. Effect of Immobilizating Adsorption on the Diffusion Time-Lag. *J. Polym. Sci. Part A-2* **1969**, *27*, 1811–1818. [CrossRef]
81. Şen, D.; Kalıpçılar, H.; Yilmaz, L. Development of Polycarbonate Based Zeolite 4A Filled Mixed Matrix Gas Separation Membranes. *J. Membr. Sci.* **2007**, *303*, 194–203. [CrossRef]
82. Wang, L.; Corriou, J.P.; Castel, C.; Favre, E. Transport of Gases in Glassy Polymers under Transient Conditions: Limit-Behavior Investigations in Dual-Mode Sorption Theory. *Ind. Eng. Chem. Res.* **2013**, *52*, 1089–1101. [CrossRef]
83. Paul, D.R.; Kemp, D.R. The Diffusion Time Lag in Polymer Membranes Containing Adsorptive Fillers. *J. Polym. Sci. Polym. Symp.* **1973**, *41*, 79–93. [CrossRef]
84. Paul, D.R.; Koros, W.J. Effect of Partially Immobilizating Sorption on Permeability and the Diffusion Time Lag. *J. Polym. Sci. Part B Polym. Phys.* **1976**, *14*, 675–685. [CrossRef]
85. Lok, B.M.; Messina, C.A.; Patton, R.L.; Gajek, R.T.; Cannan, T.R.; Flanigen, E.M. Silicoaluminophosphate Molecular Sieves: Another New Class of Microporous Crystalline Inorganic Solids. *J. Am. Chem. Soc.* **1984**, *106*, 6092–6093. [CrossRef]
86. Funke, H.H.; Chen, M.Z.; Prakash, A.N.; Falconer, J.L.; Noble, R.D. Separating Molecules by Size in SAPO-34 Membranes. *J. Membr. Sci.* **2014**, *456*, 185–191. [CrossRef]
87. Krishna, R.; van Baten, J.M. In Silico Screening of Zeolite Membranes for CO_2 Capture. *J. Membr. Sci.* **2010**, *360*, 323–333. [CrossRef]
88. Krishna, R.; van Baten, J.M. Segregation Effects in Adsorption of CO_2 Containing Mixtures and Their Consequences for Separation Selectivities in Cage-Type Zeolites. *Sep. Purif. Technol.* **2008**, *61*, 414–423. [CrossRef]
89. Zong, Z.; Feng, X.; Huang, Y.; Song, Z.; Zhou, R.; Zhou, S.J.; Carreon, M.A.; Yu, M.; Li, S. Highly Permeable N_2/CH_4 Separation SAPO-34 Membranes Synthesized by Diluted Gels and Increased Crystallization Temperature. *Microporous Mesoporous Mater.* **2016**, *224*, 36–42. [CrossRef]
90. Krishna, R.; van Baten, J.M. Insights into Diffusion of Gases in Zeolites Gained from Molecular Dynamics Simulations. *Microporous Mesoporous Mater.* **2008**, *109*, 91–108. [CrossRef]
91. Cussler, E.L. Membranes Containing Selective Flakes. *J. Membr. Sci.* **1990**, *52*, 275–288. [CrossRef]

92. Sheffel, J.A.; Tsapatsis, M. A Model for the Performance of Microporous Mixed Matrix Membranes with Oriented Selective Flakes. *J. Membr. Sci.* **2007**, *295*, 50–70. [CrossRef]
93. Sheffel, J.A.; Tsapatsis, M. A Semi-Empirical Approach for Predicting the Performance of Mixed Matrix Membranes Containing Selective Flakes. *J. Membr. Sci.* **2009**, *326*, 595–607. [CrossRef]
94. Mahajan, R.; Koros, W.J. Mixed Matrix Membrane Materials with Glassy Polymers. Part 1. *Polym. Eng. Sci.* **2002**, *42*, 1420–1431. [CrossRef]
95. Ward, J.K.; Koros, W.J. Crosslinkable Mixed Matrix Membranes with Surface Modified Molecular Sieves for Natural Gas Purification: I. Preparation and Experimental Results. *J. Membr. Sci.* **2011**, *377*, 75–81. [CrossRef]
96. Dutta, R.C.; Bhatia, S.K. Interfacial Barriers to Gas Transport in Zeolites: Distinguishing Internal and External Resistances. *Phys. Chem. Chem. Phys.* **2018**, *20*, 26386–26395. [CrossRef] [PubMed]
97. COMSOL, A.B. *COMSOL Multiphysics. User's Guide*; COMSOL A.B.: Stockholm, Sweden, 2011.
98. Jansen, J.; Macchione, M.; Drioli, E. On the Unusual Solvent Retention and the Effect on the Gas Transport in Perfluorinated Hyflon AD® Membranes. *J. Membr. Sci.* **2007**, *287*, 132–137. [CrossRef]
99. Yampolskii, Y.; Shantarovich, V. Positron Annihilation Lifetime Spectroscopy and Other Methods for Free Volume Evaluation in Polymers. In *Materials Science of Membranes for Gas and Vapor Separation*; Yampolskii, Y.P., Pinnau, I., Freeman, B.D., Eds.; John Wiley & Sons: Chichester, UK, 2006; Chapter 6; pp. 191–210.

Article

Modelling Mixed-Gas Sorption in Glassy Polymers for CO_2 Removal: A Sensitivity Analysis of the Dual Mode Sorption Model

Eleonora Ricci * and Maria Grazia De Angelis

Department of Civil, Chemical, Environmental and Materials Engineering, University of Bologna, 40131, Bologna, Italy; grazia.deangelis@unibo.it

* Correspondence: eleonora.ricci12@unibo.it; Tel.: +39-051-2090428

Received: 31 October 2018; Accepted: 20 December 2018; Published: 4 January 2019

Abstract: In an effort to reduce the experimental tests required to characterize the mixed-gas solubility and solubility-selectivity of materials for membrane separation processes, there is a need for reliable models which involve a minimum number of adjustable parameters. In this work, the ability of the Dual Mode Sorption (DMS) model to represent the sorption of CO_2/CH_4 mixtures in three high free volume glassy polymers, poly(trimethylsilyl propyne) (PTMSP), the first reported polymer of intrinsic microporosity (PIM-1) and tetrazole-modified PIM-1 (TZ-PIM), was tested. The sorption of gas mixtures in these materials suitable for CO_2 separation has been characterized experimentally in previous works, which showed that these systems exhibit rather marked deviations from the ideal pure-gas behavior, especially due to competitive effects. The accuracy of the DMS model in representing the non-idealities that arise during mixed-gas sorption was assessed in a wide range of temperatures, pressures and compositions, by comparing with the experimental results available. Using the parameters obtained from the best fit of pure-gas sorption isotherms, the agreement between the mixed-gas calculations and the experimental data varied greatly in the different cases inspected, especially in the case of CH_4 absorbed in mixed-gas conditions. A sensitivity analysis revealed that pure-gas data can be represented with the same accuracy by several different parameter sets, which, however, yield markedly different mixed-gas predictions, that, in some cases, agree with the experimental data only qualitatively. However, the multicomponent calculations with the DMS model yield more reliable results than the use of pure-gas data in the estimation of the solubility-selectivity of the material.

Keywords: dual mode sorption model; mixed-gas sorption; PIMs; glassy polymers

---

1. Introduction

In recent years, the use of polymers as membrane materials has attracted increased interest for several industrial applications, including gas separation for hydrogen recovery, nitrogen production, air dehydration, natural gas sweetening and biogas upgrading [1]. CO_2 is a typical contaminant to be removed from both natural gas and biomethane, in order to meet distribution pipelines specifications [2]. Despite the fact that CO_2 removal from natural gas with membranes has found industrial application since the 1980s [3], nowadays this technology has only about 10% of the market, which is dominated by solvent absorption using amines [1]. In membrane materials design research, countless structural and molecular modifications have been investigated in order to achieve a better separation performance, that would make membranes more competitive, in addition to being more energy-efficient and environmentally friendly [4–10]. However, one of the greatest challenges faced in membrane materials design is the existence of a trade-off between permeability and selectivity: for every gas pair the logarithm of the selectivity versus the logarithm of the permeability of the most permeable gas has been shown to lie below a limiting line, customarily referred to as the Robeson upper bound [11,12]. This is due to the fact that ultra-permeable materials usually display very poor selectivity, whereas highly-selective materials exhibit lower permeabilities [12]. This sets an upper limit to the efficiency that can be achieved by the operation, in case it is governed by diffusivity-selectivity [13,14].

Materials with improved performance, capable of surpassing the upper bound, have nonetheless been developed, among those are the family of Polymers of Intrinsic Microporosity (PIMs) [15–18] and Thermally Rearranged (TR) polybenzoxazoles [19–23]. Owing to a rigid backbone structure, consisting of a series of fused aromatic rings and to the presence of a shape-persistent site of contortion, the hindered chain packing of PIMs results in exceptionally high free volume, organized in a network of interconnected cavities. These materials have shown very high gas permeation rates, while maintaining acceptable selectivity values, and moreover they demonstrated great thermal and chemical stability [18].

Most experimental studies on prospective membrane materials are performed only with pure gases. While those data constitute a valuable benchmark of the materials properties, pure-gas tests are often insufficient to infer how the materials will behave in mixed-gas conditions. Mixed-gas permeation and sorption experiments have shown significant deviations from pure-gas ("ideal") behavior, both positive and negative [18,21,24–26]. In order to properly design a separation operation, it is necessary to characterize the relevant materials properties, namely its permeability and selectivity, as close as possible to the actual operating conditions, which can vary depending on the origin of the gaseous stream to be treated. Consequently, to uncover all the relevant phenomena, a broad experimental campaign, encompassing a wide range of temperatures, pressures and compositions, would be needed.

The transport of small molecules in dense polymeric membranes is described by the solution-diffusion model [27], according to which permeability (P) is the product of the solubility (S) and diffusion coefficients (D):

$$P = S \cdot D \tag{1}$$

Whether ultra-high free volume polymers can still be regarded as dense materials is an open question, however, there have been reports of successful modelling studies relying on this hypothesis [28]. Following this description, the selectivity of the polymer (perm-selectivity) $\alpha_{i,j}$, which is equal to the ratio between the permeability of the two gases, contains a solubility-selectivity $(\alpha^S_{i,j})$ and a diffusivity-selectivity factor $(\alpha^D_{i,j})$:

$$\alpha_{i,j} = \frac{P_i}{P_j} = \frac{S_i}{S_j} \cdot \frac{D_i}{D_j} = \alpha^S_{i,j} \cdot \alpha^D_{i,j} \tag{2}$$

Solubility-selectivity is expected to provide an important contribution to the overall permselectivity in high free volume glassy polymers, whereas for low and medium free volume polymers, where sieving effects are more important, the diffusivity-selectivity is expected to have a higher weight in the overall permselectivity. It would be interesting to be able to predict the mixed-gas behavior, using at most only pure-gas experimental measurement as input, in order to avoid or reduce the need for the more delicate and time-consuming mixed-gas tests.

The calculation of gas solubility in glassy polymers is customarily performed in the literature using the Dual Mode Sorption (DMS) model [29–39]. This model divides the total sorbed gas into two contributions: the molecules dissolving into the dense portion of the polymer (following Henry's law), and those saturating the microvoids of the excess free volume that characterizes the glassy state (described by a Langmuir curve). Its simplicity of use and its capability to correlate well the experimental sorption behavior in glassy polymers in most cases are the main reasons behind its success. However, this model does not allow to represent all types of sorption isotherms encountered, such as the sigmoidal shape of the sorption isotherms of alcohols in glassy polymers. There have been studies aimed at overcoming this limitation: for example, by incorporating multilayer sorption theory, a DMS based model capable of representing all the different shapes of sorption isotherms encountered was developed [40].

Another known issue with the use of this model is that the adjustable polymer-penetrant parameters of the DMS model depend on polymer history and operating conditions, thus lacking predictive ability outside their range of derivation, as discussed, for example, by Bondar at al. concerning the pressure range [41].

Alternatively, one can use an Equation of State (EoS) based approach to evaluate gas sorption equilibria. Some models successfully applied to the study of polymeric materials are those based on a Lattice Fluid (LF) representation of substances [42], or on hard sphere chain schemes, like the Statistical Associating Fluid Theory (SAFT) [43]. In the case of glassy polymers, due to their nonequilibrium condition, equilibrium models, such as an EoS, are not applicable. In these cases the Non-Equilibrium Thermodynamics for Glassy Polymers (NET-GP) approach [44] can be used instead. This approach extends equations of state to the nonequilibrium case, providing nonequilibrium expressions for the free energy of the system, by introducing an internal state variable, the polymer density, to describe the out-of-equilibrium degree of the systems. This approach has been successfully applied to the prediction of gas and vapor sorption in a variety of polymeric systems [28,45–47] and its capability to represent mixed-gas sorption equilibria in high free volume glassy polymer has been addressed in another work under preparation [48].

Alternatively, atomistic simulations can be employed for the prediction of sorption isotherms. Monte Carlo simulations in the Grand Canonical ensemble [49] can be performed to this aim, thanks to insertion moves that allow the polymeric system to exchange gas particles with an infinite bath until it reaches the equilibrium concentration corresponding to a given value of the chemical potential. Prediction of sorption isotherms can also be performed by post-processing of Molecular Dynamics (MD) or Monte Carlo (MC) trajectories, using the Widom test particle insertion method [50]: the intermolecular interaction energy felt by a molecule inserted in a random position in the polymer phase is related to the excess chemical potential of the penetrant inside the polymer and, in turn, to its solubility coefficient. In dense systems or in presence of large penetrant molecules, the probability of successful insertion moves decreases significantly and therefore the estimate of solubility through Widom insertions becomes less reliable. Other strategies include the use of gradual insertion of the penetrant molecules [51], or the use of particle deletion moves instead of particle insertions (Staged Particle Deletion [52], Direct Particle Deletion [53]). Atomistic simulation techniques have been increasingly applied in recent years to the study of gas transport properties of microporous polymers [54], and of PIMs in particular [55–58]. Sorption of CO_2 in PIM-1 was first simulated by Heuchel et al. [59] employing the Gusev-Suter Transition State Theory [60,61]. Even though the simulated solubility coefficients were significantly higher than the experimental ones, their work paved road to the application of molecular modelling techniques to this class of materials. Fang et al. [62,63] applied the Widom test particle insertion method [50] to predict CO_2 solubility in PIM-1 and their results were in close agreement with the experimental ones. Recently, Kupgan et al. [64] predicted CO_2 sorption in PIM-1 up to 50 bar, employing a scheme combining Grand Canonical Monte Carlo (GCMC) and Molecular Dynamics simulations devised by Hölck et al. [65], while Frentrup et al. [66] performed Nonequilibium Molecular Dynamics simulations for the direct simulation of He and CO_2 permeability through a thin membrane of PIM-1, which was in good qualitative agreement with experimental data.

Fewer modelling studies deal with the analysis of mixed gas sorption effects. Recently Rizzuto et al. [67] have coupled GCMC atomistic simulations and Ideal Adsorbed Solution Theory (IAST) [68] to investigate the mixed-gas permeation properties of CO_2/N_2 mixtures in Thermally Rearranged polymers. The simulations underestimated pure-gas sorption of both gases, however their results displayed the competitive effects between the gases expected in the case of glassy polymers, which affect greatly the solubility of the less condensable gas in the mixture. Neyertz and Brown [69] performed large-scale MD simulations of air separation with an ultra-thin polyimide membrane surrounded by an explicit gas reservoir, which allowed them to determine gas solubility, diffusivity and O_2/N_2 selectivity in multicomponent conditions, comparing favorably with experimental results. For this gas couple, the modelling study predicted a multicomponent solubility-selectivity comparable to the ideal one, that is calculated as the ratio of pure-gas solubility coefficients.

With the development of more accurate potentials, new algorithms for the generation of amorphous polymeric structures and efficient equilibration protocols, the reliability of the predictions yielded by atomistic techniques has drastically improved over the years [54,70]. Moreover, these techniques have the potential to be employed for screening purposes on existing materials, as well as on yet to be synthesized ones, as demonstrated, for instance, by Hart et al. [57] and Larsen et al. [71] for the case of CO_2/CH_4 separation with PIMs. However the extremely high computational effort required by atomistic approaches, combined to the system-specificity of several methods, remains a drawback to their application to the study of the separation properties of polymers, even though multiscale strategies, involving systematic coarse-graining and equilibration of high molecular weight models at the coarse-grained level and subsequent back-mapping to the atomistic detail have been implemented successfully to study a variety of properties of polymeric systems with reduced machine-time [72–75].

The present work is aimed at modelling gas solubility in high free volume glassy polymers both in pure- and mixed-gas conditions using the multicomponent version of the DMS model [76]. In particular, the sorption of CO_2/CH_4 mixtures in PTMSP, PIM-1 and TZ-PIM at several compositions and temperatures was studied, using experimental data presented in previous works to validate the results of the calculations [24–26,48]. The characterization of mixed-gas sorption is still quite limited and these materials are among the very few for which these experiments were performed. PTMSP, being the most permeable dense polymer [77], is a natural reference point to assess the separation performance of high free volume materials, as is PIM-1, which was the first material of the PIM class to be reported [15]. TZ-PIM constitutes an attempt at improving the selectivity of PIM-1 towards CO_2 by incorporating more CO_2-philic groups into its structure, demonstrating that post-polymerization modification techniques with controlled conversion rates represent a viable way of tuning the separation properties of these innovative materials. In this case the nitrile groups were substituted by tetrazole groups [78], but Satilmis et al. [79] showed that it is also possible to reduce them to primary amines, obtaining a material termed amine-PIM-1, with intermediate features between PIM-1 and TZ-PIM.

Lanč et al. [80] recently performed a critical analysis of the difference between gas solubility coefficients determined directly, with sorption experiments, or indirectly, from the time-lag of permeation. They investigated several high free volume glassy polymers, including PTMSP and PIM-1, concluding that the underlying approximation of a linear concentration profile across the membrane, assumed in the time-lag analysis, is a nonnegligible source of error in the indirect determination of S, but it can be mitigated by the calculation of concentration profiles using the thermodynamic Fick's law instead [81]. The authors also remarked the importance of sorption studies in uncovering fundamental aspect of gas transport in membrane materials.

Indeed, experimental measurements of mixed-gas sorption [24–26,48] allowed to understand that the competition between CO_2 and CH_4 plays a strong role in the multicomponent sorption behavior. Furthermore, the data indicate that the pure-gas solubility does not provide a good estimate of the real behavior of the mixture. In particular, pure-gas data would indicate that the main membrane parameters, like the solubility-selectivity, are a strong function of the gas mixture composition, while experimentally it is observed that the data depend very weakly on such variable. Additionally, for a set of glassy polymers comprising poly(2,6-dimethyl-1,4-phenylene oxide) (PPO), PTMSP, PIM-1 and Matrimid®, it was shown that departure functions, expressing the deviations between the multicomponent properties and the corresponding ideal values, estimated with pure component properties, obey generalized trends which resemble those observed in liquid solutions [82]. The ability of the Dual Mode Sorption model to represent these physical phenomena, as well as its quantitative accuracy in the prediction of solubility and solubility-selectivity were assessed, by comparing the results of the calculation with experimental data available for the materials considered [24–26,48], whose repeating units are shown in Figure 1. Moreover, a sensitivity analysis was carried out, in order to verify the robustness of the calculation and the reliability of the prediction in absence of experimental data for validation.

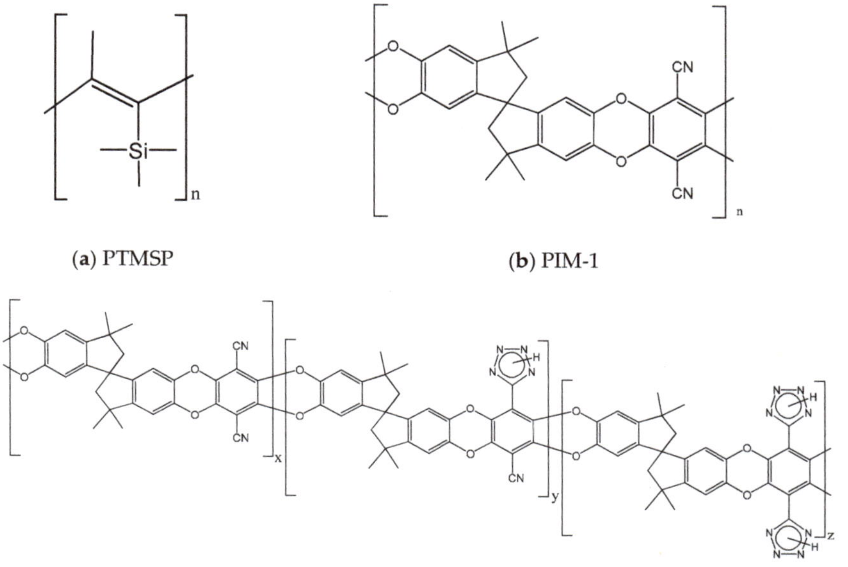

(a) PTMSP

(b) PIM-1

(c) TZ-PIM

Figure 1. Repeating units of the polymers considered in this study: (a) poly(trimethylsilyl propyne) (PTMSP); (b) the first reported polymer of intrinsic microporosity (PIM-1); (c) tetrazole-modified PIM-1 (TZ-PIM).

2. Dual Mode Sorption Model

The existence of a sorption mechanism particular to polymers in the glassy state was first postulated by Meares [35,36]. The indication that polymers below the glass transition temperature contain a distribution of microvoids frozen into their structure [35] suggested that those region of reduced density could act as preferential sorption sites. Using this concept and observing that the sorption isotherms of organic vapors in ethyl cellulose showed a curvature concave to the pressure axis, that was not witnessed in the case of rubbery materials, and, furthermore, that for these systems rather high negative values of the heat of solution were measured, Barrer et al. [37] proposed the existence of two concurrent mechanisms of sorption: dissolution and "hole-filling".

Therefore, the Dual Mode Sorption (DMS) model [29–39] postulates the existence of two different gas populations inside glassy polymers, at equilibrium with one another. One is dissolved in the dense portion of the material and can be described by Henry's law, while the other saturates the nonequilibrium excess free volume of the polymer and is described by a Langmuir curve. In this schematization, the total sorbed gas as a function of gas fugacity can be expressed as a sum of these two contributions [31,37]:

$$c_i = k_{D,i} f_i + \frac{C'_{H,i} b_i f_i}{1 + b_i f_i} \quad (3)$$

The parameter $k_{D,i}$ is Henry's law constant, while b_i is Langmuir affinity constant, which represents the ratio of the rate constants of sorption and desorption of penetrants in the microvoids and, therefore, it quantifies the tendency of a given penetrant to sorb according to the Langmuir mode.$C'_{H,i}$ is the Langmuir capacity constant, which characterizes the sorption capacity of a glassy polymer for a given penetrant in the low-pressure region. This latter parameter correlates with changes in polymer density associated with formation history or annealing treatments [83,84] and has been shown to disappear at the glass transition temperature (T_g) of the polymer [85]. For every gas-polymer pair and temperature analyzed, the three parameters are retrieved through a nonlinear least-square best fit of pure-gas sorption data.

The extension to multicomponent sorption of this model [76] is based on phenomenological arguments, suggested by the theory of competitive sorption of gases on catalysts, which exhibit a Langmuir behavior: since the amount of unrelaxed free volume in a polymer is fixed and limited (swelling is not taken into account by the model), the various penetrants will compete to occupy it and, as a consequence, the sorbed concentration is expected to decrease with respect to the pure-gas case.

It is assumed that the extent of the competitive effect is controlled by the relative values of the product of the affinity constant and partial pressure (or fugacity) of each penetrant. Under the hypothesis that the affinity parameter b, Henry's constant k_D and the molar density of a component sorbed inside the Langmuir sites are independent of the presence of other penetrants, the expression for the concentration (c) of component i in presence of a second component j is given by Equation (4):

$$c_i = k_{D,i} f_i + \frac{C'_{H,i} b_i f_i}{1 + b_i f_i + b_j f_j} \tag{4}$$

The characteristic gas-polymer parameters of the model are retrieved at each temperature from a least-square fit of a pure-gas isotherm using Equation (3). These parameters are subsequently used to predict the concentration of each gas in mixed-gas conditions at several compositions, making use of Equation (4) [86]. Therefore, in Equation (4), all parameters are the same as found in Equation (3).

It is also commonplace to write Equations (3) and (4) using the partial pressure of each gas instead of its fugacity. When high pressures are considered, such as in the present study, the approximation of ideal-gas behavior is not valid. Therefore, the fugacity is generally considered instead of partial pressure, since it constitutes a more appropriate measure of the gas chemical potential, which is the driving force for gas sorption in the polymer. Moreover, when mixtures are concerned, two gases like CH_4 and CO_2 show different departures from ideality, meaning that they can have the same partial pressure, but rather different fugacity. Even though the accuracy of the pure-gas data representation with the DMS model using either variable is the essentially the same, the values of the parameters obtained using pressure or fugacity are clearly different [87], therefore it should always be specified which variable was used in the regression, in order to enable a meaningful comparison between different parameter sets. In the context of mixed-gas sorption measurements, results are more often reported using gas fugacity, to account for the different degree of non-ideality of the components in the gas phase, therefore this was the natural choice of variable for this study as well. The accuracy of the results does not depend on this choice: it was verified that using pressure-based parameters or fugacity-based parameters yielded the same results in the mixed-gas sorption calculations. The same observation was reported also by Sanders et al. [87,88] in their studies on mixed-gas sorption of binary mixtures in poly(methyl methacrylate) (PMMA), and by Story et al. [89], in their work on mixed-gas sorption in PPO. They found that, the use of pressure-based or fugacity-based DMS parameters in the calculation of mixed-gas sorption yielded very similar results, only slightly more accurate in some of the cases when fugacity was used instead of partial pressure.

The evaluation of the solubility-selectivity is also of interest (Equation (5)). This performance indicator can be calculated using the solubility coefficients of the pure gases, as it is often done when mixture data are not available (ideal case), or, more accurately, using the solubility coefficients obtained in mixed-gas sorption tests/calculations (multicomponent case).

$$\alpha^S_{CO_2/CH_4} = \frac{S_{CO_2}}{S_{CH_4}} = \frac{c_{CO_2}/f_{CO_2}}{c_{CH_4}/f_{CH_4}} \tag{5}$$

Since a fugacity-based representation was adopted, the solubility coefficients (S) are defined as the ratio of gas concentration (c) and gas fugacity (f). The fugacity of the gases at various pressures was calculated with the Peng-Robinson equation of state (EoS) [90], both in pure- and mixed-gas condition evaluations. In the mixed-gas case, the binary parameter $k_{CO_2/CH_4} = 0.09$ [91] has been used in the Peng-Robinson EoS.

The DMS model does not account for the fact that the polymer matrix, unlike rigid porous materials, can swell when sorbing penetrants. Therefore, possible synergistic effects due to second-component induced swelling are not accounted for by such approach. However, ultra-high free volume glassy polymers have a limited tendency to swell, and the experimental data collected so far on mixed-gas sorption of CO_2/CH_4 mixtures indicate that such effects are not predominant in these materials, at least for pressures of CO_2 below 30 bar. On the other hand, in these conditions, it was observed that the prevailing multicomponent effect is the one associated with competition during sorption. Therefore, in the cases examined here, the DMS model is expected to provide a reliable estimation of the data [87].

3. Results

3.1. Pure-Gas Sorption Analysis

DMS parameters for the materials analyzed, obtained from a least-square fitting procedure with the Generalized Reduced Gradient (GRG) method [92], using concentration vs. fugacity data, are reported in Table 1. At each temperature, a different parameter set was obtained, and no constraints were applied to enforce a temperature dependence. In the last columns of the table, the Standard Error of the Estimate (*SEE*) is reported, as goodness-of-fit indicator. *SEE* was used instead of the correlation coefficient (R^2), because the underlying assumptions in the definition of R^2 are not valid in the case of a nonlinear regression model, such as the DMS model [93–95]. The following definition was used in its calculation:

$$SEE \equiv \sqrt{\frac{\sum_i \left(y_{i,exp} - y_{i,calc}\right)^2}{n-p}} \tag{6}$$

Table 1. Dual Mode Sorption model parameters (fugacity-based) for CO_2 and CH_4 sorption in PTMSP, PIM-1 and TZ-PIM, obtained by a least-square fit on data from Refs. [24–26,48] to Equation (3).

	CO_2					
	T (°C)	k_D $\left(\frac{cm^3_{STP}}{cm^3_{pol}bar}\right)$	C'_H $\left(\frac{cm^3_{STP}}{cm^3_{pol}}\right)$	b (bar^{-1})	SEE $\left(\frac{cm^3_{STP}}{cm^3_{pol}}\right)$	$\overline{SEE}_{mix}$ $\left(\frac{cm^3_{STP}}{cm^3_{pol}bar}\right)$
PTMSP	35	1.973	95.06	0.051	0.67	0.94
PIM-1	25	4.046	90.04	0.710	1.38	2.06
	35	2.890	94.83	0.388	0.92	3.90
	50	1.596	89.30	0.290	0.80	2.08
TZ-PIM	25	4.127	70.58	1.127	1.28	9.77
	35	1.982	89.53	0.378	1.57	6.66
	50	0.903	92.42	0.263	1.02	11.00
	CH_4					
	T (°C)	k_D $\left(\frac{cm^3_{STP}}{cm^3_{pol}bar}\right)$	C'_H $\left(\frac{cm^3_{STP}}{cm^3_{pol}}\right)$	b (bar^{-1})	SEE $\left(\frac{cm^3_{STP}}{cm^3_{pol}}\right)$	$\overline{SEE}_{mix}$ $\left(\frac{cm^3_{STP}}{cm^3_{pol}bar}\right)$
PTMSP	35	0.616	57.77	0.049	0.50	1.78
PIM-1	25	0.651	78.83	0.136	0.70	4.89
	35	0.541	75.87	0.106	0.51	2.69
	50	0.543	57.90	0.105	0.89	1.70
TZ-PIM	25	1.400	48.09	0.214	0.55	2.54
	35	0.378	67.12	0.087	0.97	1.40
	50	0.350	51.41	0.101	0.20	4.73

In the definition of *SEE* (Equation (6)), $y_{i,exp}$ are the experimental points, $y_{i,calc}$ are the corresponding values calculated with the model, n is the number of experimental points used in the regression and p is the number of parameters employed by the model. SEE is expressed in concentration units (as y) and lower values indicate a better agreement between experimental and calculated values. For the mixed-gas prediction, the reported value $\overline{SEE}_{mix}$ is the average deviation from three sorption isotherms at different composition calculated with the same pure-gas parameter sets.

The results are shown in Figures 2–4. The model provides an excellent fit to all the pure-gas experimental data sets. Typically, more condensable penetrants, like CO_2 in the present case, exhibit larger affinity constants, and this was indeed observed in the parameters retrieved. In addition, it would be expected that the presence of the tetrazole CO_2-philic groups in TZ-PIM would translate into higher affinity constants for CO_2 sorption, compared to PIM-1. However, this correlation of the parameter with the chemistry of the materials was not observed at all three temperatures, but only in the parameter set for the 25 °C case. This issue might relate to the parametrization route adopted, and it will be further discussed in Section 4.1.

Generally, k_D, C'_H and b are expected to decrease as temperature increases [76,96,97], consistently with their physical meaning. In the case of k_D and b, this trend was verified in all the cases inspected here, while for C'_H the expected trend was observed only in one case (CH_4 in PIM-1), while in the other cases the values fluctuated more. If the regression at each temperature is performed independently, fluctuations of the parameters have to be expected. This was noted also by Stevens et al. [98] in their analysis of Dual Mode Sorption model parameters for CO_2, CH_4 and N_2 in HAB-6FDA polyimide and its Thermally Rearranged analogues: when an unconstrained regression was performed independently at each temperature, the expected trends were followed only in some of the cases considered. In order to obtain a consistent parameter set, they imposed temperature dependence during the regression. The effect of these constraints on the mixed-gas sorption prediction will be examined in Section 4.1.

It has been reported that the DMS parameters are sensitive to the pressure range over which they are regressed [41], in particular b tends to decrease and C'_H to increase, if a broader regression range is considered, and, therefore, extrapolation outside the derivation range should be avoided. In this study, the whole isotherms were used in the regression and the pressure range was the same (0–35 bar) in all cases considered.

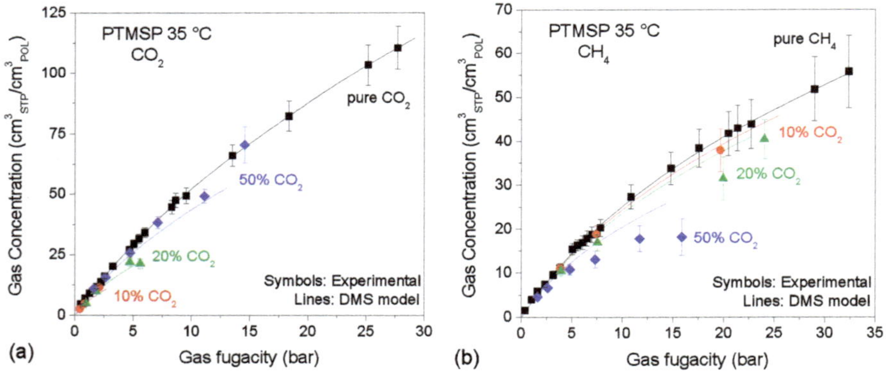

Figure 2. Sorption isotherms of CO_2 (**a**) and CH_4 (**b**) at 35 °C in PTMSP, in pure- and mixed-gas conditions (Black squares: pure-gas; Red circles: 10% CO_2 mixture; Green triangles: 20% CO_2 mixture; Blue diamonds: 50% CO_2 mixture). Experimental data from [24]. Solid lines represent Dual Mode Sorption (DMS) model predictions.

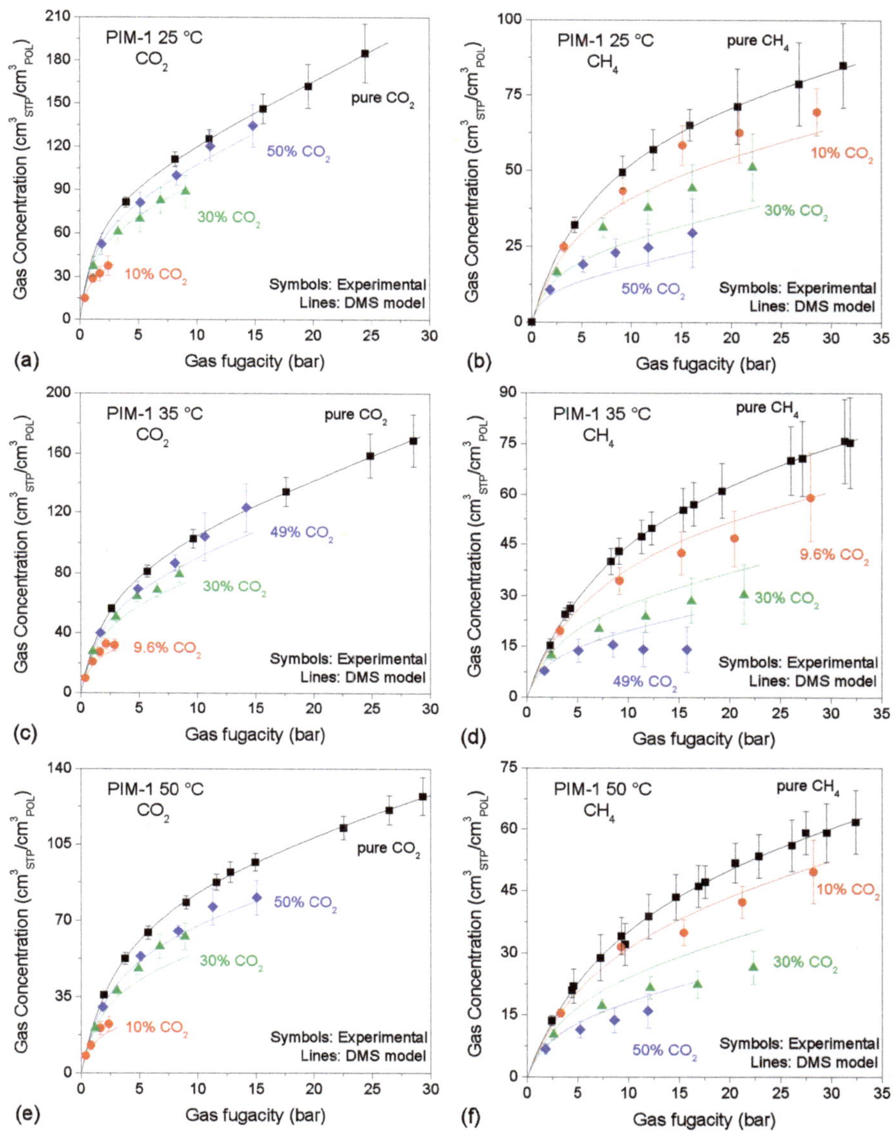

Figure 3. Sorption isotherms of CO_2 and CH_4 at 25 °C (**a**,**b**); 35 °C (**c**,**d**); 50 °C (**e**,**f**) in PIM-1, in pure- and mixed-gas conditions (Black squares: pure-gas; Red circles: ~10% CO_2 mixture; Green triangles: ~30% CO_2 mixture; Blue diamonds: ~50% CO_2 mixture). Experimental data from [26]. Solid lines are DMS model predictions.

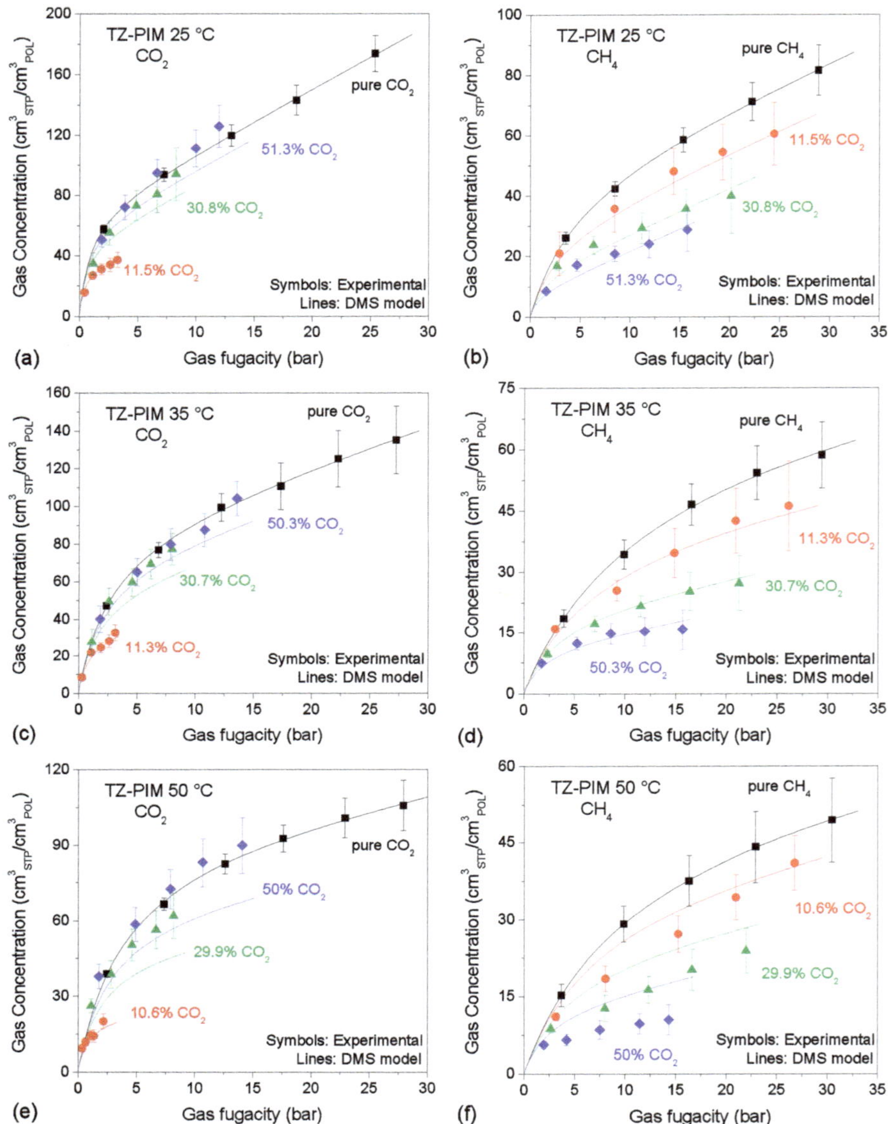

Figure 4. Sorption isotherms of CO_2 and CH_4 at 25 °C (**a**,**b**); 35 °C (**c**,**d**); 50 °C (**e**,**f**) in TZ-PIM, in pure and mixed-gas conditions (Black squares: pure gas; Red circles: ~10% CO_2 mixture; Green triangles: ~30% CO_2 mixture; Blue diamonds: ~50% CO_2 mixture). Experimental data from [48]. Solid lines are DMS model predictions.

3.2. Mixed-Gas Sorption: PTMSP

Figure 2 shows the experimental sorption data of CO_2/CH_4 mixtures (10/20/50 mol.% CO_2) in PTMSP at 35 °C [24] together with the results of mixed-gas sorption calculations with the DMS parameters reported in Table 1.

The predictions are in very good agreement with the experimental data in the case of CO_2, while in the case of CH_4 at high pressure the model overestimates the concentration for the 30:70 and 50:50 mixtures, with a maximum relative deviation of 20% and 35% respectively. Nonetheless, the model captures the fact that there is competition between the gases during sorption, but also that it is less pronounced in this polymer than in the other materials analyzed here, even at high values of the fugacity of the second component.

3.3. Mixed-Gas Sorption: PIM-1

Figure 3 shows the experimental sorption data of CO_2/CH_4 mixtures (~10/30/50 mol.% CO_2) in PIM-1 at 25, 35, 50 °C [25,26], together with the results of mixed-gas sorption calculations with the DMS model. It can be seen that, in the case of CO_2, the prediction is accurate at the lowest temperature, with average relative deviations below 5%. The average relative deviations, however, are increased to 10% at 35 °C and 50 °C.

On the other hand, in the case of CH_4, the accuracy is lower and its trend with temperature is opposite with respect to the case of CO_2. At 25 °C the concentration is significantly underestimated at all compositions (the average relative deviation is 19%), while at 35 °C it is overestimated by a similar extent (the average relative deviations is 18%). At 50 °C CH_4 sorption is still overestimated by the model, but the prediction is slightly more satisfactory, with average relative deviations of 15%. The deviation between the experimental data and the model predictions is greater than the experimental confidence intervals in several cases, therefore it does not seem to be explained fully by the uncertainty in the mixed-gas sorption measurements. Generally, for all temperatures analyzed, the lowest deviations are seen for both gases in the mixture case in which they are more abundant (50% CO_2 and 90% CH_4 respectively).

Not much can be done *a priori* to improve the quantitative accuracy of the mixed-gas prediction, because the parametrization at each temperature is independent and relies only on the accuracy of the pure-gas sorption measurements. However the effect of using a different parametrization route will be discussed in a later section.

3.4. Mixed-Gas Sorption: TZ-PIM

In Figure 4, the experimental sorption data of CO_2/CH_4 mixtures (~10/30/50 mol.% CO_2) in TZ-PIM at 25, 35, 50 °C [48], together with the results of mixed-gas sorption calculations with the DMS model are reported. In the case of TZ-PIM, the prediction of CO_2 sorption is more accurate at 25 °C and 35 °C (10% average relative deviations), while it worsens at 50 °C, where the model would seem to underestimate CO_2 concentration both in the 30% CO_2 and in the 50% CO_2 mixtures, with average relative deviations with respect to the experimental data of 26% and 23%, which are greater than the experimental confidence intervals.

In the case of CH_4, at 25 °C and 35 °C DMS predictions show very good agreement with the experimental data, with average relative deviations below 5% at all compositions. Conversely, at 50 °C the model significantly overestimates the CH_4 concentration, by as much as 32% on average.

3.5. Solubility-selectivity

Ideal and multicomponent solubility-selectivities were evaluated using Equation (5). At a fixed value of the total pressure of the mixed-gas feed, the corresponding fugacity of each gas in the mixture and multicomponent concentration values were used to obtain the multicomponent solubility-selectivity. The ideal solubility-selectivity was evaluated using the same fugacity values as in the multicomponent case, but with the corresponding concentration values taken from the pure-gas sorption isotherms.

The obtained trends are compared with the experimental data in Figure 5. For the sake of brevity, results of the comparison are shown only for the 35 °C case, but all the general observations that follow were true also for the results at 25 °C and 50 °C.

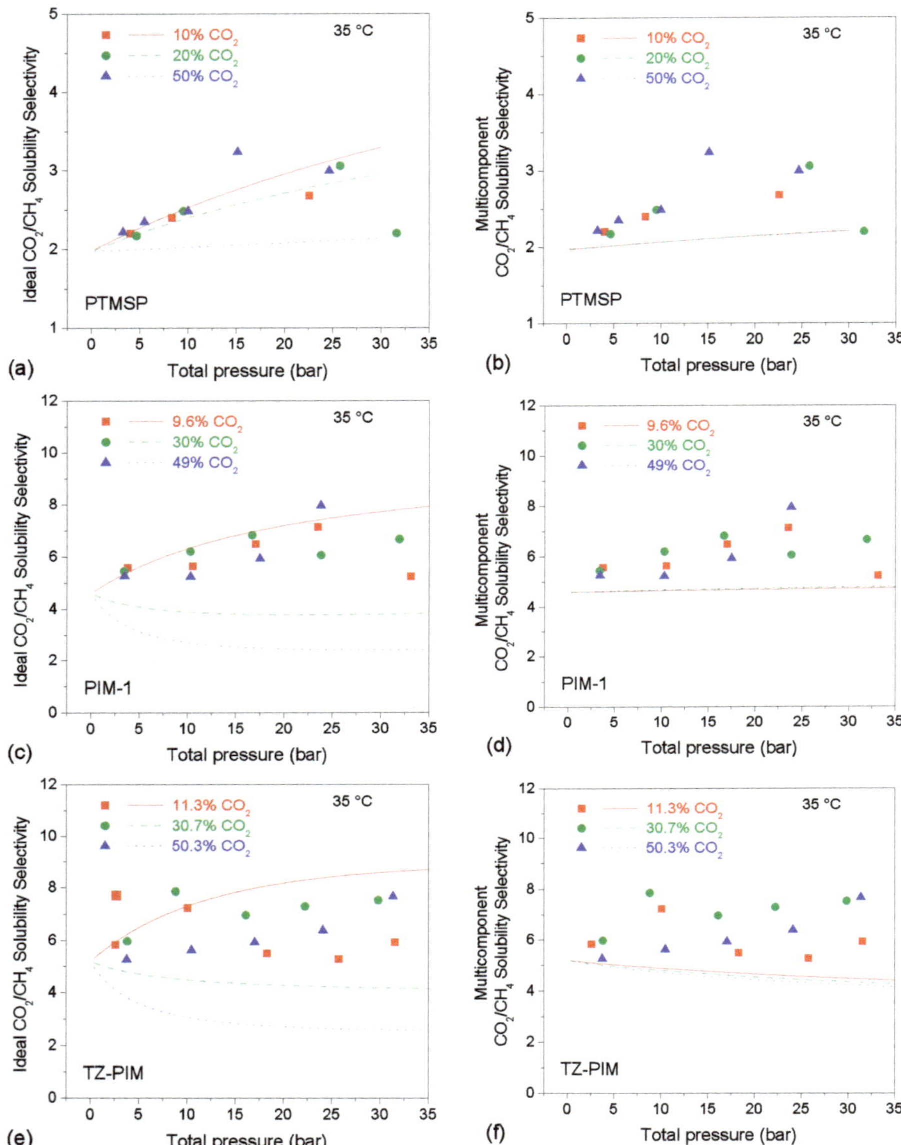

Figure 5. CO_2/CH_4 solubility-selectivity in PTMSP, PIM-1, and TZ-PIM at 35 °C at various mixture compositions. Points are experimental values [24,25,48], lines represent calculations with the DMS model. In the left column, i.e., (**a**,**c**,**e**), the lines represent ideal solubility-selectivity values, calculated with pure-gas concentrations. In the right column, i.e., (**b**,**d**,**f**), the lines represent the calculated multicomponent solubility-selectivity values.

A common remark to all cases inspected is that the mixed-gas calculations show a very different dependence on mixture composition and total pressure with respect to the ideal case calculation. In particular, mixed-gas calculations generally show a much weaker dependence than the ideal ones versus both pressure and composition. It seems, therefore, that the competitive effect, accounted for in the mixed-gas calculations, tends to stabilize the calculated solubility-selectivity with respect to fluctuations in the gas pressure and composition. The physical reason beyond this behavior, that is also confirmed by experiments, is not completely clear.

In particular, for PTMSP, the calculated values are close to the experimental ones and the trends predicted by the model exhibit almost no dependence on the gas mixture composition, with the three curves collapsing onto one another, whereas the experimental data are more scattered, and resemble more the results of the ideal-case calculation, in the lower CO_2 content cases (10–20% CO_2) and low pressure-range, where indeed the gas phase is closer to and ideal one.

Similarly, the gas composition dependence of solubility-selectivity is negligible in the mixed-gas calculation for PIM-1, and there is almost no dependence on pressure as well. In the case of TZ-PIM, the calculated values in the mixed-gas case show a very modest concentration and pressure dependence, although slightly more marked than in the other cases.

The calculated values for PIM-1 and TZ-PIM slightly underestimate the solubility-selectivity, but they would be preferable than simpler ideal-case estimates (left column of Figure 5), which, on average, could lead to larger errors. Indeed, in the evaluation of the selectivity, the experimental error of both gas concentrations is combined and, therefore, this parameter inevitably has a higher uncertainty. For this reason, it is not straightforward to infer pressure and gas mixture composition dependencies from the experimental data, due to large fluctuations and absence of monotonous trends. Nonetheless, it is clear that the calculations performed with mixed-gas concentrations yield significantly more accurate results than using the corresponding pure-gas values.

4. Sensitivity Analysis

4.1. Pure-Gas Sorption Isotherms Fitting Method

Reasons for the deviation of the DMS model predictions from the experimental data were identified originally by Koros [76] in the possible presence of non-negligible penetrant-penetrant specific interactions, or as consequence of swelling and plasticization effects, which are not accounted for in the model and that would make the parameters concentration-dependent, or require the introduction of additional terms and adjustable parameters. There have been extensions and modifications to the DMS model to include this aspects with the introduction of additional parameters [40,99,100] but the original version is still the most used one.

Since the DMS model parametrization is carried out independently at each temperature, it is striking that the accuracy of the mixed-gas prediction reported in Figures 3 and 4 varies so much between different temperatures. In order to address this issue, a different parametrization route was tried and, subsequently, a sensitivity analysis of the mixed-gas calculation to the parameter set was performed.

New parameter sets were obtained (Table 2) by taking into account during the nonlinear least-square optimization the experimental error associated with each experimental point, by minimizing χ^2 defined as follows [94]:

$$\chi^2 = \sum_{i=1}^{N} \frac{1}{\sigma_i^2} \left[c_i - \left(k_{D,i} f_i + \frac{C'_{H,i} b_i f_i}{1 + b_i f_i} \right) \right]^2 \tag{7}$$

here σ_i represents the confidence interval associated with the experimental value of the concentration c_i, N is the total number of experimental points, f_i is the gas fugacity and $k_{D,i}$, $C'_{H,i}$, b_i are the DMS parameters for the polymer-i penetrant couple. Since very often only data at one reference temperature are available, each data set was treated independently, without additional constraints to impose a temperature dependence to the parameters.

As expected, slightly different parameter sets from the ones reported in Table 1 were obtained. It was observed that an increase in the value of $C'_{H,i}$ was always accompanied by a decrease in the values of $k_{D,i}$ and b_i, and vice versa. In the electronic supplementary information (ESI) file, the comparison of the mixed-gas predictions obtained with the parameter sets from Tables 1 and 2 is reported in Figures S1–S3, where also the experimental confidence intervals are included for reference. It is remarkable that pure-gas representations are almost indistinguishable (as it can be noted also by the very similar values of Standard Error associated to the two parameter sets), even though some of the values of the parameters used differ by as much as 30%.

The mixed-gas calculations performed with the parameters reported in Table 2 provided a modest improvement in the accuracy of the prediction in some of the cases analyzed (CO_2 in TZ-PIM at 35 and 50 °C, CH_4 in TZ-PIM at 50 °C, CO_2 in PIM-1 at 35 °C, CH_4 in PIM-1 at 25 and 35 °C, CH_4 in PTMSP at 35 °C), whereas in other cases they produced slightly less accurate results (CO_2 in TZ-PIM at 25 °C, CH_4 in TZ-PIM at 25 and 35 °C, CO_2 in PIM-1 at 25 and 50 °C, CH_4 in PIM-1 at 50 °C, CO_2 in PTMSP at 35 °C). A systematic trend was not detected, at times the average accuracy was increased for both gases at the same temperature (TZ-PIM at 50 °C case and PIM-1 at 35 °C case), other times only for one of the two gases at the same temperature (TZ-PIM at 35 °C case, PIM-1 25 °C case and PTMSP 35 °C case) and in other cases for none (TZ-PIM 25 °C case and PIM-1 50 °C case). Moreover, in some instances, the results were more accurate at certain compositions but worse at others. On the whole, the discrepancies between the accuracy of the multicomponent calculations in different conditions were not eliminated by taking into account the experimental error during the parametrization.

Table 2. Dual Mode Sorption model parameters (fugacity-based) for CO_2 and CH_4 sorption in PTMSP, PIM-1, TZ-PIM, obtained by a least-square fit on data from Refs. [24–26,48] to Equation (7).

	T (°C)	k_D $\left(\frac{cm^3_{STP}}{cm^3_{pol}bar}\right)$	C'_H $\left(\frac{cm^3_{STP}}{cm^3_{pol}}\right)$	b (bar^{-1})	SEE $\left(\frac{cm^3_{STP}}{cm^3_{pol}}\right)$	$\overline{SEE}_{mix}$ $\left(\frac{cm^3_{STP}}{cm^3_{pol}bar}\right)$
CO_2						
PTMSP	35	2.373	69.26	0.067	0.73	0.97
PIM-1	25	3.664	100.25	0.506	1.60	3.72
	35	3.039	90.42	0.428	1.17	3.43
	50	1.666	86.84	0.306	0.84	2.57
TZ-PIM	25	4.023	72.88	1.019	1.44	11.21
	35	2.168	84.61	0.420	1.82	6.16
	50	1.150	84.84	0.303	1.46	10.11
CH_4						
	T (°C)	k_D $\left(\frac{cm^3_{STP}}{cm^3_{pol}bar}\right)$	C'_H $\left(\frac{cm^3_{STP}}{cm^3_{pol}}\right)$	b (bar^{-1})	SEE $\left(\frac{cm^3_{STP}}{cm^3_{pol}}\right)$	$\overline{SEE}_{mix}$ $\left(\frac{cm^3_{STP}}{cm^3_{pol}bar}\right)$
PTMSP	35	0.611	58.75	0.048	0.52	0.92
PIM-1	25	0.672	78.56	0.137	0.72	3.00
	35	0.401	82.72	0.097	0.60	1.68
	50	0.684	50.99	0.124	0.94	2.08
TZ-PIM	25	1.526	43.65	0.250	0.76	2.97
	35	0.282	73.42	0.079	0.98	1.61
	50	0.364	50.63	0.103	0.20	4.18

To improve the internal consistency of the parameters, a multi-temperature parametrization scheme was tested. For each gas, new parameters were obtained by considering the data at all temperatures simultaneously and constraining the parameters to follow the expected temperature dependence. In particular, the temperature dependence of k_D and b is described by a van't Hoff relation [96]:

$$k_D = k_{D0}\, e^{-\frac{\Delta H_D}{RT}} \tag{8}$$

$$b = b_0 \, e^{-\frac{\Delta H_b}{RT}} \tag{9}$$

In Equations (8) and (9) ΔH_D and ΔH_b are the enthalpies of sorption for Henry and Langmuir modes, R is the gas constant and T is the temperature. The pre-exponential factors k_{D0} and b_0, together with ΔH_D and ΔH_b were treated as adjustable coefficients. For C'_H, no functional temperature dependence was imposed, but the values were constrained to diminish with increasing temperature. In this way, to obtain the parameters for each gas-polymer couple at three temperatures, only 7 adjustable coefficients were used, instead of 9.

The parameter sets obtained are reported in Table 3. It is noteworthy that, in this parameter set, the values of Langmuir affinity constant for the couple CO_2-PIM-1 are always lower than the corresponding ones for CO_2-TZ-PIM at each temperature, as it would be expected given the chemical difference between the two materials.

Table 3. Dual Mode Sorption model parameters (fugacity-based) for CO_2 and CH_4 sorption in PIM-1, TZ-PIM, obtained by a least-square fit on data from Refs. [25,26,48], imposing the temperature dependence constraints expressed by Equations (8) and (9).

			CO_2			
	T (°C)	k_D $\left(\frac{cm^3_{STP}}{cm^3_{pol} bar}\right)$	C'_H $\left(\frac{cm^3_{STP}}{cm^3_{pol}}\right)$	b (bar^{-1})	SEE $\left(\frac{cm^3_{STP}}{cm^3_{pol}}\right)$	$\overline{SEE}_{mix}$ $\left(\frac{cm^3_{STP}}{cm^3_{pol} bar}\right)$
PIM-1	25	4.150	90.56	0.638	2.43	1.86
	35	2.883	89.02	0.470	4.30	3.73
	50	1.742	85.63	0.308	1.11	1.44
TZ-PIM	25	3.625	82.25	0.676	3.54	13.43
	35	2.473	77.23	0.511	3.13	10.04
	50	1.457	76.94	0.347	2.92	11.31
			CH_4			
	T (°C)	k_D $\left(\frac{cm^3_{STP}}{cm^3_{pol} bar}\right)$	C'_H $\left(\frac{cm^3_{STP}}{cm^3_{pol}}\right)$	b (bar^{-1})	SEE $\left(\frac{cm^3_{STP}}{cm^3_{pol}}\right)$	$\overline{SEE}_{mix}$ $\left(\frac{cm^3_{STP}}{cm^3_{pol} bar}\right)$
PIM-1	25	0.759	78.51	0.130	1.80	3.74
	35	0.561	72.98	0.113	0.73	2.02
	50	0.369	66.17	0.093	1.00	1.12
TZ-PIM	25	1.275	49.29	0.206	2.73	2.41
	35	0.865	42.49	0.171	1.99	3.87
	50	0.506	41.58	0.133	1.23	4.46

The comparison of the mixed-gas predictions obtained with the parameter sets from Tables 1 and 3 is reported in Figures S4 and S5 in the ESI file. To satisfy the temperature dependence constraint, a slightly less accurate representation of pure-gas sorption is generally observed, especially at higher pressure. The largest deviations are observed in the cases of CO_2 and CH_4 sorption in TZ-PIM at 25 and 50 °C, for CO_2 sorption in PIM-1 at 35 °C and for CH_4 sorption in PIM-1 at 25 °C.

The mixed-gas sorption calculations performed with the parameters reported in Table 3 provided a slight improvement in the accuracy of the prediction of CO_2 and CH_4 sorption in PIM-1 at all temperatures. On the other hand, in the case of TZ-PIM, the mixed-gas calculation yielded comparable results for both gases at 50 °C, but less accurate predictions for both gases in the 35 °C case. At 25 °C, the results for CH_4 sorption in TZ-PIM were slightly better compared to those obtained with the parameter set without a consistent temperature dependence, while those for CO_2 were slightly worse. Therefore, on the whole, this parametrization route presented improvements, in terms of internal consistency of the parameter set and their physical interpretation, but it still didn't eliminate the variability in the accuracy of the mixed-gas sorption results.

In order to address the issue systematically, a sensitivity analysis was carried out: the effect of CO_2 and CH_4 parametrization on the multicomponent calculation was studied separately, analyzing how the mixed-gas sorption results are affected by variations of $C'_{H,i}$, $k_{D,i}$, b_i while b_j is kept fixed, and, subsequently, the effect of changes in the value of b_j were also taken into account.

4.2. Confidence Intervals of the DMS Model Parameters

A comprehensive search in the parameters space was conducted, using a grid method, in order to identify a range of DMS model parameter values that allow to obtain equally satisfactory representations of the pure-gas data. Once such a range was estimated, it was tested whether different parameter sets, within those confidence intervals, could lead to better mixed-gas predictions than the ones obtained with the best-fit sets.

The results are presented in the following for the case of CH_4 solubility in PIM-1. Results analogous to the ones presented in this section were obtained also for the other cases. They are not shown, for the sake of brevity, but they can be found in the ESI file (Figures S6 and S7 complete the analysis of CH_4 sorption in PIM-1 at 35 °C and 50 °C, Figures S8–S12 and Tables S1 and S2 show the results for the case of CO_2 sorption in PIM-1 and the outcome of the calculations for CO_2 and CH_4 sorption in TZ-PIM are reported in Tables S3 and S4 and in Figures S13–S20).

Figure 6 shows a contour plot of the Standard Error in the calculation of pure CH_4 sorption in PIM-1 at 25 °C, obtained varying b and C'_H while holding k_D constant at its best fit value: 0.651 $cm^3{}_{STP}/cm^3{}_{pol}$. Each line represents a locus of constant *SEE* in the Langmuir parameter space, for a fixed value of Henry's constant.

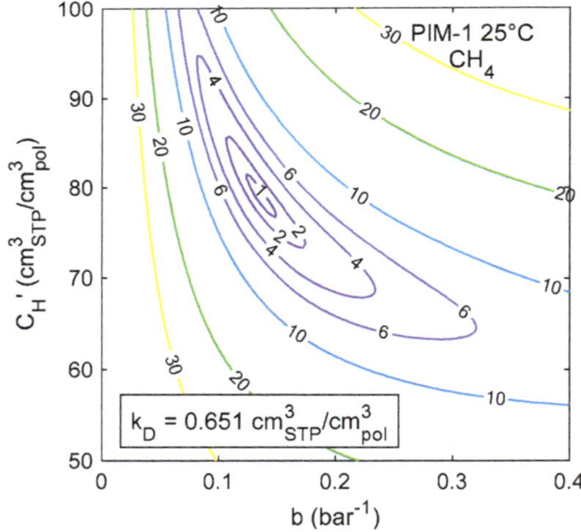

Figure 6. Contour plot of the Standard Error of the Estimate (SEE) of CH_4 sorption in PIM-1 at 25 °C, obtained varying the Langmuir sorption parameters, with a fixed value of the Henry's constant.

When k_D is allowed to vary as well, surfaces at constant error in the three-dimensional parameter space are obtained. As a criterion to delimit the confidence intervals, a maximum value for the standard error was selected. This value was chosen as SEE_{max} = 1 $cm^3{}_{STP}/cm^3{}_{pol}$ at 25 °C, which corresponds to an average relative deviation of 1.5%. The same relative deviation is attained with SEE_{max} = 0.9 $cm^3{}_{STP}/cm^3{}_{pol}$ at 35 °C and SEE_{max} = 0.8 $cm^3{}_{STP}/cm^3{}_{pol}$ at 50 °C. In Figure 7a a 3D plot is presented, in which the three colored regions correspond to domains in the parameter space where $SEE < SEE_{max}$ for CH_4 sorption in PIM-1 at the three different temperatures. Therefore, each point within the colored region corresponds to a parameter set that satisfies the accuracy criterion.

The sorption isotherms obtained with all the parameter sets included in the colored regions of part (a) of the figure are represented together in Figure 7b, and one can see that there is indeed a small, but detectable, variation in the representation of the experimental data using either of the parameter sets. This variability, however, is always lower than the experimental uncertainty of the data.

The upper and lower limits in each direction of the isosurfaces reported in Figure 7a can be used to attribute confidence intervals to the DMS parameters. Confidence intervals for nonlinear regression functions are often asymmetrical and this was observed also by other authors for the DMS model parameters [21,98].

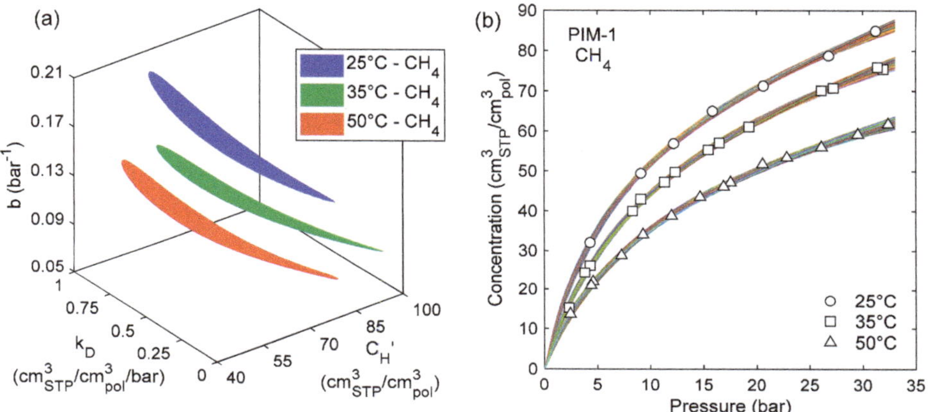

Figure 7. (**a**) Surfaces enclosing the range where DMS parameter sets yield $SEE < SEE_{max}$ in the prediction of CH_4 sorption in PIM-1 at three different temperatures; (**b**) CH_4 sorption isotherms in PIM-1 at 25, 35 and 50 °C, calculated with all the parameter sets enclosed by the corresponding colored regions in the plot on the left.

Confidence intervals of the DMS parameters for CH_4 sorption in PIM-1 are reported in Table 4. Clearly, not all the combinations of parameters within their respective confidence interval would give a valid set, otherwise the confidence region in the 3D parameter space would be represented by a parallelepiped. However, for all values included in the confidence interval of one parameter, it would be possible to find values of the other two parameters such that the accuracy criterion is satisfied. As a consequence, when using only one of the parameters, like in the case of b_j in Equation (4), all values belonging to its confidence interval should be considered acceptable.

Table 4. Confidence intervals of the fugacity-based DMS parameters yielding and average relative deviation < 1.5% in the calculation of CH_4 sorption in PIM-1 at three different temperatures.

T (°C)	k_{D,CH_4} $\left(\frac{cm^3_{STP}}{cm^3_{pol}bar}\right)$	C'_{H,CH_4} $\left(\frac{cm^3_{STP}}{cm^3_{pol}}\right)$	b_{CH_4} (bar^{-1})
25	$0.651\,^{+0.336}_{-0.290}$	$78.83\,^{+15.67}_{-15.28}$	$0.136\,^{+0.059}_{-0.031}$
35	$0.541\,^{+0.316}_{-0.408}$	$75.87\,^{+23.78}_{-15.93}$	$0.106\,^{+0.042}_{-0.029}$
50	$0.543\,^{+0.237}_{-0.443}$	$57.90\,^{+25.60}_{-12.74}$	$0.105\,^{+0.049}_{-0.035}$

4.3. Evaluation of Mixed-Gas Sorption

All the parameter sets that satisfied the condition $SEE < SEE_{max}$ in the pure-gas sorption representation, were used to calculate mixed-gas sorption isotherms, using the best-fit values reported in Table 1 for b_{CO_2}.

To quantify the accuracy of the mixed-gas prediction ($\overline{SEE}_{mix}$), the average SEE of isotherms at three concentrations (10/30/50 mol.% CO_2) for each temperature was used, and then the lowest and the highest results were selected, in order to identify the best and worst predictions, labelled respectively Set 1 and Set 2. The parameter sets that correspond to these two extreme cases and their SEE values are summarized in Table 5. The calculated sorption isotherms for each case are shown in Figure 8.

Table 5. DMS model fugacity-based parameter sets used in the calculation of mixed-gas sorption of CH_4 in PIM-1 reported in Figure 8.

	T (°C)	k_{D,CH_4} $\left(\frac{cm^3_{STP}}{cm^3_{pol}bar}\right)$	C'_{H,CH_4} $\left(\frac{cm^3_{STP}}{cm^3_{pol}}\right)$	b_{CH_4} (bar^{-1})	SEE_{pure} $\left(\frac{cm^3_{STP}}{cm^3_{pol}bar}\right)$	$\overline{SEE}_{mix}$ $\left(\frac{cm^3_{STP}}{cm^3_{pol}bar}\right)$
Set 1	25	0.945	65.08	0.186	0.999	3.98
	35	0.119	101.81	0.074	0.891	1.15
	50	0.125	82.00	0.070	0.797	1.35
Set 2	25	0.316	96.99	0.102	0.996	11.16
	35	0.869	59.26	0.152	0.899	4.95
	50	0.767	45.62	0.155	0.792	3.67

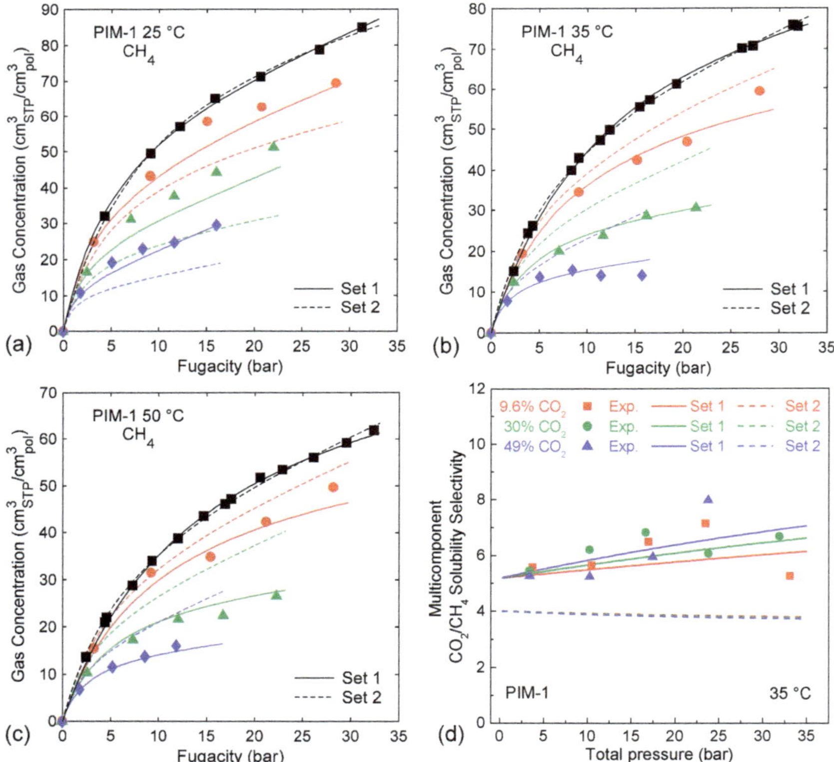

Figure 8. Dual Mode Sorption model mixed-gas predictions of CH_4 sorption in PIM-1 at 25 °C (**a**); 35 °C (**b**); 50 °C (**c**) obtained with the two parameter sets reported in Table 5; In (**d**) the solubility-selectivity calculated with the two sets at 35 °C is compared. Solid lines are obtained with Set 1, dashed ones with Set 2.

At each temperature, both parameter sets have a remarkably similar accuracy in the representation of pure-gas data, but they yield a significantly different prediction of the mixed-gas behavior. For instance, the solid lines in Figure 8 (obtained with Set 1) show a very good agreement with the experimental data, whereas the dashed ones (obtained with Set 2) are even less accurate than the initial result obtained with the best-fit parameter set.

Allowing also for experimental error, the two pure-gas representations at each temperature (black lines in Figure 8) are deemed to be equivalent and no reason for choosing one over the other could be suggested. Therefore, in absence of mixed-gas experimental data to validate against, the confidence in the accuracy of the calculation would be weakened.

The same variability was observed also in the case of CH_4 sorption in TZ-PIM, while for CO_2 sorption in both PIM-1 and TZ-PIM, the uncertainty in the mixed-gas predictions was generally lower (see Figures S9 and S14 in the ESI file).

This is reflected also in the evaluation of the solubility-selectivity: comparing the results obtained with Set 1 and Set 2 (Figure 8d) it is possible to see that not only the results obtained with Set 1 are in much better agreement with the experimental data, but also that the pressure and concentration dependence predicted by the two sets is significantly different, and in one case it is consistent with the general trend of the experimental points, while in the other case it is the opposite.

By looking at the relationship between the deviations in the pure-gas data fitting and the mixed-gas prediction, the level of uncertainty that is inherent in the calculation becomes more apparent. In Figure 9 the deviation in the pure-gas data representation (SEE_{pure}) is related to the error in the mixed-gas prediction ($\overline{SEE}_{mix}$). Each point in the plot represents a calculation performed with a different parameter set, among those meeting the accuracy criterion (Figure 7a).

It can be seen that, when moving to the right, i.e., further away from the best-fit parameter set and therefore towards slightly higher deviations in the pure-gas data correlation, a small variation of SEE_{pure} results in a much wider range of possible outcomes for $\overline{SEE}_{mix}$, thus the reliability of the calculation becomes less predictable. The interval of the values assumed by $\overline{SEE}_{mix}$ expands both towards higher and lower values, so that better mixed-gas predictions can indeed be found with slightly "less accurate" pure-gas data representations, and not with the best-fit parameter set.

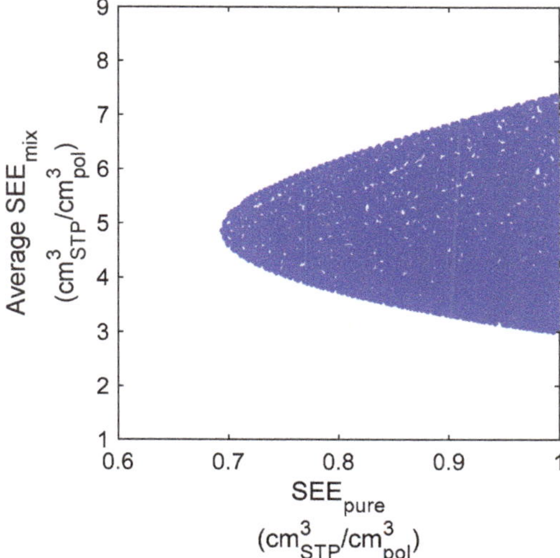

Figure 9. Accuracy range of the mixed-gas prediction (*y*-axis) for CH_4 in PIM-1 at 25 °C corresponding to a given accuracy in the pure-gas data representation (*x*-axis). Each point represents the result obtained with a different parameter set among those enclosed in the colored region of Figure 7a.

4.4. Effect of b_{CO_2}

The mixed-gas sorption results depend also strongly on the value of the Langmuir affinity constant of the second component (b_{CO_2}), which is found in the Dual Mode Sorption model expression for sorbed concentration in the multicomponent case. Following the same procedure adopted for CH_4, the region in the DMS parameter space of CO_2 sorption in PIM-1 at 25 °C satisfying the condition $SEE < SEE_{max}$ was identified. It is reported in Figure S6 of the ESI file, alongside confidence regions at 35 °C and 50 °C. From this analysis, it was possible to identify the confidence interval over which b_{CO_2} could vary $\left(b_{CO_2}^{25\,°C} = 0.710\ ^{+0.272}_{-0.239}\right)$ and then study the influence of its variations on the CH_4 mixed-gas sorption prediction.

In Figure 10, the same region of acceptable parameter sets for CH_4 sorption in PIM-1 at 25 °C found in Figure 7a is represented, but, in these plots, a color scale indicates the average accuracy in the mixed-gas prediction corresponding to each point in the plot. The calculation is repeated for different values of b_{CO_2}, chosen to span its entire $SEE < SEE_{max}$ region. It can be seen that a greater accuracy of the mixed-gas predictions is not attained with the best-fit value of b_{CO_2} reported in Table 1 (0.710 bar^{-1}), but with a lower one instead (top-left corner of Figure 10), whereas with higher values it becomes increasingly difficult to have good predictions at all (bottom-right corner of Figure 10). Once more, this was not a general trend. For example, in the cases of CH_4 sorption in PIM-1 at 35 and 50 °C, the more accurate mixed-gas predictions could be attained with values of b_{CO_2} higher than the best fit one, as it can be observed in Figures S6 and S7 of the ESI file.

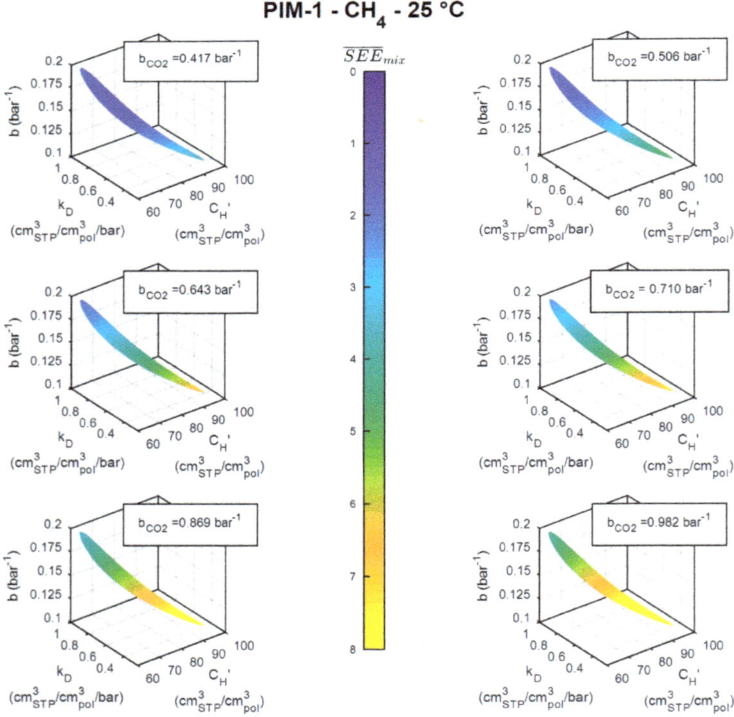

Figure 10. Isosurfaces in the DMS parameter space of CH_4 sorption in PIM-1 at 25 °C corresponding to $SEE_{pure} < SEE_{max}$, colored according to the average SEE_{mix} obtained with different values within the confidence interval of b_{CO_2}.

5. Discussion

Due to the form of Equations (3) and (4), the parameters C'_H and b are strongly coupled and, therefore, a deviation of either of them can be compensated by a corresponding deviation of the other, yielding a similar overall quality of the fit. The same remark was made also by Gleason et al. [21] in their analysis of Dual Mode parameters for mixed-gas permeation of CO_2/CH_4 in Thermally Rearranged HAB-6FDA. In order to improve the accuracy of the calculation, they chose to incorporate mixed-gas data into the fitting procedure used to retrieve the DMS parameters. Raharjo et al. [97] studied sorption of CH_4-nC_4H_{10} mixtures in PTMSP and they noticed a tendency of the DMS model to systematically overestimate CH_4 concentration in mixed-gas conditions. They subsequently re-parametrized the model, including the mixture data as well, obtaining different parameter sets from those retrieved considering only pure-gas data. In both cases [21,97] the representation of the mixture behavior was superior when the multicomponent data was included during the parametrization, but the procedure is clearly no longer predictive.

In order to reduce the uncertainty in the regression of the DMS parameters, Wang et al. [101] suggested to obtain Henry's constant independently, through the analysis of the temperature dependence of the solubility coefficient above T_g, and then retrieve only C'_H and b from the best-fit of the sorption data. This approach yielded different sets from those obtained in a simultaneous regression of all three parameters and, even though those sets had lower values of the goodness-of-fit indicator, they showed improved self-consistency and the expected temperature dependence. This method, however, was not applicable to the materials studied here, and in general for glassy polymers with very high T_g, for which gas solubility data above T_g are not available.

Comparing the results displayed here for mixed-gas CH_4 sorption in PIM-1 to those of CH_4 sorption in TZ-PIM and also to those of CO_2 sorption in PIM-1 and TZ-PIM, it is not straightforward to identify a general trend and therefore draw guidelines to mitigate the issue. The parameter set obtained by imposing a temperature dependence yielded the most reliable results, therefore, this parametrization route should be followed whenever possible, if the intended use of the parameters is that of performing predictive mixed-gas sorption calculations. If data at only one temperature are available and the quantitative accuracy of the mixed-gas sorption is necessitated, caution in the use of this model is advised.

In general, the prediction was either satisfactory for all compositions or for none: a low average SEE_{mix} was always the consequence of a similar representation of all three mixed-gas sorption isotherms. Therefore, if one could validate the parameter set adopted at least against experimental data at one composition, it should be possible to calculate the behavior at other compositions with greater confidence.

6. Conclusions

The sorption of CO_2/CH_4 mixtures in three high free volume glassy polymers, PTMSP, PIM-1 and TZ-PIM, was modelled with the multicomponent extension of the Dual Mode Sorption model. The three model parameters were retrieved from the best fit of pure-gas sorption isotherms and yielded an excellent representation of the experimental data. Multicomponent calculations provided a good qualitative picture of sorption in mixed-gas conditions, displaying the reduction in the solubility that is observed experimentally, due to competition with the second gas present in the mixture, and how this effect is more pronounced for the less soluble gas (CH_4). Moreover, a reasonable estimate of solubility-selectivity was obtained. The quantitative agreement with the experimental data in multicomponent conditions, however, varied greatly between the cases inspected.

A sensitivity analysis revealed that a small uncertainty in the pure-gas data, which would translate into a different parametrization, is greatly amplified in the mixed-gas calculation, which is much less robust than expected. Great variability is also introduced by the affinity constant of the second component, b_j, which could also assume slightly different values due to the experimental uncertainty.

In the absence of experimental mixed-gas data to validate against, it is difficult to estimate a priori the quantitative accuracy of the mixed-gas prediction.

In conclusion, given its simplicity and immediacy of use, this model is a useful tool for a first estimate of the mixed-gas effects, but great care should be used when quantitative accuracy is of interest, resorting eventually to other, yet more complex, models instead [48].

Supplementary Materials: An Electronic Supplementary Information file is available online at http://www.mdpi.com/2077-0375/9/1/8/s1, containing additional Figures and Tables. Figure S1: Effect of including the experimental error during the DMS model parametrization on the prediction of mixed-gas sorption of CO_2 and CH_4 in PTMSP, Figure S2: Effect of including the experimental error during the DMS model parametrization on the prediction of mixed-gas sorption of CO_2 and CH_4 in PIM-1, Figure S3: Effect of including the experimental error during the DMS model parametrization on the prediction of mixed-gas sorption of CO_2 and CH_4 in TZ-PIM, Figure S4: Effect of constraining the temperature dependence of the parameters during the DMS model parametrization on the prediction of mixed-gas sorption of CO_2 and CH_4 in PIM-1, Figure S5: Effect of constraining the temperature dependence of the parameters during the DMS model parametrization on the prediction of mixed-gas sorption of CO_2 and CH_4 in TZ-PIM, Figure S6: Effect of b_{CO_2} on the calculated mixed-gas sorption of CH_4 in PIM-1 at 35 °C, Figure S7: Effect of b_{CO_2} on the calculated mixed-gas sorption of CH_4 in PIM-1 at 50 °C, Figure S8: Confidence intervals of CO_2/PIM-1 DMS model parameters, Figure S9: Range of variation of the DMS model predictions of CO_2 sorption in PIM-1 in multicomponent conditions, Figure S10: Effect of b_{CH_4} on the calculated mixed-gas sorption of CO_2 in PIM-1 at 25 °C, Figure S11: Effect of b_{CH_4} on the calculated mixed-gas sorption of CO_2 in PIM-1 at 35 °C, Figure S12: Effect of b_{CH_4} on the calculated mixed-gas sorption of CO_2 in PIM-1 at 50 °C, Figure S13: Confidence intervals of CO_2/TZ-PIM and CH_4/TZ-PIM DMS model parameters, Figure S14: Range of variation of the DMS model predictions of CO_2 and CH_4 sorption in TZ-PIM in multicomponent conditions, Figure S15: Effect of b_{CO_2} on the calculated mixed gas sorption of CH_4 in TZ-PIM at 25 °C, Figure S16: Effect of b_{CO_2} on the calculated mixed-gas sorption of CH_4 in TZ-PIM at 35 °C, Figure S17: Effect of b_{CO_2} on the calculated mixed-gas sorption of CH_4 in TZ-PIM at 50 °C, Figure S18: Effect of b_{CH_4} on the calculated mixed-gas sorption of CO_2 in TZ-PIM at 25 °C, Figure S19: Effect of b_{CH_4} on the calculated mixed gas sorption of CO_2 in TZ-PIM at 35 °C, Figure S20: Effect of b_{CH_4} on the calculated mixed-gas sorption of CO_2 in TZ-PIM at 50 °C, Table S1: Confidence intervals of CO_2/PIM-1 DMS model parameters, Table S2: DMS parameter sets yielding the most and least accurate predictions of CO_2 sorption in PIM-1 in multicomponent conditions, Table S3: Confidence intervals of CO_2/TZ-PIM and CH_4/TZ-PIM DMS model parameters, Table S4: DMS parameter sets yielding the most and least accurate predictions of CO_2 and CH_4 sorption in TZ-PIM in multicomponent conditions.

Author Contributions: Conceptualization, E.R. and M.G.D.A.; Formal analysis, E.R.; Visualization, E.R.; Writing—original draft, E.R.; Supervision, M.G.D.A.; Writing—review & editing M.G.D.A.

Funding: This research received no external funding.

Conflicts of Interest: The authors declare no conflict of interest.

References

1. Galizia, M.; Chi, W.S.; Smith, Z.P.; Merkel, T.C.; Baker, R.W.; Freeman, B.D. 50th Anniversary Perspective: Polymers and Mixed Matrix Membranes for Gas and Vapor Separation: A Review and Prospective Opportunities. *Macromolecules* **2017**, *50*, 7809–7843. [CrossRef]
2. Baker, R.W.; Lokhandwala, K. Natural gas processing with membranes: An overview. *Ind. Eng. Chem. Res.* **2008**, *47*, 2109–2121. [CrossRef]
3. Baker, R.W.; Low, B.T. Gas Separation Membrane Materials: A Perspective. *Macomolecules* **2014**, *47*, 6999–7013. [CrossRef]
4. Li, C.; Meckler, S.M.; Smith, Z.P.; Bachman, J.E.; Maserati, L.; Long, J.R.; Helms, B.A. Engineered Transport in Microporous Materials and Membranes for Clean Energy Technologies. *Adv. Mater.* **2018**, *30*, 1704953. [CrossRef] [PubMed]
5. Luo, S.; Zhang, Q.; Zhu, L.; Lin, H.; Kazanowska, B.A.; Doherty, C.M.; Hill, A.J.; Gao, P.; Guo, R. Highly Selective and Permeable Microporous Polymer Membranes for Hydrogen Purification and CO_2 Removal from Natural Gas. *Chem. Mater.* **2018**, *30*, 5322–5332. [CrossRef]
6. Low, B.T.; Wang, Y.; Chung, T. Polymeric Membranes for Energy Applications. *Encycl. Polym. Sci. Technol.* **2013**, 1–37. [CrossRef]
7. Kim, S.; Lee, Y.M. Rigid and microporous polymers for gas separation membranes. *Prog. Polym. Sci.* **2015**, *43*, 1–32. [CrossRef]
8. Kim, W.; Nair, S. Membranes from nanoporous 1D and 2D materials: A review of opportunities, developments, and challenges. *Chem. Eng. Sci.* **2013**, *104*, 908–924. [CrossRef]

9. Zhang, C.; Cao, B.; Li, P. Thermal oxidative crosslinking of phenolphthalein-based cardo polyimides with enhanced gas permeability and selectivity. *J. Membr. Sci.* **2018**, *546*, 90–99. [CrossRef]
10. Zhang, C.; Fu, L.; Tian, Z.; Cao, B.; Li, P. Post-crosslinking of triptycene-based Tröger's base polymers with enhanced natural gas separation performance. *J. Membr. Sci.* **2018**, *556*, 277–284. [CrossRef]
11. Robeson, L.M. Correlation of separation factor versus permeability for polymeric membranes. *J. Membr. Sci.* **1991**, *62*, 165–185. [CrossRef]
12. Robeson, L.M. The upper bound revisited. *J. Membr. Sci.* **2008**, *320*, 390–400. [CrossRef]
13. Park, H.B.; Kamcev, J.; Robeson, L.M.; Elimelech, M.; Freeman, B.D. Maximizing the right stuff: The trade-off between membrane permeability and selectivity. *Science* **2017**, *365*, 1138–1148. [CrossRef] [PubMed]
14. Lin, H.; Freeman, B.D. Materials selection guidelines for membranes that remove CO_2 from gas mixtures. *J. Mol. Struct.* **2005**, *739*, 57–74. [CrossRef]
15. Budd, P.M.; Ghanem, B.S.; Makhseed, S.; McKeown, N.B.; Msayib, K.J.; Tattershall, C.E. Polymers of intrinsic microporosity (PIMs): Robust, solution-processable, organic nanoporous materials. *Chem. Commun.* **2004**, *230*. [CrossRef] [PubMed]
16. Budd, P.M.; Msayib, K.J.; Tattershall, C.E.; Ghanem, B.S.; Reynolds, K.J.; McKeown, N.B.; Fritsch, D. Gas separation membranes from polymers of intrinsic microporosity. *J. Membr. Sci.* **2005**, *251*, 263–269. [CrossRef]
17. McKeown, N.B.; Budd, P.M. Exploitation of intrinsic microporosity in polymer-based materials. *Macromolecules* **2010**, *43*, 5163–5176. [CrossRef]
18. Swaidan, R.; Ghanem, B.S.; Litwiller, E.; Pinnau, I. Pure- and mixed-gas CO_2/CH_4 separation properties of PIM-1 and an amidoxime-functionalized PIM-1. *J. Membr. Sci.* **2014**, *457*, 95–102. [CrossRef]
19. Park, H.B.; Jung, C.H.; Lee, Y.M.; Hill, A.J.; Pas, S.J. Polymers with Cavities Tuned for Fast Selective Transport of Small Molecules and Ions. *Science* **2007**, *318*, 254–258. [CrossRef]
20. Park, H.B.; Han, S.H.; Jung, C.H.; Lee, Y.M.; Hill, A.J. Thermally rearranged (TR) polymer membranes for CO_2 separation. *J. Membr. Sci.* **2010**, *359*, 11–24. [CrossRef]
21. Gleason, K.L.; Smith, Z.P.; Liu, Q.; Paul, D.R.; Freeman, B.D. Pure- and mixed-gas permeation of CO_2 and CH_4 in thermally rearranged polymers based on 3,3′-dihydroxy-4,4′-diamino-biphenyl (HAB) and 2,2′-bis-(3,4-dicarboxyphenyl) hexafluoropropane dianhydride (6FDA). *J. Membr. Sci.* **2015**, *475*, 204–214. [CrossRef]
22. Smith, Z.P.; Sanders, D.F.; Ribeiro, C.P.; Guo, R.; Freeman, B.D.; Paul, D.R.; McGrath, J.E.; Swinnea, S. Gas sorption and characterization of thermally rearranged polyimides based on 3,3′-dihydroxy-4,4′-diamino-biphenyl (HAB) and 2,2′-bis-(3,4-dicarboxyphenyl) hexafluoropropane dianhydride (6FDA). *J. Membr. Sci.* **2012**, *415–416*, 558–567. [CrossRef]
23. Sanders, D.F.; Smith, Z.P.; Ribeiro, C.P.; Guo, R.; McGrath, J.E.; Paul, D.R.; Freeman, B.D. Gas permeability, diffusivity, and free volume of thermally rearranged polymers based on 3,3′-dihydroxy-4,4′-diamino-biphenyl (HAB) and 2,2′-bis-(3,4-dicarboxyphenyl) hexafluoropropane dianhydride (6FDA). *J. Membr. Sci.* **2012**, *409*, 232–241. [CrossRef]
24. Vopička, O.; De Angelis, M.G.; Sarti, G.C. Mixed gas sorption in glassy polymeric membranes: I. CO_2/CH_4 and n-C4/CH_4 mixtures sorption in poly(1-trimethylsilyl-1-propyne) (PTMSP). *J. Membr. Sci.* **2013**, *449*, 97–108. [CrossRef]
25. Vopička, O.; De Angelis, M.G.; Du, N.; Li, N.; Guiver, M.D.; Sarti, G.C. Mixed gas sorption in glassy polymeric membranes: II. CO_2/CH_4 mixtures in a polymer of intrinsic microporosity (PIM-1). *J. Membr. Sci.* **2014**, *459*, 264–276. [CrossRef]
26. Gemeda, A.E.; De Angelis, M.G.; Du, N.; Li, N.; Guiver, M.D.; Sarti, G.C. Mixed gas sorption in glassy polymeric membranes. III. CO_2/CH_4 mixtures in a polymer of intrinsic microporosity (PIM-1): Effect of temperature. *J. Membr. Sci.* **2017**, *524*, 746–757. [CrossRef]
27. Wijmans, J.G.; Baker, R.W. The solution-diffusion model: A review. *J. Membr. Sci.* **1995**, *107*, 1–21. [CrossRef]
28. Minelli, M.; Friess, K.; Vopička, O.; De Angelis, M.G. Modeling gas and vapor sorption in a polymer of intrinsic microporosity (PIM-1). *Fluid Phase Equilib.* **2013**, *347*, 35–44. [CrossRef]
29. Vieth, W.R.; Tam, P.M.; Michaels, A.S. Dual sorption mechanisms in glassy polystyrene. *J. Colloid Interface Sci.* **1966**, *22*, 360–370. [CrossRef]
30. Koros, W.J.; Paul, D.R. Design considerations for measurement of gas sorption in polymers by pressure decay. *J. Polym. Sci. Polym. Phys. Ed.* **1976**, *14*, 1903–1907. [CrossRef]

31. Michaels, A.S.; Vieth, W.R.; Barrie, J.A. Diffusion of gases in polyethylene terephthalate. *J. Appl. Phys.* **1963**, *34*, 13–20. [CrossRef]
32. Koros, W.J.; Paul, D.R. Carbon dioxide sorption and transport in polycarbonate. *J. Polym. Sci. Polym. Phys.* **1976**, *14*, 687–702. [CrossRef]
33. Paul, D.R.; Koros, W.J. Effect of partially immobilizing sorption on permeability and the diffusion time lag. *J. Polym. Sci. Polym. Phys. Ed.* **1976**, *14*, 675–685. [CrossRef]
34. Fredrickson, G.H.; Helfand, E. Dual-Mode Transport of Penetrants in Glassy Polymers. *Macromolecules* **1985**, *18*, 2201–2207. [CrossRef]
35. Meares, P. The Diffusion of Gases Through Polyvinyl Acetate. *J. Am. Chem. Soc.* **1954**, *76*, 3416–3422. [CrossRef]
36. Meares, P. The solubilities of gases in polyvinyl acetate. *Trans. Faraday Soc.* **1957**, *54*, 40–46. [CrossRef]
37. Barrer, R.M.; Barrie, J.A.; Slater, J. Sorption and Diffusion in Ethyl Cellulose. Part III. Comparison between Ethyl Cellulose and Rubber. *J. Polym. Sci.* **1958**, *27*, 177–197. [CrossRef]
38. Vieth, W.R.; Howell, J.M.; Hsieh, J.H. Dual sorption theory. *J. Membr. Sci.* **1976**, *1*, 177–220. [CrossRef]
39. Vieth, W.R.; Alcalay, H.H.; Frabetti, A.J. Solution of Gases in Oriented Poly(ethylene Terephthalate). *J. Appl. Polym. Sci.* **1964**, *8*, 2125–2138. [CrossRef]
40. Feng, H. Modeling of vapor sorption in glassy polymers using a new dual mode sorption model based on multilayer sorption theory. *Polymer* **2007**, *48*, 2988–3002. [CrossRef]
41. Bondar, V.I.; Kamiya, Y.; Yampol'skii, Y.P. On pressure dependence of the parameters of the dual-mode sorption model. *J. Polym. Sci. Part B Polym. Phys.* **1996**, *34*, 369–378. [CrossRef]
42. Lacombe, R.H.; Sanchez, I.C. Statistical Thermodynamics of Fluid Mixtures. *J. Phys. Chem.* **1976**, *80*, 2568–2580. [CrossRef]
43. Chapman, W.G.; Gubbins, K.E.; Jackson, G.; Razosz, M. SAFT: Equation-of-state solution model for associating fluids. *Fluid Phase Equilib.* **1989**, *52*, 31–38. [CrossRef]
44. Doghieri, F.; Sarti, G.C. Nonequilibrium Lattice Fluids: A Predictive Model for the Solubility in Glassy Polymers. *Macromolecules* **1996**, *29*, 7885–7896. [CrossRef]
45. Minelli, M.; Doghieri, F. Predictive model for gas and vapor solubility and swelling in glassy polymers I: Application to different polymer/penetrant systems. *Fluid Phase Equilib.* **2014**, *381*, 1–11. [CrossRef]
46. Doghieri, F.; Quinzi, M.; Rethwisch, D.G.; Sarti, G.C. Predicting Gas Solubility in Glassy Polymers through Nonequilibrium EOS. In *Advanced Materials for Membrane Separations*; American Chemical Society: Washington, DC, USA, 2004; pp. 74–90.
47. Minelli, M.; Campagnoli, S.; De Angelis, M.G.; Doghieri, F.; Sarti, G.C. Predictive model for the solubility of fluid mixtures in glassy polymers. *Macromolecules* **2011**, *44*, 4852–4862. [CrossRef]
48. Ricci, E.; Gemeda, A.E.; Du, N.; Li, N.; De Angelis, M.G.; Guiver, M.D.; Sarti, G.C. Sorption of the CO_2/CH_4 mixture in TZ-PIM, PIM-1 and PTMSP: Experimental data and NELF-model based analysis of competitive sorption and its impact on the selectivity. *J. Membr. Sci.* **2018**. submitted for publication.
49. Allen, M.P.; Tildesley, D.J. *Computer Simulation of Liquids*; Clarendon Press: Wotton-under-Edge, UK, 1989; ISBN 9780198556459.
50. Widom, B. Some Topics in the Theory of Fluids. *J. Chem. Phys.* **1963**, *39*, 2808. [CrossRef]
51. Ramdin, M.; Balaji, S.P.; Vicent-Luna, J.M.; Gutiérrez-Sevillano, J.J.; Calero, S.; De Loos, T.W.; Vlugt, T.J.H. Solubility of the precombustion gases CO_2, CH_4, CO, H_2, N_2, and H_2S in the ionic liquid [bmim][Tf2N] from Monte Carlo simulations. *J. Phys. Chem. C* **2014**, *118*, 23599–23604. [CrossRef]
52. Boulougouris, G.C.; Economou, I.G.; Theodorou, D.N. On the calculation of the chemical potential using the particle deletion scheme. *Mol. Phys.* **1999**, *96*, 905–913. [CrossRef]
53. Siegert, M.R.; Heuchel, M.; Hofmann, D. A generalized direct-particle-deletion scheme for the calculation of chemical potential and solubilities of small- and medium-sized molecules in amorphous polymers. *J. Comput. Chem.* **2007**, *28*, 877–889. [CrossRef] [PubMed]
54. Kupgan, G.; Abbott, L.J.; Hart, K.E.; Colina, C.M. Modeling Amorphous Microporous Polymers for CO_2 Capture and Separations. *Chem. Rev.* **2018**, *118*, 5488–5538. [CrossRef] [PubMed]
55. Tocci, E.; De Lorenzo, L.; Bernardo, P.; Clarizia, G.; Bazzarelli, F.; McKeown, N.B.; Carta, M.; Malpass-Evans, R.; Friess, K.; Pilnáček, K.; et al. Molecular modeling and gas permeation properties of a polymer of intrinsic microporosity composed of ethanoanthracene and Tröger's base units. *Macromolecules* **2014**, *47*, 7900–7916. [CrossRef]

56. Hart, K.E.; Colina, C.M. Estimating gas permeability and permselectivity of microporous polymers. *J. Membr. Sci.* **2014**, *468*, 259–268. [CrossRef]
57. Hart, K.E.; Abbott, L.J.; McKeown, N.B.; Colina, C.M. Toward effective CO_2/CH_4 separations by sulfur-containing PIMs via predictive molecular simulations. *Macromolecules* **2013**, *46*, 5371–5380. [CrossRef]
58. Rose, I.; Bezzu, C.G.; Carta, M.; Comesaña-Gándara, B.; Lasseuguette, E.; Ferrari, M.C.; Bernardo, P.; Clarizia, G.; Fuoco, A.; Jansen, J.C.; et al. Polymer ultrapermeability from the inefficient packing of 2D chains. *Nat. Mater.* **2017**, 1–39. [CrossRef]
59. Heuchel, M.; Fritsch, D.; Budd, P.M.; McKeown, N.B.; Hofmann, D. Atomistic packing model and free volume distribution of a polymer with intrinsic microporosity (PIM-1). *J. Membr. Sci.* **2008**, *318*, 84–99. [CrossRef]
60. Gusev, A.A.; Arizzi, S.; Suter, U.W.; Moll, D.J. Dynamics of light gases in rigid matrices of dense polymers. *J. Chem. Phys.* **1993**, *99*, 2221–2227. [CrossRef]
61. Gusev, A.A.; Suter, U.W. Dynamics of small molecules in dense polymers subject to thermal motion. *J. Chem. Phys.* **1993**, *99*, 2228. [CrossRef]
62. Fang, W.; Zhang, L.; Jiang, J. Polymers of intrinsic microporosity for gas permeation: A molecular simulation study. *Mol. Simul.* **2010**, *36*, 992–1003. [CrossRef]
63. Fang, W.; Zhang, L.; Jiang, J. Gas Permeation and Separation in Functionalized Polymers of Intrinsic Microporosity: A Combination of Molecular Simulations and Ab Initio Calculations. *J. Phys. Chem. C* **2011**, *115*, 14123–14130. [CrossRef]
64. Kupgan, G.; Demidov, A.G.; Colina, C.M. Plasticization behavior in polymers of intrinsic microporosity (PIM-1): A simulation study from combined Monte Carlo and molecular dynamics. *J. Membr. Sci.* **2018**, *565*, 95–103. [CrossRef]
65. Hölck, O.; Böhning, M.; Heuchel, M.; Siegert, M.R.; Hofmann, D. Gas sorption isotherms in swelling glassy polymers—Detailed atomistic simulations. *J. Membr. Sci.* **2013**, *428*, 523–532. [CrossRef]
66. Frentrup, H.; Hart, K.E.; Colina, C.M.; Müller, E.A. In silico determination of gas permeabilities by non-equilibrium molecular dynamics: CO_2 and He through PIM-1. *Membranes* **2015**, *5*, 99–119. [CrossRef]
67. Rizzuto, C.; Caravella, A.; Brunetti, A.; Park, C.H.; Lee, Y.M.; Drioli, E.; Barbieri, G.; Tocci, E. Sorption and Diffusion of CO_2/N_2 in gas mixture in thermally-rearranged polymeric membranes: A molecular investigation. *J. Membr. Sci.* **2017**, *528*, 135–146. [CrossRef]
68. Myers, A.L.; Prausnitz, J.M. Thermodynamics of mixed-gas adsorption. *AIChE J.* **1965**, *11*, 121–127. [CrossRef]
69. Neyertz, S.; Brown, D. Air Sorption and Separation by Polymer Films at the Molecular Level. *Macromolecules* **2018**, *51*, 7077–7092. [CrossRef]
70. Theodorou, D.N. Principles of Molecular Simulation of Gas Transport in Polymers. In *Materials Science of Membranes for Gas and Vapor Separation*; John Wiley & Sons: Hoboken, NJ, USA, 2006; ISBN 047085345X.
71. Larsen, G.S.; Hart, K.E.; Colina, C.M. Predictive simulations of the structural and adsorptive properties for PIM-1 variations. *Mol. Simul.* **2014**, *40*, 599–609. [CrossRef]
72. Spyriouni, T.; Boulougouris, G.C.; Theodorou, D.N. Prediction of sorption of CO_2 in glassy atactic polystyrene at elevated pressures through a new computational scheme. *Macromolecules* **2009**, *42*, 1759–1769. [CrossRef]
73. Vogiatzis, G.G.; Theodorou, D.N. Multiscale Molecular Simulations of Polymer-Matrix Nanocomposites. *Arch. Comput. Methods Eng.* **2017**. [CrossRef]
74. Mathioudakis, I.; Vogiatzis, G.G.; Tzoumanekas, C.; Theodorou, D.N. Molecular modeling and simulation of atactic polystyrene/amorphous silica nanocomposites. *J. Phys. Conf. Ser.* **2016**, *738*, 012021. [CrossRef]
75. Mathioudakis, I.G.; Vogiatzis, G.G.; Tzoumanekas, C.; Theodorou, D.N. Multiscale simulations of PS–SiO_2 nanocomposites: From melt to glassy state. *Soft Matter* **2016**, *12*, 7585–7605. [CrossRef]
76. Koros, W.J. Model for sorption of mixed gases in glassy polymers. *J. Polym. Sci. Polym. Phys. Ed.* **1980**, *18*, 981–992. [CrossRef]
77. George, G.; Bhoria, N.; Alhallaq, S.; Abdala, A.; Mittal, V. Polymer membranes for acid gas removal from natural gas. *Sep. Purif. Technol.* **2016**, *158*, 333–356. [CrossRef]
78. Du, N.; Park, H.B.; Robertson, G.P.; Dal-Cin, M.M.; Visser, T.; Scoles, L.; Guiver, M.D. Polymer nanosieve membranes for CO_2-capture applications. *Nat. Mater.* **2011**, *10*, 372–375. [CrossRef] [PubMed]
79. Satilmis, B.; Lan, M.; Fuoco, A.; Rizzuto, C.; Tocci, E.; Bernardo, P.; Clarizia, G.; Esposito, E.; Monteleone, M.; Dendisová, M.; et al. Temperature and pressure dependence of gas permeation in amine-modified PIM-1. *J. Membr. Sci.* **2018**, *555*, 483–496. [CrossRef]

80. Lanč, M.; Pilnáček, K.; Mason, C.R.; Budd, P.M.; Rogan, Y.; Malpass-Evans, R.; Carta, M.; Gándara, B.C.; McKeown, N.B.; Jansen, J.C.; et al. Gas sorption in polymers of intrinsic microporosity: The difference between solubility coefficients determined via time-lag and direct sorption experiments. *J. Membr. Sci.* **2018**, *570–571*, 522–536. [CrossRef]
81. Pilnáček, K.; Vopička, O.; Lanč, M.; Dendisová, M.; Zgažar, M.; Budd, P.M.; Carta, M.; Malpass-Evans, R.; McKeown, N.B.; Friess, K. Aging of polymers of intrinsic microporosity tracked by methanol vapour permeation. *J. Membr. Sci.* **2016**, *520*, 895–906. [CrossRef]
82. Ricci, E.; Minelli, M.; De Angelis, M.G. A multiscale approach to predict the mixed gas separation performance of glassy polymeric membranes for CO_2 capture: The case of CO_2/CH_4 mixture in Matrimid®. *J. Membr. Sci.* **2017**, *539*. [CrossRef]
83. Chan, A.H.; Koros, W.J.; Paul, D.R. Analysis of hydrocarbon gas sorption and transport in ethyl cellulose using the dual sorption/partial immobilization models. *J. Membr. Sci.* **1978**, *3*, 117–130. [CrossRef]
84. Kanehashi, S.; Nagai, K. Analysis of dual-mode model parameters for gas sorption in glassy polymers. *J. Membr. Sci.* **2005**, *253*, 117–138. [CrossRef]
85. Koros, W.J.; Paul, D.R. CO_2 sorption in poly(ethylene terephthalate) above and below the glass transition. *J. Polym. Sci. Part B Polym. Phys.* **1978**, *16*, 1947–1963. [CrossRef]
86. Koros, W.J.; Sanders, E.S. Multicomponent gas sorption in glassy polymers. *J. Polym. Sci. Polym. Symp.* **1985**, 141–149. [CrossRef]
87. Sanders, E.S.; Koros, W.J. Sorption of CO_2, C_2H_4, N_2O and their Binary Mixtures in Poly(methyl methacrylate). *J. Polym. Sci. B* **1986**, *188*, 175–188. [CrossRef]
88. Sanders, E.S.; Koros, W.J.; Hopfenberg, H.B.; Stannett, V.T. Mixed gas sorption in glassy polymers: Equipment design considerations and preliminary results. *J. Membr. Sci.* **1983**, *13*, 161–174. [CrossRef]
89. Story, B.J.; Koros, W.J. Sorption of CO_2/CH_4 mixtures in poly(phenylene oxide) and a carboxylated derivative. *J. Appl. Polym. Sci.* **1991**, *42*, 2613–2626. [CrossRef]
90. Peng, D.; Robinson, D.B. A New Two-Constant Equation of State. *Ind. Eng. Chem. Fundam.* **1976**, *15*, 59–64. [CrossRef]
91. Sandler, S.I. *Chemical, Biochemical, and Engineering Thermodynamics*, 5th ed.; John Wiley & Sons: Hoboken, NJ, USA, 2017.
92. Abadie, J.; Carpentier, J. Generalization of the Wolfe reduced gradient method to the case of nonlinear constraints. In *Optimization*; Fletcher, R., Ed.; Academic Press: London, UK, 1969; pp. 37–49.
93. Yamane, T. *Statistics: An Introductory Analysis*; Harper & Row, John Weatherhill: New York, NY, USA; Evanston, IL, USA; London, UK; Tokyo, Japan, 1973; ISBN 0060473134.
94. Bevington, P.R.; Robinson, D.K. *Data Reduction and Error Analysis for the Physical Sciences*, 3rd ed.; McGraw-Hill: Boston, MA, USA, 2003.
95. Myers, R.H. *Classical and Modern Regression with Applications*; Duxbury Press: Belmont, CA, USA, 1990.
96. Koros, W.J.; Paul, R.; Huvard, G.S. Energetics of gas sorption in glassy polymers. *Polymer* **1979**, *20*, 956–960. [CrossRef]
97. Raharjo, R.D.; Freeman, B.D.; Sanders, E.S. Pure and mixed gas CH_4 and n-C_4H_{10} sorption and dilation in poly(1-trimethylsilyl-1-propyne). *Polymer* **2007**, *48*, 6097–6114. [CrossRef]
98. Stevens, K.A.; Smith, Z.P.; Gleason, K.L.; Galizia, M.; Paul, D.R.; Freeman, B.D. Influence of temperature on gas solubility in thermally rearranged (TR) polymers. *J. Membr. Sci.* **2017**, *533*, 75–83. [CrossRef]
99. Kamiya, Y.; Hirose, T.; Mizoguchi, K.; Naito, Y. Gravimetric study of high-pressure sorption of gases in polymers. *J. Polym. Sci. Part B Polym. Phys.* **1986**, *24*, 1525–1539. [CrossRef]
100. Chiou, J.S.; Maeda, Y.; Paul, D.R. Gas and vapor sorption in polymers just below Tg. *J. Appl. Polym. Sci.* **1985**, *30*, 4019–4029. [CrossRef]
101. Wang, J.S.; Kamiya, Y. A method of validation and parameter evaluation for dual-mode sorption model. *J. Membr. Sci.* **1999**, *154*, 25–32. [CrossRef]

Review

Models for Facilitated Transport Membranes: A Review

Riccardo Rea, Maria Grazia De Angelis and Marco Giacinti Baschetti *

Dipartimento di Ingegneria Civile, Chimica, Ambientale e dei Materiali (DICAM), Università di Bologna, Via Terracini 28, 40131 Bologna, Italy; riccardo.rea3@unibo.it (R.R.); grazia.deangelis@unibo.it (M.G.D.A.)
* Correspondence: marco.giacinti@unibo.it

Received: 9 December 2018; Accepted: 15 January 2019; Published: 2 February 2019

Abstract: Facilitated transport membranes are particularly promising in different separations, as they are potentially able to overcome the trade-off behavior usually encountered in solution-diffusion membranes. The reaction activated transport is a process in which several mechanisms take place simultaneously, and requires a rigorous theoretical analysis, which unfortunately is often neglected in current studies more focused on material development. In this work, we selected and reviewed the main mathematical models introduced to describe mobile and fixed facilitated transport systems in steady state conditions, in order to provide the reader with an overview of the existing mathematical tools. An analytical solution to the mass transport problem cannot be achieved, even when considering simple reaction schemes such as that between oxygen (solute) and hemoglobin (carrier) ($A + C \rightleftarrows AC$), that was thoroughly studied by the first works dealing with this type of biological facilitated transport. Therefore, modeling studies provided approximate analytical solutions and comparison against experimental observations and exact numerical calculations. The derivation, the main assumptions, and approximations of such modeling approaches is briefly presented to assess their applicability, precision, and flexibility in describing and understanding mobile and fixed site carriers facilitated transport membranes. The goal is to establish which mathematical tools are more suitable to support and guide the development and design of new facilitated transport systems and materials. Among the models presented, in particular, those from Teramoto and from Morales-Cabrera et al. seem the more flexible and general ones for the mobile carrier case, while the formalization made by Noble and coauthors appears the most complete in the case of fixed site carrier membranes.

Keywords: facilitated transport; permeability models; membrane separation; CO_2 capture; gas separation; modeling

---

1. Introduction

In membrane separation processes, the term "facilitated transport" (FT) refers to a coupled transport mechanism composed of two main effects: pure physical diffusion and reversible chemical reaction between the target solute (A) and the carrier (C). At the feed side of the membrane, the solute reacts and the reaction product (AC), also referred to as "complex" in the usual terminology of FT, diffuses across the thickness. At the downstream side, the solute is released by the inverse chemical reaction. The unreacted solute, as well as the other dissolved species (that do not interact chemically with the carrier) merely diffuse in the system according to Fick's law. This carrier-mediated transport leads to a superior performance in terms of solute flux and separation ability of the membranes [1,2].

Based on the mobility of the carrier species, there are two main families of carrier-mediated membranes: the mobile carrier (MC) systems and the fixed site carrier (FSC) ones. In the first case the carrier is free to diffuse across the membrane (supported or immobilized liquid membranes), in the second one the carrier agent is fixed, or bounded, to the polymeric backbone. In this latter case, the complex diffuses following a more complex mechanism than the one in the mobile carrier systems. This point will be further explained in the section referring to transport in FSC membranes.

In 1940, Osterhout et al. [3], firstly investigated systematically the molecular carrier mechanism acting on the transport of potassium and sodium ions in human body, opening a new frontier. After that, in the early 1960s, Scholander [4] and Wittemberg [5] recognized, separately, the facilitation effect that hemoglobin (Hb) and myoglobin (Mg) had on the oxygen transport in blood. They concluded that Hb and Mg react reversibly with the oxygen and diffuse across a liquid film as oxy-Hb or oxy-Mb while, at the opposite side, the oxygen is released and the carrier re-formed.

Starting from those results, many subsequent studies dealt with facilitated transport in membranes. In particular, many works were devoted to the understanding of the principles behind the mechanism, on the key parameters governing it, as well as on the effect of such parameters on the separation performances [6–9].

The compactness, energy-efficiency, and ease of operation of membrane systems make facilitated transport membranes one of the most competitive and promising technologies in industrial separations [10–12]. In this context, great interest is focused on some typical separations of the chemical industry.

Olefin and paraffin separation by facilitated transport membranes (FTM) are reported since the 1980s. In this field, the commonly used carrier is a solution of silver nitrate, usually supported on a polymer matrix, that reacts selectively with the pi-electrons of olefins. In this field of application, Teramoto et al. [13] reported in 1986 results obtained for the ethylene/ethane separation, while Ho and co-worker, in 1994, studied the butene and isobutene separation from butane and propane, respectively [14]. One year before, Funke et al. [15] used hydrated Nafion as polymeric network in enhanced cis and trans 2-butene transport. A review of olefin/paraffin separation using FTM has been published in 2012 by Faiz and Li [16]. More recent works dealt with the use of N-methyl pyrrolidone coupled with $AgNO_3$ salt as liquid membranes, supported on tetrafluoroethylene (PTFE), by Azizi et al. [17] to separate propylene from propane in 2015. Supported ionic liquid membranes have been recently used by Zarca and co-workers [18] to enhance the selectivity of propylene/propane. FSC systems based on polymer incorporating silver nanoparticles are under study in olefin separation, to overcome the chemical stability issues of the silver ion [19–23].

Metal complexes have been extensively studied in FSC membranes to facilitate the transport of oxygen. In the 1986–87 two different papers were published, Okahata et al. [24] and Nishide et al. [25], about the facilitation effect in oxygen separation membranes made by cobalt complex, CoPIm, chained to poly-butyl methacrylate polymer. Two years later [26] Ohyanagi and co-workers investigated the change of polymer matrix and metal complexes on the oxygen transport by using FePIm as fixed carrier in poly-methyl methacrylate membrane. Poly-carbonate/COSALEN cobalt complex systems was used in 1996 by Chen and Lai [27], picket-fence Cobalt porphyrin was englobed in different polymer matrix by Shentu and Nischide in 2003 [28]. Preethi et al. [29] complexed the poly vinyl imidazole with cobalt-pthalocyanine in 2006 and studied the effects on the reversible complexation with the oxygen. Three years ago, in 2016, Choi et al. [30] prepared hollow fibers made by poly-ether sulfone and poly-dimethyl siloxane and cobalt-tetraphenylporphyrin (CoPTT) and analyzed the possibility to use these systems to separate oxygen in post-combustion as flue gas pre-treatment to remove CO_2 before the scrubber unit.

A large number of studies deals with the CO_2 separation from other gases, such as N_2 (post-combustion), CH_4 (biomethane and natural gas upgrading), H_2 (steam reforming purification) [31–33]. The massive anthropogenic emissions of the last decades are changing dramatically the environment in terms of global temperature raise, oceans acidification, as well as Arctic and Antarctic defrost and sea level rise [34]. Facilitated transport membranes could be an interesting, promising, ecofriendly alternative to the common solutions, such as amino based-separations unity. Guha et al. [35] in 1990 used an aqueous solution of diethanol amine (DEA) immobilized in poly-propylene microporous support, Davis and Sandall also used DEA aqueous solution but immobilized in poly-ethylene glycol, in 1993 [36]. Teramoto, in 1996, used poly-vinyl diene difluoride as supporting polymer for DEA and monoethanol amine (MEA) solutions [37]. FSC membranes have been extensively tested in the CO_2 separation in recent years. Matsuyama et al. used poly-ethylene imine blended with poly-vinyl alcohol [38], and Ho and co-workers studied mobile-fixed hybrid carrier systems made by poly-allyl amine/poly-vinyl alcohol blend with amino acid and potassium hydroxide in 2006 and 2008, [39,40]. Deng et al. [41] presented their results about the facilitated transport of carbon dioxide in poly-vinyl amine and poly-vinyl alcohol blend membranes in 2009. Poly-vinyl amine membranes were also tested in a pilot scale plant in 2013 by the same research group [42] showing stability of performance over four months of test and another pilot scale test was conducted in 2017 and reported here, [43]. Nowadays, concerning FTM for CO_2 separations, the hybrid systems containing nanofillers [44–46] as well as mobile and fixed sites carrier, such as ionic liquids or amino acid salts or polymers containing suitable pendant groups, [47–52] are investigated to increase membrane stability and at the same time to obtain high selectivity and permeability.

Despite the high number of data reported in literature, especially in the context of FSC systems, there is a general lack of knowledge of the physics underlying the phenomenon. While a great interest is posed on the experimental measurements, the modeling efforts often reduce to finding a fitting curve that interpolates the experimental data. As shown in the next section, the facilitated transport problem, except in some special cases, usually cannot be solved analytically. For the reaction scheme under study in this review, many authors used numerical techniques to solve the differential mass transfer problem. Kutchai et al. [53] used the quasi linearization method to provide the numerical solution and to demonstrate the effect of the physical and chemical parameters on the transport. Ward [54] gave a quantitative explanation of nitric oxide facilitated transport in ferrous chloride solution using Galerkin's method [55]. Kemena et al. [7] confirmed the presence of an optimal equilibrium constant value, previously hypothesized by Schultz et al. [1], which allows to obtain the maximum facilitation factor as a function of the dimensionless parameters governing the problem. Kirkoppru-Dindi and Noble [56] extended that approach for the multiple sites carrier case. The methods of orthogonal collocation on finite elements, introduced by Carey and Finlayson [57], was used by Jain and Shultz [58] to predict experimental results coming from real cases such as carbon monoxide/hemoglobin solution, carbon dioxide/bicarbonate solution and nitric oxide/ferrous chloride solution. Barbero and Manzanares [59], by the boundary element method procedure of Ramachandran [60], calculated the results for the facilitation factor over the entire range comprise between the chemical equilibrium regime to the pure physical diffusion transport. Basaran et al. [8] extended the work of Kemena et al. [7] to study how the diffusivity ratio between the carrier and the reaction product affects the facilitation. Ebadi Amooghin et al. [61] modeled the CO_2 facilitated transport in a poly-vinyl alchol (PVA)-amines membrane by the finite element method. Because of the lack of exact analytical solutions, the reaction-augmented transport problem still represents an interesting topic in mathematical and computational studies in the differential nonlinear systems field [59,62–67].

The deep investigation by a phenomenological model of this complex transport mechanism, however, could greatly help in developing new and more performing materials, in recognizing the best condition of use as well as in providing useful information for the overall process optimization. The purpose of this review is therefore to provide an insight on the modeling works that, in the last sixty years, various authors developed to model facilitated transport systems. A series of analytical, approximate models are reviewed in this work, starting from the one presented by Friedlander and Keller in 1965 [68] to the one of Zarca et al. [18] proposed in 2017.

For this purpose, the models described in the present work have been divided in two different sections, Mobile Carrier and Fixed Carrier, each of one organized based on the publish date of the single contribution, from the past to the present.

Before that, the common mathematical background is introduced which allows to describe the mass transfer problem in these kind of systems. In this concern, we chose to focus the review on the most used reaction mechanism in the literature, that is:

$$A + C \rightleftarrows AC$$

Mathematical Background

The carrier-mediated facilitated transport of a gas in a membrane consists of simple diffusion transport, coupled with one or more reversible chemical reactions which allow to transport the target species also in the form of a complex. Such a process is not easy to describe mathematically, because in most cases the differential transport problem is nonlinear and no exact analytical solution exists.

In order to approach the problem, many authors have developed analytical solutions assuming some simplification or approximation such as the attainment of the local equilibrium condition, equal diffusivities coefficients for carrier and complex, and excess of carrier in respect to the solute. These solutions and models proposed will be further investigated and discussed in the next sections.

Here we present the general formulation of the facilitated transport problem in terms of mathematical differential equations expressing the mass balance for the species involved. The treatment will be based on a single chemical reaction in which three species co-exist: solute, carrier, and their reaction product (also called complex).

In a rectangular geometry, for each species i involved, we can write the continuity equation:

$$\frac{DC_i}{Dt} = \left(\frac{\partial C_i}{\partial t} + v_x\frac{\partial C_i}{\partial x} + v_y\frac{\partial C_i}{\partial y} + v_z\frac{\partial C_i}{\partial z}\right) = -\left(\frac{\partial J_{ix}}{\partial x} + \frac{\partial J_{iy}}{\partial y} + \frac{\partial J_{iz}}{\partial z}\right) - r_i + r_i' \quad (1)$$

where:

C = concentration
J = flux
r = dissipative term
r' = generative term
v = velocity
x, y, z = cartesian directions
t = time

to be considered with consistent units.

From Equation (1), considering the one dimensional problem (x direction) in steady state condition, without convective contribution, we have:

$$-\frac{dJ_i}{dx} - r_i + r_i' = 0 \quad (2)$$

To complete the description of the problem, a reaction scheme that describes the chemical interaction between the carrier and the solute species has to be considered. In the present work, the reaction scheme used originally to depict the oxygen facilitated transport by the hemoglobin or myoglobin has been used. Indeed, the facilitated transport modeling was introduced, in the 1960s, to give a theoretical explanation of oxygen enhanced diffusion in hemoglobin and myoglobin solutions experimentally observed [4,5,69–71]. Moreover, even in cases where the modeling approaches were not addressed directly to the oxygen transport, the latter case was used as benchmark to test and compare the obtained solutions.

The reaction scheme mentioned, already shown in the introduction, is depicted in Equation (3):

$$A + C \rightleftarrows AC \quad (3)$$

where A, C and AC indicate the solute, the carrier and the carrier complex species respectively. Following this reaction scheme Equation (2) may be rewritten for each component as follows:

$$-\frac{dJ_A}{dx} - k_f C_A C_C + k_r C_{AC} = 0 \tag{4}$$

$$-\frac{dJ_C}{dx} - k_f C_A C_C + k_r C_{AC} = 0 \tag{5}$$

$$-\frac{dJ_{AC}}{dx} + k_f C_A C_C - k_r C_{AC} = 0 \tag{6}$$

where the forward reaction rate is assumed to depend on both the carrier and solute concentration, while the backward reaction has a linear dependence on the reaction product concentration, C_{AC}. Both kinetic constants, k_f and k_r, are considered concentration-independent.

The boundary conditions (BC) needed to close the mathematical problem are obtained by assuming that the only species that can enter and leave freely the membrane is the solute. Carrier and carrier complex are considered to remain confined in the membrane, so that no flux of carrier nor reaction product across the boundaries is allowed. In addition, the solute concentration at the interface is generally considered to be in equilibrium with the gas bulk phase, even if this constraint can be removed, as done by Noble et al. [72] which also considered the existence of an external mass transfer resistance.

Based on what discussed, the following boundary conditions holds:

$$x = 0 \quad C_A = C_A^0 \quad {J_{AC}}^0 = {J_C}^0 = 0 \tag{7}$$

$$x = L \quad C_A = C_A^L \quad {J_{AC}}^L = {J_C}^L = 0 \tag{8}$$

where L is the membrane thickness. The previous BCs state that the carrier and the reaction product are considered as nonvolatile species and are confined inside the condensed phase.

Since neither the carrier nor the complex can leave the membrane, one has the following integral constraint:

$$\int_0^L (C_C + C_{AC})dx = C_T L \tag{9}$$

where C_T is the initial carrier concentration, before it starts to react with the solute.

Now, if we consider that the diffusion mechanism follows Fick's law for each component, we have:

$$J_i = -D_i \frac{dC_i}{dx} \tag{10}$$

And Equations (4)–(6) can be written as:

$$D_A \frac{d^2 C_A}{dx^2} - k_f C_A C_C + k_r C_{AC} = 0 \tag{11}$$

$$D_C \frac{d^2 C_C}{dx^2} - k_f C_A C_C + k_r C_{AC} = 0 \tag{12}$$

$$D_{AC} \frac{d^2 C_{AC}}{dx^2} + k_f C_A C_C - k_r C_{AC} = 0 \tag{13}$$

And the boundary conditions can now be presented in the form:

$$x = 0 \quad C_A = C_A^0 \quad \frac{dC_{AC}}{dx} = \frac{dC_C}{dx} = 0 \tag{14}$$

$$x = L \quad C_A = C_A^L \quad \frac{dC_{AC}}{dx} = \frac{dC_C}{dx} = 0 \tag{15}$$

By the sum of Equations (11) and (13), and integrating twice, it is obtained:

$$D_A C_A + D_{AC} C_{AC} = C_1 x + C_2 \tag{16}$$

where C_1 and C_2 are integration constants, and C_1 in particular represents the total solute flux.

Similarly, adding Equations (12) and (13), we have, for the carrier:

$$D_C C_C + D_{AC} C_{AC} = C_3 x + C_4 \tag{17}$$

where now C_3 represents the total carrier flux, however by using the boundary Conditions (14) or (15) we can see that such flux is 0 at the boundaries and therefore, being C_3 a constant, it is zero everywhere.

In the case of mobile carrier, then, by considering the carrier and complex diffusivities equal ($D_C = D_{AC} = D$), the Equation (17) can be rewritten as:

$$C_C + C_{AC} = C_5 \tag{18}$$

where $C_5 = C_4/D$.

Moreover, in this case, with the integral Constraint (9) it can be shown that at any point in the membrane:

$$C_C + C_{AC} = C_5 = C_T \tag{19}$$

In the case of fixed carrier, on the other hand, the diffusion coefficient D_C can be considered as zero while D_{AC} now is an effective diffusion coefficient associated to a sort of jumping mechanism of A from one fixed site to the other.

The solute, in the facilitated transport process, is transferred from one boundary to the other by two different mechanisms, the pure diffusion in unreacted state, and the diffusion as a complexed species. Once on the downstream side of the membrane, the reverse reaction takes place and the solute is released. Hence, the total solute flux is expressed as:

$$J_A = -D_A \frac{dC_A}{dx} - D_{AC} \frac{dC_{AC}}{dx} \tag{20}$$

The ratio between the total flux of A inside the membrane, Equation (20), and the flux given by the pure diffusion mechanism leads to the definition of the facilitation factor F:

$$F = \frac{-D_A \frac{dC_A}{dx} - D_{AC} \frac{dC_{AC}}{dx}}{-D_A \frac{dC_A}{dx}} \tag{21}$$

The facilitation factor is a direct measure of the effect of the reaction on the transport. It easy to see that, if no chemical reaction occurs or the facilitation factor is equal to one, we fall in the pure physical diffusion case.

The set of equations given above, Equations (11)–(13), together with the boundary Conditions (14) and (15), compose a system of nonlinear differential equations that cannot be exactly solved by analytical methods.

In some particular conditions, however, the problem can be easily solved and the analytical solution found. This is the case when two asymptotic conditions are satisfied, that are: (i) chemical equilibrium among the species throughout the membrane thickness (fast chemical reaction) and (ii) pure physical diffusion (negligible effect of chemical reaction). Olander [73] was the first who solved the facilitation by using the chemical equilibrium assumption. In this regime, the overall reaction rate is zero, and the equilibrium constant of the reaction, in terms of concentrations, is given by:

$$K_{eq} = \frac{C_{AC}}{C_A C_C} \tag{22}$$

In the mobile carrier case, if we consider the case of equal carrier and complex diffusivities, using Equation (22) for the chemical equilibrium, the facilitation factor assumes the form of Equation (23):

$$F = 1 + \frac{K_{eq} C_T}{\left(1 + K_{eq} C_A^0\right)\left(1 + K_{eq} C_A^L\right)} \frac{D_{AC}}{D_A} \tag{23}$$

The assumption of equal carrier and complex diffusivity was used in most of the models regarding mobile carrier systems reported in this review. This approximation could be explained by considering that the first models for mobile carrier systems were developed to explain how oxygen is transported in hemoglobin solutions. Hemoglobin (carrier) and oxyhemoglobin (reaction product) are very similar in molar volume so this approximation is reasonable for these biological systems [74]. Moreover, this approximation simplifies the integral constraint given in Equation (9) and leads to the Equation (19) for the carrier conservation.

For the fixed carrier case, in which the carrier has zero mobility, this expression, Equation (23), in general it is no longer true also in the equilibrium assumption, but holds in some special cases, as pointed out in the referring section.

For the mobile carrier system the above equation, or the equilibrium condition, represents a good approximation for some real cases of study, as mentioned by Ward [54], Olander [73], Schultz et al. [1,2], and Fatt and La Force [75]. Nevertheless, the assumption of chemical equilibrium leads to an expression for the complex and carrier concentration that cannot satisfy the boundary conditions. In fact, the reaction product is correlated to the solute concentration by:

$$C_{AC} = \frac{K_{eq} C_T C_A}{1 + K_{eq} C_A} \tag{24}$$

Differentiating (24) in the x-direction, we have:

$$\frac{dC_{AC}}{dx} = \frac{K_{eq} C_T}{\left(1 + K_{eq} C_A\right)^2} \frac{dC_A}{dx} \tag{25}$$

that does not vanish at the boundaries, violating the boundary conditions, since the solute flux is not zero there. Despite this inconsistency, the chemical equilibrium approximation represents a useful limiting case for the modeling of facilitated transport mechanism. Indeed such approximated approach, as shown in the next section, will be used, together with the pure diffusion one, as a starting point for several approximate analytical solution procedures.

2. Mobile Carrier Systems

Mobile carrier systems are the most used ones, especially in the carbon capture field. The liquid phase containing the carrier agent ensures high diffusivities of the reaction product, showing better flux performance in respect to the fixed carrier [76,77]. Moreover, in hybrid systems, the fixed sites facilitation effect is usually lower than the mobile one, so that the membranes performance, in terms of flux and selectivity, are more affected by the presence of mobile agents [78]. On the other hand, these systems can undergo chemical degradation, evaporation or wash out of carrier solutions, with consequent loss of separation ability [79]. To avoid these issues, the use of less volatile carrier solutions and/or the use of poly-electrolytes as supporting membranes, that increase the interaction between carrier and polymer phase, represent two methods to improve the membranes stability properties [80–82].

2.1. Models for Mobile Carrier Facilitated Transport Membranes

In this section a review of several analytical approximate solutions to the facilitated transport problem will be illustrated for the mobile carrier case, from the one of Friedlander and Keller (1965), to that of Morales and Cabrera (2002).

The mass balances, the boundary conditions, and the other mathematical considerations which will be considered hereafter are those already introduced in the mathematical background. For the cases in which they are different from the usual form, we will report the appropriate formulations.

2.1.1. Friedlander and Keller, 1965. Mass Transfer in Reacting System Near Equilibrium: Use of the Affinity Function

In 1965 Friedlander and Keller [68] proposed a solution for the facilitated transport of the solute across a liquid film using the affinity function for reacting system near chemical equilibrium. The reaction rate was assumed linear with activity, and the result was employed in the transport problem across a liquid reactive film. It should be noticed that the present model was derived assuming small variations of (low) concentration across the layer and, for this reason, it is not suitable for systems in which the concentration varies significantly across the thickness.

Near the chemical equilibrium condition, it is possible to linearize the reaction term of the mass flux by means of a Taylor series expansion around zero. By truncating the Taylor expansion at the first order term, the authors approximate the solution for the facilitated factor as:

$$F = \frac{1}{1 - \frac{\frac{1}{m_A}}{\sum_i \frac{1}{m_i}} E(\eta)} \tag{26}$$

In Equation (26), with m_i we indicate the mobility parameter of the i-th component.

$$\lambda = \left[B \left(\sum_i \frac{v_i^2}{m_i} \right) \right]^{-1/2} \tag{27}$$

$$\eta = \frac{L}{\lambda} \tag{28}$$

$$E(\eta) = 1 - \frac{2}{\eta} \left(\frac{cosh(\eta) - 1}{sinh(\eta)} \right) \tag{29}$$

η indicates a dimensionless length defined as the ratio between the membrane thickness and a characteristic length λ, given in Equation (27), which is a measure of the ratio between the reaction and the pure diffusive rate transport effects. In the definition of λ, v_i are the stoichiometric coefficients, B is the ratio among the forward reaction rate in the equilibrium condition and RT, where R is the universal gas constant and T the temperature. In dilute conditions, the mobility parameter can be easily expressed as:

$$m_i = \frac{C_i D_i}{RT} \tag{30}$$

and the solution for the facilitation factor is calculated in a straightforward manner.

From Equation (26), it is clear that the reaction always enhances the flux, even though the magnitude of the enhancement is function of the $E(\eta)$ parameter, Equation (29). Two limiting cases were identified: for $\eta > 100$, $E(\eta)$ is close to unity, and the assumption chemical equilibrium throughout the membrane is reasonable. On the other hand, values of η lower than 1 lead to the pure diffusional case in which the effect of the chemical reaction can be neglected. Furthermore, despite the approximation of near equilibrium condition is assumed throughout the system, the departure from chemical equilibrium is greater close to the boundaries, while it is negligible in the core of the membrane. Indeed, by varying the value of L in Equation (28), one can estimate the minimum thickness required to validate the equilibrium approximation in the actual system conditions.

2.1.2. Blumenthal and Katchalsky, 1969. The Effect of Carrier Association–Dissociation Rate on Membrane Permeation

In 1969 Blumenthal and Katchalsky [83] also provided an approximate analytical solution for systems close to chemical equilibrium. In their work they assumed equal diffusion coefficients for the carrier and the reaction product. In this model, however, the local concentration is allowed to slightly deviate from the equilibrium value, by a departure factor that is function of position. With this assumption, the reaction rate can be expressed as:

$$J_{react} = k_f (C_A{}^{eq} + \delta_A)(C_C{}^{eq} + \delta_C) - k_r (C_{AC}{}^{eq} + \delta_{AC}) \tag{31}$$

where $C_A{}^{eq}, C_C{}^{eq}, C_{AC}{}^{eq}$ are the concentration of the various species in the middle of the membrane, considered in equilibrium condition, and taken equal to their mean value across the thickness. Neglecting the second order departure term, Equation (31) becomes:

$$J_{react} = k_f C_A{}^{eq} \delta_C + k_f C_{AC}{}^{eq} \delta_A - k_r \delta_{AC} \tag{32}$$

Differentiating twice the Equation (32) and using Equations (11)–(13) to make the proper substitution, it is possible to write down the following equations:

$$\frac{d^2 J_{react}}{dx^2} = \frac{1}{\lambda^2} J_{react} \tag{33}$$

$$\lambda = \left[\frac{k_f C_A{}^{eq} + k_r}{D_{AC}} + \frac{k_r C_C{}^{eq}}{D_A} \right]^{-1/2} \tag{34}$$

The parameter λ, expressed by Equation (34), is in principle similar to the characteristic length introduced by Friedlander and Keller [68] and discussed above.

Finally, the authors found the following solution for the reaction rate as function of the distance from the upstream boundary:

$$J_{react} = J_{react}^0 \left[cosh\left(\frac{x}{\lambda}\right) - coth\left(\frac{L}{2\lambda}\right) sinh\left(\frac{x}{\lambda}\right) \right] \tag{35}$$

J_{react}^0 is the reaction rate evaluated at solute, carrier and complex concentrations at the upstream membrane side.

From Equation (35) it follows that:

$$J_{react} = -J_{react}^0 \quad x = L$$

$$J_{react} = 0 \quad x = L/2$$

In other words Equation (35) states that at the center of the slab the chemical equilibrium region is reached, as the net reaction rate is zero. Moreover, the higher the ratio $\frac{L}{2\lambda}$, the larger the size of the (near) equilibrium region in the membrane. As one can see in Figure 1a, below, by increasing the value of λ, the equilibrium region becomes smaller, finally being represented by a single point, located in the middle.

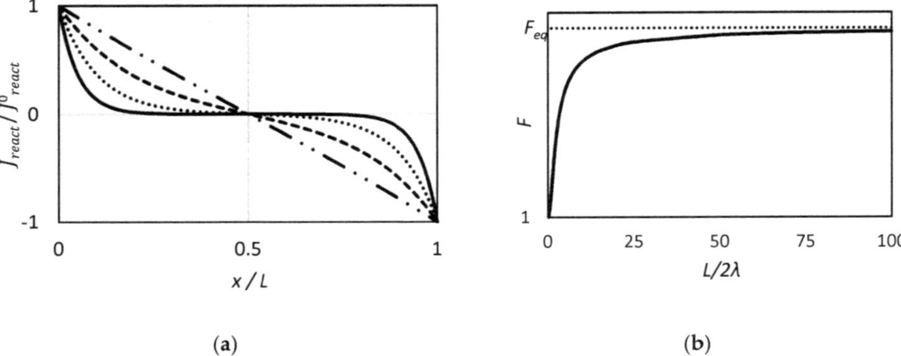

Figure 1. Effect of ratio between membrane thickness (L) and characteristic length (λ) on the transport in the model of Blumenthal and Katchalsky. (**a**) Size of equilibrium core in the membranes for different λ values (Continuous lines λ = 0.05, dotted lines λ = 0.1, broken lines λ = 0.2, broken dotted lines λ = 1); (**b**) Facilitation factor as function of λ. F_{eq} represents Equation (40).

The total flux of permeate is expressed as the sum of two contributions, the diffusion in free state and the diffusion in carrier-complex form, according to Equation (20). While usually the concentration of the free species is known at the boundaries, Equations (7) and (8), the reaction product concentration at the boundaries needs to be estimated to calculate the total flux. Blumenthal and Katchalsky derived the analytical solution to calculate the latter, and the solution for the total solute flux becomes:

$$J_A = D_A \frac{\Delta C_A}{L} + D_{AC} \frac{K_{eq}^{-1} C_T}{\left(1+\frac{1}{\chi}\right)\left(K_{eq}^{-1}+C_A{}^{eq}\right)\left(K_{eq}^{-1}+C_A{}^{eq}+\frac{D_{AC} K_{eq}^{-1} C_T}{D_A\left(K_{eq}^{-1}+C_A{}^{eq}\right)(1+\chi)}\right)} \frac{\Delta C_A}{L} \tag{36}$$

or, in terms of facilitation factor:

$$F = 1 + \frac{D_{AC}}{D_A} \frac{K_{eq}^{-1} C_T}{\left(1+\frac{1}{\chi}\right)\left(K_{eq}^{-1}+C_A{}^{eq}\right)\left(K_{eq}^{-1}+C_A{}^{eq}+\frac{D_{AC} K_{eq}^{-1} C_T}{D_A\left(K_{eq}^{-1}+C_A{}^{eq}\right)(1+\chi)}\right)} \tag{37}$$

where

$$\chi = \frac{L}{2\lambda} coth\left(\frac{L}{2\lambda}\right) - 1 \tag{38}$$

If the membrane thickness is such that $L/2\lambda \gg 1$, Equation (36) can be further simplified. Below we show the flux of solute in these cases for which the local equilibrium condition holds:

$$J_A = D_A \frac{\Delta C_A}{L} + \frac{D_{AC} K_{eq}^{-1} C_T}{\left(K_{eq}^{-1}+C_A{}^{eq}\right)^2} \frac{\Delta C_A}{L} \tag{39}$$

Thus, the facilitation factor is given by:

$$F = 1 + \frac{D_{AC}}{D_A} \frac{K_{eq}^{-1} C_T}{\left(K_{eq}^{-1}+C_A{}^{eq}\right)^2} \tag{40}$$

The differences among Equation (40) and the equilibrium solution reported in Equation (23) come from the approximations used by the authors; they considered a small solute concentration difference between the two boundaries, so that both the upstream and downstream side concentration can be approximated with the mean value in equilibrium condition.

For the opposite case, $L/2\lambda \ll 1$, the parameter χ tends to zero and the latter term in Equation (36) is negligible with respect to the first one, the pure physical diffusion contribution. Obviously in this case the facilitation factor is equal to one, by definition.

In conclusion, the approach of Blumenthal and Katchalsky, although being derived for the near equilibrium conditions, was able to predict the two different limiting case of pure physical diffusion and chemical equilibrium. The ratio between the membrane thickness and the characteristic length, λ, discriminates between the two asymptotic cases.

2.1.3. Goddard et al., 1969. On Membrane Diffusion with Near Chemical Equilibrium Reaction

A different approach was used by Goddard, Shultz and Basset [84] to solve the steady state facilitated transport in liquid membranes. They studied the problem of a fast reaction by using the so called "method of matched asymptotic expansion", that was already used in many boundary layers problem of transport. By using this method, the spatial domain is divided in three layers: two thin layers close to the boundaries and one thick layer in the middle of the slab. For each layer, a solution is found in terms of power expansion in ϵ terms, which is once again a measure of the importance of reaction terms with respect to the pure diffusion. It is assumed that the middle of the slab is at chemical equilibrium, while in the outer thin layers the reaction and diffusion rate have the same order of magnitude. Since the given solutions are different approximations of the same function, there is a region (overlapping region) in which they are simultaneously valid, and the approximate functions match.

In order to take into account the fast reaction regime, the differential equations, boundary conditions and stoichiometric constraint, Equations (9), (11), (12), (13), (14) and (15) are rewritten by the authors in dimensionless form as follows:

$$\epsilon^2 \frac{d^2\psi_i}{dy^2} - \mu_i\theta = 0 \quad i = A, C, AC \tag{41}$$

$$y = 0 \quad \psi_A = \psi_A^0 \quad \frac{d\psi_C}{dy} = \frac{d\psi_{AC}}{dy} = 0 \tag{42}$$

$$y = 1 \quad \psi_A = \psi_A^L \quad \frac{d\psi_C}{dy} = \frac{d\psi_{AC}}{dy} = 0 \tag{43}$$

$$\int_0^1 (\psi_B + \psi_C) dy = 1 \tag{44}$$

where:

$$\psi_i = \frac{C_i}{C_T} \tag{45}$$

$$\epsilon^2 = \frac{D_A C_T}{L^2 J^*_{react}} \tag{46}$$

$$\theta = \frac{J_{react}}{J^*_{react}} \tag{47}$$

$$J^*_{react} = k_f C_T^2 \tag{48}$$

$$\mu_i = \frac{D_A}{D_i} \tag{49}$$

and y indicates the dimensionless distance from the upstream boundary, x/L. Note that ϵ is qualitatively the ratio between the factor λ, given previously, in Equation (34) and the thickness L and that, as said above, it represents a measure of the ratio between the diffusion and the reaction rate.

For the equilibrium core, a power series expansion expresses the concentration profile as function of ϵ:

$$\psi_i = \tau_i^{(0)} + \epsilon\tau_i^{(1)}(y) + \epsilon^2\tau_i^{(2)}(y) + 0\left(\epsilon^3\right) \quad 0(\epsilon) < y < 1 - 0(\epsilon) \tag{50}$$

In Equation (50), the expansion coefficients $\tau_i{}^j$ are independent from ϵ. τ_i^0 reflects the dimensionless equilibrium core concentrations, while the terms multiplied by ϵ represent the departure from the zero reaction rate state, i.e., higher order correction terms of the expansion.

For the reactive zone near the boundaries ($0 \leq y < 0(\epsilon)$; $1 - 0(\epsilon) < y \leq 1$) the following expansions were used:

$$\psi_i = b_i^{(0)}(\overline{y}) + \epsilon b_i^{(1)}(\overline{y}) + \epsilon^2 b_i^{(2)}(\overline{y}) + 0\left(\epsilon^3\right) \quad 0 \leq y < 0(\epsilon) \tag{51}$$

$$\psi_i = c_i^{(0)}(\underline{y}) + \epsilon c_i^{(1)}(\underline{y}) + \epsilon^2 c_i^{(2)}(\underline{y}) + 0\left(\epsilon^3\right) \quad 1 - 0(\epsilon) < y \leq 1 \tag{52}$$

where:

$$\overline{y} = \frac{y}{\epsilon}$$

$$\underline{y} = \frac{(1-y)}{\epsilon}$$

The two "strained" coordinates $\overline{y}$ and $\underline{y}$ are used to define the limits of the boundary regions, starting from the upstream and downstream boundaries to the internal core respectively.

Even in this case, the core equilibrium length is strictly correlated to a parameter that relates the diffusion and reaction contribution to the transport, ϵ. The lower the value of ϵ, the thinner the boundary regions and, thus, the thicker the equilibrium core. According to the asymptotic matching technique [85], the internal core solution (50) has to match, for $y \to 0(\epsilon)$ and $y \to 1 - 0(\epsilon)$, with the solution for the reactive boundaries, Equations (51) and (52), for $\overline{y} \to \infty$ and $\underline{y} \to \infty$ respectively.

The final solution was presented, for the case in which $D_C = D_{AC}$, in terms of ratio between actual facilitation factor and facilitation factor in the equilibrium regime, F/F_{eq} truncated to the first term of the expansion:

$$\frac{F}{F_{eq}} = 1 - \left[G(\overline{Z}) + G(\overline{Z})\right]\sigma^{1/2}\epsilon \tag{53}$$

where F_{eq} is the one derived in the mathematical background and hereafter again shown:

$$F_{eq} = 1 + \frac{K_{eq}C_T}{\left(1 + K_{eq}C_A^0\right)\left(1 + K_{eq}C_A^L\right)}\frac{D_{AC}}{D_A}$$

The functions required to calculate the facilitated factors, $G(Z)$, as function of $\overline{Z}$, $\overline{Z}$ and σ are reported in the Supplementary Materials (SM).

This model showed a good agreement with the numerical calculations of Kutchai et al. [53] regarding the facilitated transport of oxygen in hemoglobin solutions for three different membrane thicknesses, but the model becomes less accurate as thickness decreases. Moreover, in some situations, when the facilitated equilibrium factor is too large, the present method, truncated at first order in ϵ, is unable to give quantitative agreement. In those cases, the authors suggest to use the present solution as a tool to estimate qualitatively the departure from equilibrium of the system of interest.

2.1.4. Kreuzer and Hoofd, 1970. Facilitated Diffusion of Oxygen in the Presence of Hemoglobin

In 1970, Kreuzer and Hoofd [86] derived a set of analytical equations which allow to calculate the total permeate flux and to describe the concentration profiles in facilitated transport systems by considering the same diffusivity for carrier and complex. In this work, the high solute concentration side is located at $x = L$.

The solution was still found considering a zone close to the boundaries where there is no chemical equilibrium. However, according to the authors, such region is reasonably small and the carrier concentration therein should not differ much from its boundary values, due to the surface impermeability Conditions (14) and (15). This approximation will be removed in a subsequent work by the same authors in 1972 [87], discussed in the following. In the middle of the slab, on the other hand, the near equilibrium condition was employed to derive the solution.

As mentioned, in the region close to the boundaries it is reasonable to make some assumptions. For x close to zero, i.e., in the low solute concentration side, the carrier concentration is approximately constant and very similar to its boundary value. On the opposite side, the same assumption holds. By using these considerations, the system of differential equations governing the problem can be solved for the concentration and, ultimately, for the total flux.

Equations (54) and (55) report the solutions derived for the solute and complex concentrations for $x \to 0$.

$$C_A = k_r\left(\frac{J_A x + B}{\alpha^2 D_A D_{AC}}\right) - D_{AC}E_1 e^{-\alpha x} \quad x \cong 0 \tag{54}$$

$$C_{AC} = k_f C_C^0\left(\frac{J_A x + B}{\alpha^2 D_A D_{AC}}\right) - D_A E_1 e^{-\alpha x} \quad x \cong 0 \tag{55}$$

where α, defined below, is a parameter similar in meaning to λ^{-1} in Equation (34).

$$\alpha = \sqrt{\frac{k_f C_C^0}{D_A} + \frac{k_r}{D_{AC}}} \tag{56}$$

Similarly, for the region close to other boundary, we have:

$$C_A = k_r\left(\frac{J_A x + B}{\beta^2 D_A D}\right) + D_{AC}H_1 e^{-\beta(L-x)} \quad x \cong L \tag{57}$$

$$C_{AC} = k_f C_C^0\left(\frac{J_A x + B}{\alpha^2 D_A D_{AC}}\right) - D_A H_1 e^{-\beta(L-x)} \quad x \cong L \tag{58}$$

where β is the same as α but calculated at the opposite limiting surface:

$$\beta = \sqrt{\frac{k_f C_C^L}{D_A} + \frac{k_r}{D_{AC}}} \tag{59}$$

The constants H_1, E_1, and B are reported in the supporting materials.

By differentiating twice Equations (54) and (57) it is easy to recognize an exponential decay (growth) of solute concentration with the distance from the boundary located at $x = L$ ($x = 0$). This suggests the presence of an equilibrium region inside the membrane where the concentration could be considered constant.

With the assumption of local equilibrium in the middle of the slab, the following equations allow to evaluate the boundaries concentration of the reaction product, C_{AC}^0 and C_{AC}^L:

$$C_{AC}^0 = \frac{k_f C_T (C_A^0 + E)}{k_r + k_f (C_A^0 + E)} + \frac{D_A}{D_{AC}}E \tag{60}$$

$$C_{AC}^L = \frac{k_f C_T (C_A^L - H)}{k_r + k_f (C_A^L - H)} - \frac{D_A}{D_{AC}}H \tag{61}$$

where $E = D_{AC}E_1$, $H = D_{AC}H_1$.

For the total flux, after some algebra, it is possible to derive the following formulas:

$$J_A L = \left(C_A^L - H - C_A^0 - E\right)\left\{D_A + \frac{D_{AC}k_f k_r C_T}{\left(k_r + k_f C_A^L - k_f H\right)\left(k_r + k_f C_A^0 + k_f E\right)}\right\} \tag{62}$$

$$J_A = \alpha D_A E \left\{ 1 + \frac{D_A \left(k_r + k_f C_A^0 + k_f E\right)^2}{D_{AC}\left(k_f k_r C_T\right)} \right\} \tag{63}$$

$$J_A = \beta D_A H \left\{ 1 + \frac{D_A \left(k_r + k_f C_A^L - k_f H\right)^2}{D_{AC}\left(k_f k_r C_T\right)} \right\} \tag{64}$$

The set of five Equations (62), (63) and (64) plus the two ones for E_1 and H_1 (see Supplementary Materials), in five unknowns, α, β, H, E, J_A, must be solved via a numerical procedure, in order to calculate the flux for fixed values of the other physical and chemical parameters. By the knowledge of these parameters, it is also possible to calculate the concentration profiles of the species inside the thickness. The equations needed to do that are reported in the original paper, to which we address the reader [86].

If we neglect the departure from the equilibrium condition, Equation (62) collapses on the equilibrium solution for the solute flux, given in Equation (23) in terms of facilitation factor, as one can expect.

The authors tested their calculations against experimental data coming from different authors [5,69,70].

In Figure 2 we report the results of the present model when compared with the experimental data of Wittemberg [5] relative to the system of oxygen–hemoglobin. The flux reported in the y-axis is relative to the reactive component of the total solute flux. As we can see, the model agrees very well with the measured data of oxygen in presence of hemoglobin solution impregnated on a Millipore membrane. Moreover, the present approach is able to recognize that the solute concentration gradient is equal for both the downstream and upstream boundary region, suggesting that the pure physical diffusion is a good approximation in this area. However, if tested again the oxygen–myoglobin system, the model is less satisfactory [87].

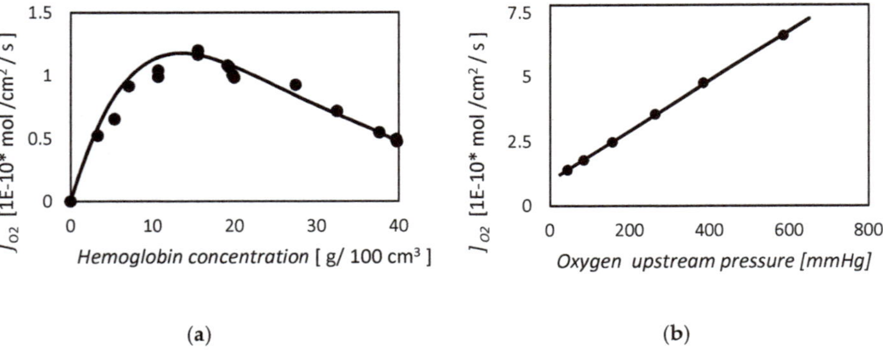

Figure 2. Model of Kreuzer and Hoofd versus experimental data by Wittemberg [5]. (**a**) Oxygen facilitated flux in function of hemoglobin concentration; (**b**) oxygen facilitated flux in function of the oxygen pressure at the feed side. In both figures the points are the experimental data, the line reports the model prediction. The flux reported is the reactive component of the total flux.

To improve the prediction performance, after two years, in 1972, Kreuzer and Hoofd published a new analytical approach that was an improvement of the method discussed above [86] by using better approximations in the boundary layers.

2.1.5. Kreuzer and Hoofd, 1972, Factors Influencing Facilitated Diffusion of Oxygen in the Presence of Hemoglobin and Myoglobin

In the work of 1972, the previous description of the membrane region was retained together with the equal carrier and complex diffusivities approximation [87]. However, the assumption of constant carrier concentrations near the boundaries was removed, leading to a better, albeit more complicated, analytical description of the process. As in the previous case, $x = L$ is the high solute concentration side. Here the boundaries concentrations were evaluated as the difference between the equilibrium contributions and a departure term, function of the distance from the membrane surfaces.

The solute concentration close to boundaries is thus given by:

$$C_A = C_A^{eq} - \Delta(x) \tag{65}$$

where the departure function, $\Delta(x)$, vanishes far from surfaces.

The carrier and complex concentrations are expressed as follows:

$$C_C = C_C^{eq} - \frac{D_A}{D_{AC}}\Delta(x) \tag{66}$$

$$C_{AC} = C_{AC}^{eq} + \frac{D_A}{D_{AC}}\Delta(x) \tag{67}$$

The equilibrium concentrations appearing in Equations (66) and (67) are those for carrier, solute and complex evaluated at the two boundaries, between square brackets in Equations (70) and (71), and not the equilibrium ones in the internal core, where the near equilibrium hypothesis is assumed. In other words, the equilibrium concentrations are position-dependent. Inserting Equations (65) to (67) in Equation (1) and using the above mentioned assumptions, the resulting second order differential equation can be solved near the boundaries in term of $\Delta(x)$.

The correction term, for $x \to 0$ and $x \to L$, was found to be equal to:

$$\Delta(x) = \frac{6\alpha_0^2 D_{AC} q_0 e^{-\alpha_0 x}}{k_f(1+q_0 e^{-\alpha_0 x})^2} \qquad x \to 0 \tag{68}$$

$$\Delta(x) = \frac{-6\alpha_1^2 D_{AC} q_1 e^{-\alpha_1(L-x)}}{k_f\left(1-q_1 e^{-\alpha_1(L-x)}\right)^2} \qquad x \to L \tag{69}$$

where:

$$\alpha_1 = \sqrt{\frac{k_f\left[C_C^{eq}\right]_{x=L}}{D_A} + \frac{k_r + k_f\left[C_A^{eq}\right]_{x=L}}{D_{AC}}} \tag{70}$$

$$\alpha_0 = \sqrt{\frac{k_f\left[C_C^{eq}\right]_{x=0}}{D_A} + \frac{k_r + k_f\left[C_A^{eq}\right]_{x=0}}{D_{AC}}} \tag{71}$$

The impermeability condition at the boundary for carrier and complex still holds, Equations (14) and (15).

For $x = 0$ and $x = L$, considering the previous results, we have the following expressions for the flux:

$$\frac{J_A}{D_A}\left[\frac{dC_{AC}^{eq}}{dC_A^{eq}}\right]_{x=0} = \frac{6\alpha_0^3 q_0(1-q_0)}{k_f(1+q_0)^3}\left(D_A + D_{AC}\left[\frac{dC_{AC}^{eq}}{dC_A^{eq}}\right]_{x=0}\right) \qquad x \to 0 \tag{72}$$

$$\frac{J_A}{D_A}\left[\frac{dC_{AC}^{eq}}{dC_A^{eq}}\right]_{x=L} = \frac{6\alpha_1^3 q_1(1+q_1)}{k_f(1-q_1)^3}\left(D_A + D_{AC}\left[\frac{dC_{AC}^{eq}}{dC_A^{eq}}\right]_{x=L}\right) \qquad x \to L \tag{73}$$

and for the solute concentration at two boundaries:

$$C_A^0 = \left[C_A^{eq}\right]_{x=0} - \frac{6\alpha_0^2 D q_0}{k_f(1+q_0)^2} \quad x \to 0 \tag{74}$$

$$C_A^L = \left[C_A^{eq}\right]_{x=L} - \frac{6\alpha_1^2 D q_1}{k_f(1-q_1)^2} \quad x \to L \tag{75}$$

The last six Equations (70)–(75) together with the exact solution for the total flux, Equation (76) (integral form of Equation (20)), compose the system of seven equations in the seven unknowns α_0, α_1, $q_0, q_1, \left[C_A^{eq}\right]_{x=0}, \left[C_A^{eq}\right]_{x=L}, J_A$ to be solved, again, with numerical techniques.

$$J_A = D_A \frac{(C_A^L - C_A^0)}{L} + D_{AC} \frac{(C_{AC}^L - C_{AC}^0)}{L} \tag{76}$$

Once the system is solved, the concentration profiles may be determined solving the following equations for C_{AC}^{eq} and C_A^{eq}:

$$J_A x + B = D_A C_A^{eq} + D_{AC} C_{AC}^{eq} \tag{77}$$

$$B = D_A \left[C_A^{eq}\right]_{x=0} + D_{AC} \left[C_{AC}^{eq}\right]_{x=0} \tag{78}$$

Note that in Equations (77) and (78) only equilibrium concentrations appear. As shown in the mathematical background section, they are reciprocally correlated, so that the last two equations can be solved obtaining the species profiles. The model was used by the authors to highlight the influence of characteristic parameters, such as complex diffusivity, backward reaction rate, oxygen back-pressure and actual membrane thickness, on the facilitation factor. However, despite the better agreement reached, in some cases the model predictions remain poor. As pointed out by Jain and Shultz [58], the solutions provided by Kreuzer and Hoofd, even if accurate in describing some systems, becomes progressively less accurate as the Damköhler number drops (as the 'thin' film case studied by Smith et al. [88], reported below in this work). In other words, this kind of approach in not suitable for systems in the low facilitation factor regime.

In 1979 [89] the same authors published a revised version of the above approach in which a correction function was used, similarly to Δ in Equations (68) and (69). The correction function was found by solving the problem on the whole volume, instead that on the boundaries only. This fact should improve the quality of the solution. However, the solution procedure has not been extensively explained by the authors, and for such reason it has not been reported in this review.

2.1.6. Yung and Probstein, 1973. Similarity Considerations in Facilitated Transport

A systematic study of the characteristic dimensionless parameters affecting the facilitated transport mechanism was performed by Yung and Probstein in 1973 [90]. For downstream solute concentration equal to zero, they identified three characteristic groups and calculated the numerical solution by varying their values. For some special cases, they also provided analytical solutions which represent the asymptotic behavior of those systems. The approximate solution, hereafter presented, is written in terms of the dimensionless parameters defined below:

$$\kappa = C_A^0 K_{eq} \tag{79}$$

$$\epsilon = \frac{D_{AC}}{k_r L^2} \tag{80}$$

$$\delta = \frac{D_A}{\left(k_f C_T\right) L^2} \tag{81}$$

And the so called 'disequilibrium parameter' η is defined as:

$$\eta = \frac{\left(\frac{C_C}{C_C^0} - 1\right)}{\kappa} \tag{82}$$

The first parameter presented, κ, in Equation (79), represents the dimensionless equilibrium constant normalized by solute concentration in the upstream side of the membrane. In Equation (80), ϵ is an inverse Damköhler number, i.e., a ratio between the characteristic time of the inverse reaction and the characteristic time of the complex diffusion. δ is an inverse Damköhler number too, but is referred to the diffusion of solute A and to the direct reaction rate, Equation (81).η can be seen as a measure of the departure from the chemical equilibrium condition. Indeed, given the hypothesis of equal diffusivities of carrier and complex and downstream solute concentration equal to zero, it is possible to show that the carrier concentration at the downstream boundary coincides with the total carrier one. Moreover, if the equilibrium condition holds throughout the thickness, the ratio between carrier concentration at the downstream and upstream boundaries is equal to:

$$\frac{C_C{}^L}{C_C^0} = 1 + \kappa$$

$$\frac{J_A L}{C_A^0 D_A} = 1 + \Delta\frac{\eta_L}{\alpha} = F \tag{83}$$

Hence, the parameter reported in Equation (82), evaluated at $x = L$ and indicated as η_L, ranging between one, in the equilibrium condition, and zero in the absence of reaction. Equation (83) reports the facilitation factor solution as function of these characteristic parameters, where Δ is the ratio between ϵ and δ, and α is the ratio between the total carrier concentration, C_T and the concentration at the upstream boundary, C_C^0.

If no chemical reaction occurs, η_L is equal to zero and the left side of Equation (83) is equal to 1, corresponding to pure physical diffusion regime.

The mass balances were rewritten using the parameters defined above, and the exact analytical solutions were found for some special cases, i.e., for $\Delta = 0$ or $\kappa = 0$.

The parameter vanishes if the equilibrium reaction constant is << 1 or if the complex diffusivity is negligible with respect to the pure solute one. If the first condition occurs, κ also vanishes. Then, if $\Delta = 0$, the solution for η is provided by the following equation:

$$\eta = \eta_1 \int_0^{\varepsilon} \frac{1}{\eta_1} \left[\int_0^{\varepsilon'} \beta \eta_1 d\varepsilon'' \right] d\varepsilon' \tag{84}$$

where η_1, the solution to the linear Airy equation, defined in [91].

While, if $\kappa = 0$, but $\Delta \neq 0$ (i.e., for low values of C_A^0), the analytical solution is in the form of equation:

$$\eta_L = \frac{\left[1 - 2\sqrt{\tau} tanh\left(\frac{1}{2\sqrt{\tau}}\right)\right]}{\left[1 + 2\sqrt{\tau} tanh\left(\frac{1}{2\sqrt{\tau}}\right)\right]} \tag{85}$$

where:

$$\tau = \frac{\epsilon}{(1 + \Delta)}$$

It has been shown that Equation (85) provides excellent agreement with the numerical solution calculated by the same authors, and also with results from Kutchai et al. [53] and Bzdil et al. [92], in a wide range of conditions. At last, also the analytical solution for the facilitation factor, F, is presented:

$$F = 1 + \frac{2\eta_L \Delta}{2 + \kappa(\eta_L + 1)} \tag{86}$$

and tested against the exact values reported by Kutchai et al. [53] for two different sets of κ and Δ by changing ϵ. As we can observe in Figure 3, the approximate solution is in agreement with the exact results. However, for higher values of κ and Δ, as reported in Figure 3b, the discrepancies of Equation (86) become more pronounced, reaching a maximum relative percentage deviation of about 35%, while for low values of these parameters, Figure 3a, the maximum error is 5% in the range investigated.

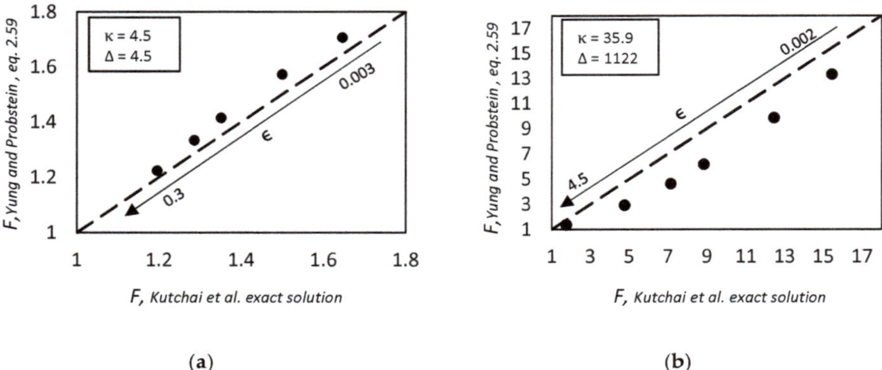

Figure 3. Comparison between facilitation factor F calculated numerically by Kutchai et al. [53] and the value calculated by Equation (86) proposed by Yung and Probstein [90] for different values of ϵ and κ and Δ fixed. (**a**) κ = 4.5, Δ = 4.5; (**b**) κ = 35.9, Δ = 1122. Dashed lines represent the parity condition.

2.1.7. Smith et al., 1973. An Analysis of Carrier Facilitated Transport

Starting from a magnitude analysis on the governing equations of facilitated transport, in 1973, Smith et al. [88] also provided an approximate analytical solution with the matching of asymptotic expansion technique.

They studied two different cases: the thin film and thick film ones, and found approximate solution for each case. By matching the asymptotic behavior of these, a global solution was found which is able to depict the whole range of thicknesses, approaching the two asymptotes of pure diffusion and chemical equilibrium, respectively. The assumption of equal carrier and complex diffusivities holds in this study.

(1) Thin Films

Based on the governing equations discussed in the mathematical background, one could ensure that the diffusional contribution is predominant over the reactive one in the case of vanishing membrane thickness. However, this condition is also met if the reaction rate is small, compared to the pure diffusion one, so the 'thin film' condition is not based on geometric properties only, but, more in general, to the Damköhler number.

For solute and carrier concentrations, and for the total solute flux, the solutions were given as a series expansion, truncated to the second term, containing a dimensionless parameter ϵ, that combines the Damköhler number and equilibrium constant. By using the definition of facilitation factor, the authors derived the following expression:

$$F = \frac{J_A L}{D_A \Delta C_A} = 1 + \epsilon\gamma_1 + \epsilon^2\gamma_2 \tag{87}$$

where:

$$\epsilon = \frac{1}{12}\frac{k_f C_T}{k_f \overline{C}_A + k_r}\left(\frac{k_r L^2}{D_A}\right) \tag{88}$$

$$\Delta C_A = C_A^0 - C_A^L \tag{89}$$

$$\overline{C}_A = \frac{C_A^0 + C_A^L}{2} \tag{90}$$

The terms appearing in the expansions, γ_1, γ_2, are function of the dimensionless length and are reported in the Supplementary Materials.

By knowledge of the physical and chemical parameters, such as the direct and reverse reaction rate, the diffusivities and the total carrier concentration, it is possible to identify the cases in which a first order approximation is accurate for the flux calculation. Indeed, from Equation (87), we see that if $\epsilon\gamma_1 \gg \epsilon^2\gamma_2$ the equation reduces to:

$$\frac{J_A L}{D_A \Delta C_A} = F = 1 + \epsilon\gamma_1 \tag{91}$$

And it is possible to consider Equation (91), instead of Equation (87), for thicknesses below a threshold value given by:

$$L \ll \sqrt{12\frac{\left(k_f \overline{C}_A + k_r\right) D_A}{k_f k_r C_T \gamma_2}} \tag{92}$$

(2) Thick Films

When considering thick films, the solution for the concentration profiles is found as a departure from the equilibrium values. For both the carrier and the solute, the equilibrium concentrations were considered not fixed, but rather function of position, indicated here as $\hat{C}_A$, $\hat{C}_C$. These are to be intended, then, as local equilibrium concentrations. Moreover, even if the local chemical equilibrium could be a reasonable assumption for thick membranes, it does not satisfy all boundary conditions, as shown in the mathematical background section. Therefore, for the departure functions the authors considered that near the surfaces, a boundary layer region is still present where the diffusional effect plays a certain role according to Goddard et al. [84], and Kreuzer and Hoofd [86,87,89].

The concentration profiles are defined as:

$$C_A = \hat{C}_A + \delta C_A(x) \tag{93}$$

$$C_C = \hat{C}_C + \delta C_C(x) \tag{94}$$

where the departure functions $\delta C_A(x)$ and $\delta C_C(x)$ are reciprocally correlated by Equation (95):

$$\delta C_C(x) = \frac{D_A}{D_{AC}}\delta C_A(x) \tag{95}$$

Near the two boundaries, as already mentioned, the solution is quite different from the one in the core of the system. Here, the perturbation function δC_A was found to be:

$$\delta C_A = \left[C_A^0 - \hat{C}_A(0)\right] e^{-\frac{x}{\lambda_0}} \quad \frac{x}{L} \to 0 \tag{96}$$

$$\delta C_A = \left[C_A^L - \hat{C}_A(L)\right] e^{-\frac{(L-x)}{\lambda_L}} \quad \frac{x}{L} \to 1 \tag{97}$$

where λ is a length scale, defined below, and the subscripts indicate the position where the function is evaluated.

$$\lambda = \left[\frac{k_f \hat{C}_A(x) + k_r}{D_{AC}} + \frac{k_f k_r C_T}{D_C\left(k_f \hat{C}_A(x) + k_r\right)}\right]^{-1/2} \tag{98}$$

λ_0 and λ_L are representative of the thickness of the two existing boundary layers near the membrane surfaces.

The system to be solved via trial and error procedure for the solute equilibrium concentrations at the boundaries, $\hat{C}_A(0)$, $\hat{C}_A(L)$, and for the total flux consists of three equations in three unknowns and is reported hereafter. For the sake of brevity, we address to the original paper for the whole mathematical derivation.

The boundary layers solutions are then used to obtain a solution for the internal zone of the membrane. In fact, the departure function δC_A above was employed to obtain a perturbation term in the core.

For the internal zone the authors derived a system of equations: hereafter we reported only the one for the flux, dimensionless. The other ones are reported in the Supplementary Materials.

$$J_A^* = \frac{F}{F^{eq}} = \{1 - N_4 G_0(0) - M_4 G_L(0)\}\left\{\frac{1 + \frac{F_{eq}-1}{[1-N_3 G_0(0)][1+M_3 G_L(0)]}}{F_{eq}}\right\} \tag{99}$$

In the above equation: $\widetilde{\lambda_L}$ and $\widetilde{\lambda_0}$ are the parameters defined by Equation (98), calculated for $\hat{C}_A(L) = C_A^L$ and $\hat{C}_A(0) = C_A^0$, F^{eq} is the facilitation factor in equilibrium conditions given in Equation (23) and G_0, G_L as well as N_3, N_4, M_3, M_4, are given in the supplementary section. Also in this case, due to the nonlinearity of the system, a numerical procedure is required to calculate the dimensionless flux.

Finally, for the core, the departure function is found:

$$\delta C_A(x)^{core} = 2\frac{k_f^2 k_r C_T J_A^2}{D_A^3 {D_{AC}}^2} \tag{100}$$

The solution method described here provides successful agreement with the numerical calculations of Kutchai et al. [53] in a large variety of cases and for both thin and thick layers.

In Figure 4 we report the model prediction of this approach compared with numerical solution of Kutchai et al. [53] for oxygen transport in hemoglobin solution and, also, with the Kreuzer and Hoofd [87] results. The present method can describe well the solute flux over a wide range of thicknesses, unlike the model Kreuzer and Hoofd [86,87] that does not describe thin films very well.

Despite the higher predictive ability, the method contains many equations and the calculation procedure could be not easy for the thick film case. However, it is worth mentioning that, if the membrane is very thin or very thick, the solution is easily obtained, using Equation (91) or the linearized form of Equation (99) (reported in the Supplementary Materials), respectively.

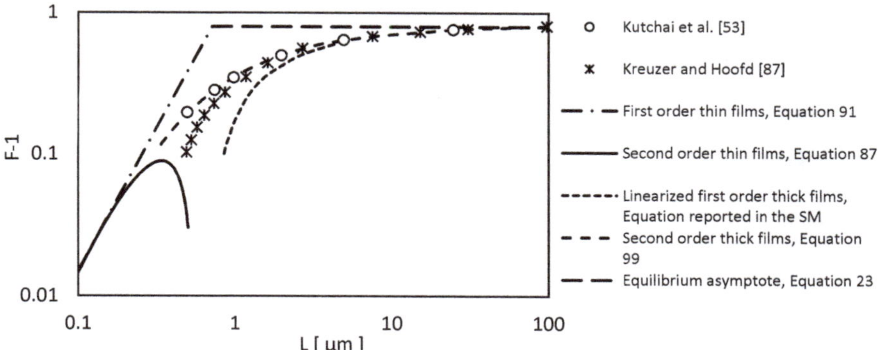

Figure 4. Reactive contribution to facilitation factor as function of membrane thickness.

2.1.8. Smith, Quinn, 1979. The Prediction of Facilitation Factors for Reaction-Augmented Membrane Transport

A straightforward method to calculate the facilitation factor was given in 1979 by Smith and Quinn [93]. Their equation was originally introduced by Donaldson and Quinn [94] as an exact solution to the facilitation transport of CO_2 in enzymatically bounded polymer in particular conditions. That work has not been discussed in this review, because the reaction scheme used deviates from the one taken as case of study here.

Considering the reaction scheme in Equation (3), the authors, by linearization of the reaction rate, showed that it is possible to obtain an analytical approximate solution that covers the entire regime of kinetic conditions, from fast to slow reaction. For zero downstream solute concentration, with the assumption of equal carrier and complex diffusivities, and considering a constant carrier concentration in the membrane, the reaction rate may be linearized and the solution for the facilitation factor is simply given by:

$$F = \frac{1+X}{1+\frac{X}{\theta}tanh\theta} \tag{101}$$

with:

$$X = \frac{D_{AC}K_{eq}C_C}{D_A} \tag{102}$$

$$\theta = \frac{1}{2}\sqrt{\frac{k_f C_C L^2}{D_A}\left(\frac{1+X}{X}\right)} \tag{103}$$

$$C_C = \frac{C_T}{1+K_{eq}C_A} \tag{104}$$

The Equation (101) can describe quite well the entire range of conditions, from near pure diffusion to near chemical equilibrium condition when an excess of carrier is present. Moreover, for slow chemical reaction rates, the solution may be approximated with the following one:

$$F \cong 1 + \frac{1}{12}\frac{k_f C_C L^2}{D_A} \tag{105}$$

Such equation is analogous to the thin film asymptotic solution described in Smith et al. [88]. Indeed, it is easy to recognize from Equations (91) and (105) that, if one assumes the carrier concentration constant and equal to C_T, and the chemical reaction is very slow, these two equations give exactly the same result.

On the other hand, the opposite case of fast reaction, or chemical equilibrium condition, leads to the following equation for the facilitation factor:

$$F \cong 1 + \frac{K_{eq} C_C D_{AC}}{D_A} \tag{106}$$

which is the proper equilibrium asymptote for the facilitation factor. By substitution of Equation (104) in (106), for zero downstream side solute concentration, Equation (23) is found.

When we consider the near equilibrium situation, or fast chemical reaction, assuming that the carrier is constant and equal to its equilibrium value, and with the solute concentration set equal to its upstream value, Equation (101) can predict facilitated transport for systems in which the downstream side concentration of solute is negligible. Another case in which the present solution can be used is when a large excess of carrier is employed, as in such case the concentration of carrier is approximatively equal to the total concentration denoted by C_T. Note that $C_C = C_T$ requires that $K_{eq} C_A^0 \ll 1$, condition that can be achieved even if the forward chemical reaction is very slow.

The discussed solution is mathematically simple, but fails for large constant equilibrium values. Indeed, it is widely recognized that, for high values of K_{eq} the facilitation effect disappears and F tends to unity due to a saturation of the carrier agent (Schultz et al. [1]). This phenomenon implies that there is a value of the equilibrium constant that maximizes the facilitation, as explained in detail by Kemena et al. [7]. On the other hand, in the solution of Smith and Quinn presented here, the facilitation factor reaches a plateau by increasing the equilibrium constant value, as shown in Figure 5.

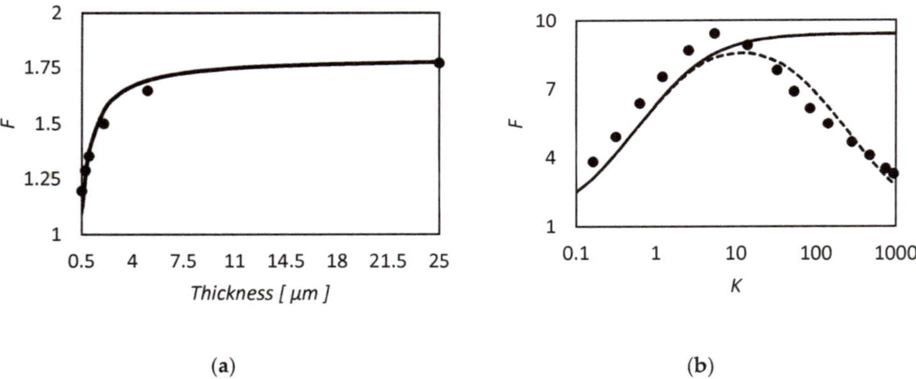

(a) (b)

Figure 5. Model prediction of facilitated factor F. (**a**) Numerical results of Kutchai et al. [53] and Equation (101) calculations; (**b**) Asymptotic behavior of Equation (101) as function of dimensionless equilibrium constant ($K = K_{eq}\ C_A{}^0$). Symbols are numerical results from Kemena et al. [7], filled line is Equation (101), dotted line is the improved model of Jeema and Noble, Equation (126).

However in the proper region of validity, the solution of Smith and Quinn represents a powerful tool to provide a prediction of F. In Figure 5a we reported as example the prediction of Equation (101) versus the numerical calculation of Kutchai et al. [53] as a function of membrane thickness. The agreement between the two calculations are excellent. On the other hand, Figure 5b shows that the model fails for higher values of equilibrium constant (filled points are the numerical results of Kemena et al. [7]).

2.1.9. Noble et al., 1986. Effect of Mass Transfer Resistance on Facilitated Transport

The external mass transfer resistance was investigated by Noble et al. in 1986 [72] with the aim to avoid the predicted asymptotic behavior of Equation (101). As in the work of Smith and Quinn [93], the case of excess carrier was considered to propose an analytical approximate method.

The solution was presented for the facilitated factor in dimensionless form, using two Sherwood numbers to account for the external resistance to mass transfer. The authors assumed equal diffusion coefficients for carrier and carrier complex and zero downstream solute bulk concentration. It has been found that, for large Sherwood numbers, the results collapse to those proposed by Smith and Quinn [93] and analyzed before.

In the present work, the boundary conditions for the solute are different from those used in previous ones. In particular, to take into account the mass transfer resistance, the boundary conditions are written as:

$$\bar{k}_0\left(\frac{C_A{}^0}{m} - C_{A,0}\right) = -D_A\frac{dC_A}{dx} \quad x = 0 \tag{107}$$

$$\bar{k}_L(C_{A,L} - 0) = -D_A\frac{dC_A}{dx} \quad x = L \tag{108}$$

while for the carrier and the carrier-complex the impermeability surfaces condition still holds.

$\bar{k}_0, \bar{k}_L$ are the mass transfer coefficients in upstream and downstream side respectively, m is the partition coefficient (ratio between external phase concentration and membrane phase concentration), $C_A{}^0$ is the external phase solute concentration in the upstream side, and $C_{A,0}$ and $C_{A,L}$ are the membrane phase concentrations at the upstream and downstream sides respectively.

The facilitation factor is then expressed as:

$$F = \frac{\left(1 + \frac{\alpha K}{1+K}\right)\left(1 + \frac{1}{Sh_0} + \frac{1}{Sh_L}\right)}{1 + \frac{\alpha K}{1+K}\frac{tanh(\lambda)}{\lambda} + \left(1 + \frac{\alpha K}{1+K}\right)\left(\frac{1}{Sh_0} + \frac{1}{Sh_L}\right)} \tag{109}$$

where the dimensionless parameters, defined below, according to Kemena et al. [7] and Folkner and Noble [95] represent: a mobility ratio between complex and solute (α), the dimensionless equilibrium constant (K), a parameter that contain the previous three (λ), the Sherwood numbers for the upstream and downstream side (Sh_0 and Sh_L).

$$\alpha = \frac{D_{AC}C_T m}{D_A C_A{}^0} \tag{110}$$

$$K = K_{eq}\frac{C_A{}^0}{m} \tag{111}$$

$$\epsilon = \frac{D_{AC}}{k_r L^2} \tag{112}$$

$$\lambda = \frac{1}{2}\sqrt{\left(\frac{1 + (\alpha + 1)K}{\epsilon(1 + K)}\right)} \tag{113}$$

$$Sh = \frac{\bar{k}L}{D_A} \tag{114}$$

It is easy to see that, considering the carrier concentration expressed by Equation (115), the solution given here is the same proposed by Smith and Quinn [93], Equation (101), if the Sherwood numbers tends to infinity, and hereafter reported in Equation (116) by using the parameters defined above in Equations (110)–(113).

$$C_C = \frac{C_T}{1 + K_{eq}\frac{C_A{}^0}{m}} \tag{115}$$

$$F = \frac{\left(1 + \frac{\alpha K}{1+K}\right)}{1 + \frac{\alpha K}{1+K}\frac{tanh(\lambda)}{\lambda}} \tag{116}$$

It is equally easy to observe that the lower the Sherwood numbers, the lower the facilitation effect. The main drawback of the external mass transfer resistance are a lower solute concentration in the membrane phase, in the upstream side, and, at the opposite boundary, a higher level of solute. These two effects lead to a reduced driving force for the transport. The present approach was compared with the numerical results calculated by Kemena et al. [7] and the agreement between the approximate solution and numerical calculation was quite good in the range investigated.

2.1.10. Basaran et al., 1989. Facilitated Transport with Unequal Carrier and Complex Diffusivities

A systematic investigation of the effect of diffusivity difference between carrier and complex on the facilitation factor was performed by Basaran et al. in 1989 [8]. Two different asymptotic solutions were found for the two limiting cases of large and small Damköhler number. Those were compared with numerical calculations made by the same authors using the Galerkin finite elements method. It was found that, in general, the ratio between the carrier-complex and the pure carrier diffusivity enhances the facilitated transport effect. They extended the work of Kemena et al. [7] by finding the optimal equilibrium values that maximizes the facilitation also for the case of $D_C/D_{AC} \neq 1$.

(1) Small Damköhler Number

For small Damköhler number, the solution for the concentration profiles and facilitation factor was found following the regular perturbation analysis, similarly to what done by Smith et al. [88].

$$P = \frac{k_r L^2}{D_A} \tag{117}$$

The concentration profiles for the solute and the carrier, Equations (118) and (119), are expressed as power expansions in the Damköhler number, defined in Equation (117) and here indicated as P.

$$\overline{C}_A = \sum_{n=0}^{\infty} P^n c_A^{(n)} \tag{118}$$

$$\overline{C}_C = \sum_{n=0}^{\infty} P^n c_C^{(n)} \tag{119}$$

$\overline{C}_A$ and $\overline{C}_C$ are dimensionless concentration, referred to the upstream side, and $c_A^{(n)}$ and $c_C^{(n)}$ are the expansion coefficients. For the calculation of the facilitation factor, the expansion was truncated at the first order for $\overline{C}_C$ while, for $\overline{C}_A$ the last term considered was the second order.

With this approach, the resulting facilitation factor was presented as:

$$F = 1 + \frac{1}{12}PKc_C^{(0)} - \frac{1}{720}\left(PKc_C^{(0)}\right)^2\left\{1 + 6\frac{ST}{PKc_C^{(0)}T}\left[R + K\left(1 + \frac{\overline{C}_A^L}{2}\right)\right] - \frac{1}{12}\frac{SK}{PT}\left(\overline{C}_A^L - 1\right)^2\right\} \tag{120}$$

where the zero-th order expansion coefficients appearing, $c_C^{(0)}$, is reported in the equation below:

$$c_C^{(0)} = \frac{T}{1 + K\frac{(1+\overline{C}_A^L)}{2}} \tag{121}$$

K is the dimensionless equilibrium constant, $K = K_{eq}C_A^0$, and $T = \frac{C_T}{C_A^0}$.

The other parameters appearing in Equation (120), R and S, are directly correlated to the ratio between the diffusivities of the species by the following relationships:

$$R = \frac{D_C}{D_{AC}} \tag{122}$$

$$S = P\frac{D_A}{D_{AC}} \tag{123}$$

From Equation (120), one can see that, keeping all the other dimensionless groups fixed, the facilitation factor decreases with increasing R.

(2) Large Damköhler Number

For the large Damköhler number case, the authors found the solution for the flux via a singular perturbation analysis.

The effect of the diffusivities ratio was investigated by linearization of the facilitation factor around $R = 1$, truncated at first order, for the case of the reaction rate is higher than the diffusion one:

$$F = 1 + \Im_1\left(1 - \left(\sigma\frac{\eta_0\eta_1\omega + \rho\sigma\delta}{\eta_C\eta_1 + \sigma\delta} + \frac{\gamma_0}{\eta_0} + \frac{\gamma_1}{\eta_1} - 2\right)\left(\frac{1-R}{R}\right)\right) \tag{124}$$

In the equation above $\Im_1$, reported below in Equation (125), is the facilitation factor in the case of $R = 1$. All the other functions in Equation (124) are reported in the Supplementary Materials.

$$\Im_1 = \frac{\sigma\delta}{\gamma_0\gamma_1}\left[-\frac{1}{2p_0}\left(p_1 + \left(p_1^2 - 4p_0p_2\right)^{1/2}\right)\right] \tag{125}$$

It was shown that:

$$\sigma\frac{\eta_0\eta_1\omega + \rho\sigma\delta}{\eta_0\eta_1 + \sigma\delta} + \frac{\gamma_0}{\eta_0} + \frac{\gamma_1}{\eta_1} - 2 \leq 0$$

Thus, also in this case, the same dependence of the facilitated factor on the ratio R is shown, as for the case of low Damköhler number.

Comparison with numerical calculations done by the authors, shows a qualitatively good agreement between the asymptotic expansion and the numerical solution in both the Damköhler's number. However, for the near diffusion limit, the approximated solution cannot describe quantitatively the results.

Figure 6 shows the two opposite limits. Near the chemical equilibrium condition, Figure 6b, the approximated solution is able to predict the diffusivity ratio effect on the facilitation qualitatively and, moreover, by increasing the Damköhler's numbers P from 1000 to 10,000, the agreement becomes quantitative too. On the contrary, Figure 6a, near the pure diffusion condition the approximated method, even if qualitatively right, can give facilitation factor lower than one, that is impossible by definition (Equation (21)). However, from these results one can conclude that if the system is near the pure physical diffusion, the ratio between carrier and complex diffusivity does not play a significant role in the facilitation.

The dependence of the optimum dimensionless equilibrium constant as function of the Damköhler number was also investigated. As a confirmation of the results already explained, it was shown that for small Damköhler number, the optimum K is not affected by the R ratio while a pronounced effect appears when the system approaches the fast reaction regime.

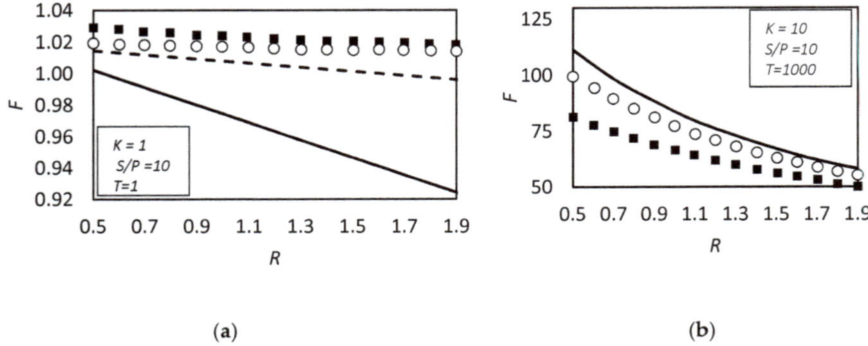

(a) (b)

Figure 6. Diffusivity ratio influence on facilitation. (**a**) Small Damköhler number. Black squares and white circles are numerical results for $P = 1$ and 0.5 respectively, full and broken lines are Equation (120) for $P = 1$ and 0.5 respectively; (**b**) Large Damköhler number. Black squares and white circles are numerical results for $P = 1000$ and 10,000 respectively. Full line is Equation (124). In both the case small but nonzero downstream concentration has been considered ($\overline{C}_A^L = 0.01$).

2.1.11. Jemaa and Noble, 1992. Improved Analytical Prediction of Facilitation Factors in Facilitated Transport

As already mentioned, the approximate model provided by Smith and Quinn [93] can describe quite accurately the facilitation effect, provided that the dimensionless equilibrium constant is not too high but it is not able to describe the gaussian shape of the facilitation factor as function of the equilibrium constant (Figure 5b). Indeed, for value of K higher than the optimal one, the facilitation factor predicted by Smith and Quinn approaches asymptotically a constant value.

To overcome this limitation, Jemaa and Noble [96] proposed an improvement of the approximate solution to take into account a small, but non zero, downstream solute concentration. However, that concentration does not correspond to the actual value, but is rather an adjustable parameter that can be varied to match the analytical approximation to the numerical solution, provided by Kemena et al. [7]. In a graphical way, they reported the adjustable concentration parameter that minimizes the error between the analytical and numerical solution for different operational conditions. A set of optimal dimensionless concentration parameters were calculated for different operative conditions.

The approximate facilitation factor, in the assumption of excess carrier and chemical equilibrium throughout the thickness, and equal diffusivity of carrier and carrier-complex, as in their previous study [72] and in agreement with Smith and Quinn [93], for this case, is:

$$F = \frac{1 + \frac{\alpha K}{(1+K)(1+\theta K)}}{1 + \frac{\alpha K}{(1+K)(1+\theta K)} \frac{tanh(\lambda)}{\lambda}} \tag{126}$$

where the nonzero downstream dimensionless concentration is represented by θ and defined as:

$$\theta = \frac{C_A^L}{C_A^0}$$

λ, the composed parameter already defined in Equation (113), is now expressed as:

$$\lambda = \frac{1}{2}\sqrt{\left(\frac{1 + (\alpha + 1)K}{\epsilon(1 + K)(1 + \theta K)}\right)} \tag{127}$$

while the other dimensionless parameters α, K and ϵ are still given by the Equations (110), (111) and (112).

Note that for θ equal to zero, Equation (126) is the same as the one reported by Smith and Quinn [93], Equation (116). In Figure 5b above we reported a comparison of the present approach with the exact results of Kemena et al. [7] and with the model results of Smith and Quinn [93] for the parameter set α = 50, ϵ = 0.1 (to use these, refer to Equation (116) instead of (101)). As shown, the present approach has a better predictive ability than the one of Smith and Quinn. Moreover, for non zero actual downstream concentration, Equation (129) could provide a good estimation for the facilitation factor, without any adjustment on the numerical results.

This method could be seen as a simple way to adjust the analytical solution without mathematical complications and provides a good prediction, even if the use of a fitting parameter makes it not fully predictive.

2.1.12. Teramoto, 1994. Approximate Solution of Facilitation Factor in Facilitated Transport

Teramoto in 1994 [74], proposed an analytical approach to describe the facilitated transport for the entire range of Damköhler number, without assuming equal diffusivities or zero solute concentration at downstream side. Starting from the two membrane surfaces, two different facilitation factors were written, the first one from the upstream boundary and the second one from the downstream side. The final solution was found via a trial and error procedure that converges when the two factors match.

For the first case, upstream boundary, it is assumed that the influx could be adequately represented by considering the carrier concentration constant and equal to its value at boundary, an approach very similar to that already used by of Kreuzer and Hoofd [86,87] and discussed above in this work. In analogy, for the second case, the outflux could be express considering the carrier concentration constant and equal to the downstream side value.

For the solute influx, the following boundary conditions were used:

$$C_A = C_A^0 \quad C_{AC} = C_{AC}^0 \quad \frac{dC_{AC}}{dx} = 0 \quad x = 0 \tag{128}$$

$$C_A = C_A^L \qquad x = L \tag{129}$$

and for the outflux:

$$C_A = C_A^0 \qquad x = 0 \tag{130}$$

$$C_A = C_A^L \quad C_{AC} = C_{AC}^L \quad \frac{dC_{AC}}{dx} = 0 \quad x = L \tag{131}$$

The mathematical treatment of Teramoto lead to the violation of some boundary conditions, as it can be seen by comparing Equations (128) to (131) to the ones reported in the mathematical background Equations (7) and (8). According to van Krevelen and Hoftijzer [97], since the solute influx could be more influenced from the carrier concentration at feed side rather than at the downstream, these violations should not dramatically influence the solution for the solute flux through the upstream surface. The same considerations hold for the downstream side.

The set of parameters defined to solve the facilitation problem is:

$$\overline{C}_A = \frac{C_A}{C_A^0} \tag{132}$$

$$\overline{C}_{AC} = \frac{C_{AC}}{C_T} \tag{133}$$

$$\overline{C}_C = \frac{C_C}{C_T} \tag{134}$$

$$q = \frac{D_{AC}C_T}{D_A C_A^0} \tag{135}$$

$$r = \frac{D_{AC}}{Dc} \tag{136}$$

$$\delta = L\sqrt{\left(\frac{k_f C_T}{D_A}\right)} \tag{137}$$

$$\gamma_0 = \delta\sqrt{\overline{C}_C^0 + \frac{1}{rqK}} \tag{138}$$

and K is the dimensionless equilibrium constant, $K = K_{eq}C_A^0$.

From the upstream point of view, the facilitation factor was found to be equal to:

$$F_0 = \gamma_0 \frac{\left[1 + \frac{1}{rqK\overline{C}_C^0}\right](1 - \overline{C}_A^L) + (cosh(\gamma_0) - 1)\left[1 - \frac{\overline{C}_{AC}^0}{K\overline{C}_C^0}\right]}{sinh(\gamma_0) + \frac{\gamma_0}{rqK\overline{C}_C^0}} \tag{139}$$

In a similar way, for the opposite boundary, $x = L$ the facilitation factor is:

$$F_L = \gamma_L \frac{\left[1 + \frac{1}{rqK\overline{C}_C^L}\right](1 - \overline{C}_A^L) + (cosh(\gamma_L) - 1)\left[\frac{\overline{C}_{AC}^L}{K\overline{C}_C^L} - \overline{C}_A^L\right]}{sinh(\gamma_L) + \frac{\gamma_0}{rqK\overline{C}_C^L}} \tag{140}$$

where:

$$\gamma_L = \delta\sqrt{\overline{C}_C^L + \frac{1}{rqK}} \tag{141}$$

In steady state condition, Equation (139) must be equal to Equation (140). A set of Equations (142), (143), and (144), can be used to express the relationship between the carrier and complex concentrations at the boundaries:

$$\overline{C}_C^L = \overline{C}_C^0 + \frac{(F - 1 + \overline{C}_A^L)}{q} \tag{142}$$

$$\overline{C}_{AC}^L = \overline{C}_{AC}^0 - \frac{(F - 1 + \overline{C}_A^L)}{rq} \tag{143}$$

$$\frac{(\overline{C}_{AC}^L + \overline{C}_{AC}^0) + (\overline{C}_C^L + \overline{C}_C^0)}{2} = 1 \tag{144}$$

Those equations were found by considering sigmoidal and reverse sigmoidal shape of complex and carrier concentrations respectively. The system can be solved to obtain the five unknowns $\overline{C}_C^L$, $\overline{C}_{AC}^L$, $\overline{C}_C^0$, $\overline{C}_{AC}^0$ and F via numerical procedure for a given set of values for q, r, K, δ, and $\overline{C}_A^L$.

Note that the sigmoidal approximation used to express the carrier continuity, as an alternative to the integral constraint given by Equation (9), leads to an algebraic equation, that is easier to use in order to solve the system.

The facilitated factor calculated with the present approximate method was found in excellent agreement, over a wide range of cases studied, with the numerical calculations made by Basaran et al. [8], Kemena et al. [7], and Jain and Schultz [58].

However, discrepancies arise between the calculated solute concentration profiles (not reported in this work) when $\delta \gg 1$. In this case, the solutions obtained starting from the different boundaries give different results, as shown in Figure 7a. To improve the concentration profiles accuracy in this case, an equilibrium core in the middle part of the membrane was introduced, and the following equations were given for the middle part of the slab:

$$(1-\overline{C}_A)+q\left(\overline{C}_C-\overline{C}_C^0\right)=Fy \tag{145}$$

$$(1-\overline{C}_A)+rq\left(\overline{C}_{AC}-\overline{C}_{AC}^0\right)=Fy \tag{146}$$

Together with the equilibrium condition, and using this approximated method to evaluate $\overline{C}_C^L, \overline{C}_{AC}^L, \overline{C}_C^0, \overline{C}_{AC}^0$ and F Equations (145) and (146) can be used to calculate the concentration profiles. With the last improvement, it is possible describe the solute concentration profile over the entire thickness. On the other hand, if δ is not too large, the concentration profile is well depicted without the equilibrium core correction, Figure 7b.

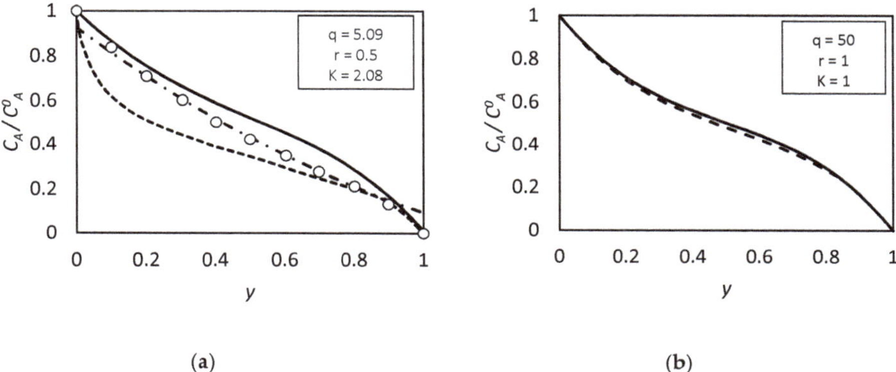

Figure 7. Solute concentration profiles. (**a**) δ = 22.5, circles are numerical solution from Jain and Shultz [58], solid line is the upstream solution, broken line is the downstream solution, dash dotted line is the equilibrium core solution; (**b**) δ = 5, solid line is the upstream solution, broken line is the downstream solution. In both cases the downstream solute concentration is zero.

2.1.13. Morales-Cabrera et al., 2002. Approximate Method for the Solution of Facilitated Transport Problems in Liquid Membranes

An improvement of the method developed by Teramoto [74] was given by Morales-Cabrera et al. in 2002 [98]. Based on the same boundaries approach, the solution, in terms of concentration profiles and facilitated flux, was proposed analyzing the solute flux only at the two boundaries which, at steady state, must obviously be the same. The present method is based on the linearization of the reaction term, using a Taylor expansion truncated at the first order, evaluated at the boundary surfaces.

With respect to the similar approach of Teramoto [74], the approximate solution of Morales-Cabrera et al. consists of a higher number of nonlinear equations, but does not involve simplifications on the carrier continuity equation, Equation (9).

Unlike the other treatments presented, here the $x = 0$ is set in the middle plane of the thickness and the system is split in two subregions. The first region goes from the upstream boundary, $x = -L/2$, to the middle plane, $x = 0$, and the second one from the middle plane, $x = 0$, to the downstream boundary surface, $x = +L/2$.

The following equations are derived for the dimensionless concentration profiles of the solute in the left region $\overline{C}_{A,L}$, and right region $\overline{C}_{A,R}$:

$$\overline{C}_{A,L}=A_L sinh(\varphi_L y)+B_L cosh(\varphi_L y)-\frac{\alpha_L}{\varphi_L{}^2}y-\frac{\beta_L}{\varphi_L{}^2}\quad -1\le y\le 0 \tag{147}$$

$$\overline{C}_{A,R} = A_R sinh(\varphi_R y) + B_R cosh(\varphi_R y) - \frac{\alpha_R}{{\varphi_R}^2} y - \frac{\beta_R}{{\varphi_R}^2} \quad 0 \le y \le 1 \tag{148}$$

φ_L *and* φ_R are two modified Damköhler numbers evaluated in the left and right region respectively, defined in Equations (150) and (151), ϕ^2 is the Damköhler number defined in Equation (152), y is the dimensionless length centered in the middle plane, Equation (149) and $A_L, A_R, B_L, B_R, \alpha_L, \alpha_R, \beta_L, \beta_R$ are constants that can be found in the Supplementary Materials.

$$y = \frac{x}{\frac{L}{2}} \tag{149}$$

$$\varphi_L^2 = \phi^2 \left[\overline{C}_C^{-1} + \frac{1}{r_C} + \frac{1}{K r_{AC}} \right] \tag{150}$$

$$\varphi_R^2 = \phi^2 \left[\overline{C}_C^1 + \frac{\overline{C}_A^1}{r_C} + \frac{1}{K r_{AC}} \right] \tag{151}$$

$$\phi^2 = \frac{k_f L^2 C_A^{-L/2}}{D_A} \tag{152}$$

In the above equations, K ins the dimensionless equilibrium constant at the upstream boundary ($KC_A^{-\frac{L}{2}}$), r_C and r_{AC} are diffusivities ratio carrier/solute and complex/solute respectively. All the concentrations appearing are dimensionless and relative at the upstream solute concentration. $\overline{J}_A$ is the dimensionless solute flux of permeate at the boundaries $y = -1$ or $y = 1$, defined as:

$$\overline{J}_A = \frac{d\overline{C}_A}{dy} \quad y = \pm 1$$

The other equations needed to close the system are the conservation of the total carrier concentration, now expressed by Equation (153), and the two integration constants, T_C and T_{AC}, coming from the solution to the differential system of mass balance equations:

$$\begin{aligned} \overline{C}_T = \tfrac{1}{2}\left(\tfrac{1}{r_C} - \tfrac{1}{r_{AC}}\right) \Big\{ & \left[\tfrac{A_L}{\varphi_L}\left(1 - \tfrac{1}{cosh(\varphi_L)}\right) + \left(1 - \tfrac{\alpha_L}{{\varphi_L}^2} + \tfrac{\beta_L}{{\varphi_L}^2}\right) \tfrac{tanh(\varphi_L)}{\varphi_L} + \tfrac{\alpha_L}{2{\varphi_L}^2} - \tfrac{\beta_L}{{\varphi_L}^2} \right] \\ & - \left[\tfrac{A_R}{\varphi_R}\left(1 - \tfrac{1}{cosh(\varphi_R)}\right) - \left(\overline{C}_A^1 + \tfrac{\alpha_R}{{\varphi_R}^2} + \tfrac{\beta_R}{{\varphi_R}^2}\right) \tfrac{tanh(\varphi_R)}{\varphi_R} + \tfrac{\alpha_R}{2{\varphi_R}^2} - \tfrac{\beta_R}{{\varphi_R}^2} \right] \Big\} - \tfrac{T_C}{r_C} + \tfrac{T_{AC}}{r_{AC}} \end{aligned} \tag{153}$$

$$T_C = 1 - \overline{J}_A + r_C \overline{C}_C^{-1} = \overline{C}_A^1 + \overline{J}_A + r_C \overline{C}_C^1 \tag{154}$$

$$T_{AC} = 1 - \overline{J}_A + r_{AC} \overline{C}_{AC}^{-1} = \overline{C}_A^1 + \overline{J}_A + r_{AC} \overline{C}_{AC}^1 \tag{155}$$

Together with the two new boundary conditions in $y = 0$, reported in Equation (156):

$$\begin{gathered} \frac{d\overline{C}_{A,L}}{dy} = \frac{d\overline{C}_{A,R}}{dy} \\ y = 0 \\ \overline{C}_{A,L} = \overline{C}_{A,R} \end{gathered} \tag{156}$$

Concentration profiles and solute flux can be calculated by solving the set of nonlinear equations derived.

Once the solute concentration profile is known, it is possible to evaluate its derivative required for the facilitation factor calculation:

$$F = 1 + \frac{2\overline{J}_A}{\overline{\overline{C_A^{-1} - C_A^1}}} \tag{157}$$

In the above equations, the superscript −1 or 1 indicates the point at which the concentration is evaluated according to the dimensionless coordinates *y*, defined in Equation (149).

In terms of facilitation factor, the results of this approximate method showed excellent agreement with the numerical solution given by Basaran [8] and Kutchai et al. [53] for different Damköhler numbers, different diffusivities ratio and downstream solute concentrations. The method is also able to predict the presence of an optimal value of equilibrium constant like the one of Teramoto [74].

Figure 8a shows how the facilitation factor is affected by the Damköhler number, defined in Equation (152), for different values of downstream solute concentration. The drop in facilitation factor due to the increase of backpressure is well described. In Figure 8b, it is also reported the same dependence of the facilitated factor to the ratio of carrier and carrier complex diffusivity. Moreover, the present approach can describe the equilibrium regime which occurs for the entire range of Damköhler numbers, in accordance with both the analytical equilibrium solution reported by Ward [54] and the Kutchai's numerical solution [53].

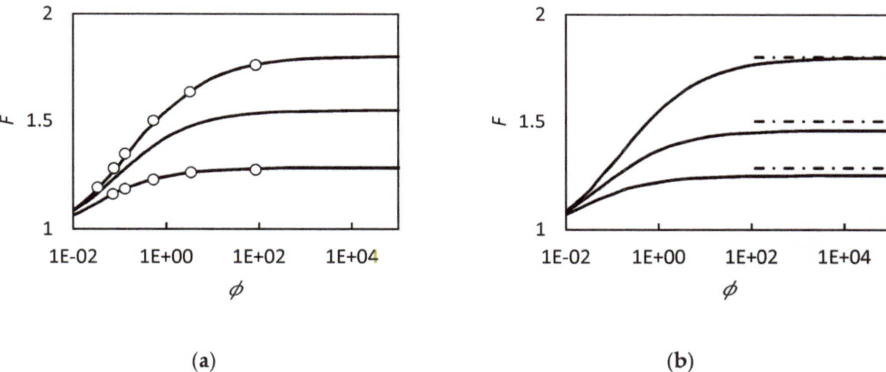

Figure 8. Facilitation factor as function of Damköhler number (Equation (157)). (**a**) Influence of the downstream solute concentration (dimensionless) for the case $D_{AC} = D_C$. From the top, $\overline{C}_A^1 = 0, 0.1, 0.4$. Circles are numerical solution from Kutchai et al. [53], lines are calculations from the present model; (**b**) diffusivities ratio effect, $r = D_{AC}/D_C$, for zero downstream concentration. From the top, $r = 1$, $r = 0.5$, and $r = 0.25$. Continuous lines are the model of Morales and Cabrera [98], dash and dotted lines are the asymptotic solution of Teramoto [74].

Even if this approach represents an improvement of the one proposed by Teramoto, it suffers in terms of concentration profiles description in the middle of the membrane, for values of a Damköhler number larger than one, in the near chemical equilibrium regime [99]. In this regime, differences between the approximate and exact solutions still remain in the middle region while, near the boundaries, the solutions agree. However, looking at the overall membrane properties, in terms of solute flux or facilitation factor, similarly to Teramoto [74], the present approach can describe properly the actual behavior of these kind of systems, despite the solution procedure is not straightforward.

3. Fixed Carrier Systems

In this family of facilitated transport systems, the carrier agent is attached to the polymer backbone. Unlike the MCS case, the carrier cannot diffuse across the membrane but, rather, can vibrate around its equilibrium position in the chain [100]. As a general consequence of that, FSC systems are characterized by lower solute fluxes than MC ones but, on the other hand, they own a better long term stability, since the carrier agent cannot leave the membrane [48,82,101,102]. This advantage make the fixed carrier systems a valid alternative to the more common mobile carrier systems.

3.1. Models for Fixed Sites Facilitated Transport Membranes

The present section will start with introduction of the dual mode (DM) transport model. This model was introduced, at the end of the 1950s, to describe the gas solubility and diffusion in glassy polymers, without any explicit chemical reaction between the polymer and solute. However, since the Langmuir sorption mechanism [103], which provides one of the terms of the DM model, could be explained by chemical reaction mechanisms [104], some authors used the DM approach for the mathematical description of FSC systems.

In particular, this was done in the first studies concerning the facilitated transport of oxygen in polymer-based metal complex membranes, as in the works by Okahata et al. [24], Nishide et al. [25,105], but also to represent the selective permeation of CO_2 in polymeric membranes having aminoethoxycarbonyl moieties (Yoshikawa et al. [106,107]). Moreover, the dual mode comprehensive transport framework derived by Barrer [108] in 1984, was used by Noble [109–111] to derive some crucial results about the fixed carrier facilitated transport. Even if the dual mode has been applied to many fixed carrier systems, with good results in terms of fitting, this theory does not provide a clear and full explanation of the transport mechanism in these facilitated membranes, in particular regarding the chemical reaction effect.

Cussler et al., [100], and Noble, [109–111], derived phenomenological models to answer that issue.

Their models, though conceptually and mathematically quite different, consider explicitly how the chemical reaction mechanism and the morphological polymer structure act on the facilitation. Cussler found that the facilitation is possible only if the carrier concentration exceeds a percolation limit, while in the works of Noble there is no such limit. One of the most interesting results of Noble's group is that, in the condition of excess carrier concentration, the facilitation factor in FSC systems obeys the same law valid for the mobile carrier case. However, even if the work suggests a useful qualitative interpretation of the facilitation, on a quantitative point of view it works well only if the system is far from the carrier saturation condition.

Kang et al. [112,113] used a physical analogy with the parallel RC alternate voltage circuits to calculate the facilitation factor in fixed carrier membranes. In their picture, the solute transport mechanism could be seen as the electron transport in RC circuit: the pure physical diffusion is the analogous of the electron transport in the resistor element, while the reaction sites are compared to the capacitor element. A single RC circuit and a series of RC circuits was taken into account and compared.

Recently, two models have been developed by Zarca et al. [18,114], in 2017. The first one regards the pure fixed sites transport and proposes a simple mathematical relation for the permeability calculation by the use of two fitting parameters and it allows to predict the percolation concentration limit, if it exists. The second is an extension developed to describe hybrid facilitated systems in which both mobile and fixed carriers act simultaneously.

3.1.1. Dual Mode Theory

The dual mode transport theory indicates a family of transport models developed by different authors. The fundamentals of such theory were derived between the late 1950s and the mid 1980s, originally to explain the excess of gas sorption in glassy polymer with respect to the value predicted by Henry's law [115], valid for the rubbery systems. It was Meares, in 1954, [116] who hypothesized that a population of microholes exists in a glassy polymer matrix, which could contain the excess of sorbed molecules. Barrer, in 1958, firstly proposed the mathematical formulation of the gas solubility in those matrices [117]. He observed that the equilibrium concentration of solutes in glassy polymers could be described as sum of two different contribution: the Henry's law [115] to consider the gas dissolved in the polymer matrix, and a Langmuir [103] term to take into account the gas adsorbed in the holes. The sum of these two contributions gives Equation (158), which is the dual mode sorption relation:

$$C_A = C_D + C_H = k_d p + \frac{CH'bp}{1+bp} \tag{158}$$

where:

C_A = solute concentration in polymer phase
C_H = solute concentration trapped in 'holes'
C_D = solute concentration dissolved
p = solute partial pressure
k_d = Henry's constant
CH' = holes saturation level concentration
b = affinity constant

Equation (158) is used in all the subsequent developments of the dual mode sorption theory to represent gas solubility. Some variations were proposed by considering the fugacity or activity rather than the pressure of the gas phase, as done for example by Petropoulos [118], but on the conceptual point of view nothing changes. For simplicity's sake, we report all the model equations in terms of concentration and pressure.

The representation of gas diffusion in such systems followed a very different destiny, and many approaches were proposed in the literature, which are recalled hereafter.

In 1963 Michaels and Vieth [119], while studying gas diffusion in poly-ethylene terephthalate, considered that the diffusion coefficient of molecules trapped in the polymer "holes" should be zero, so that the gas sorbed by Langmuir mechanism is totally immobilized in the matrix. Moreover they considered that, in every point of the polymer, there is local chemical equilibrium between dissolved and adsorbed gas.

In this case, it is possible to express the adsorbed concentration as function of the dissolved one, through Equation (159):

$$C_H = \frac{CH'MC_D}{1+MC_D} \tag{159}$$

where $M = b/k_d$.

Based on the previous hypothesis, the solute permeability should be given simply by the product between Henry's constant and the diffusion coefficient of dissolved species, Equation (160).

$$P = k_d D_d \tag{160}$$

Petropoulos in 1970 [118] pointed out that some assumptions previously used to treat the diffusion in glassy polymers were not based on theoretical justifications. He focused mainly on removing the total immobilization hypothesis for adsorbed species, providing an equation to account for the diffusivity of those molecules.

Following that idea, it has been shown that the solute overall permeability has the following form:

$$P = k_d D_d + \frac{CH'D_h}{(p_A^0 - p_A^L)} ln \frac{(1+bp_A^0)}{(1+bp_A^L)} \tag{161}$$

where D_d, D_h indicates the diffusivity of dissolved and adsorbed species respectively, p_A^0, p_A^L are the upstream and downstream solute partial pressures. Equation (161) still holds even if the local chemical equilibrium condition is not considered.

In 1976 Paul and Koros, [120], and Koros et al. [121] provided a mathematical description of transport in glassy polymers, to take into account the partial immobilization of the adsorbed solute. They assumed that a fraction of adsorbed molecules, together with the totality of dissolved ones, can freely diffuse in the polymer matrix, while the remaining part of the Langmuir's type species is frozen in the holes with zero mobility.

This leads to rewrite the Fick's law, Equation (10), for the flux as:

$$J_A = -D\frac{dC_m}{dx} \tag{162}$$

where C_m is the concentration of solute able to diffuse, and is given by the following equation:

$$C_m = C_D(1 + RC_H) \tag{163}$$

The ratio between the diffusivity coefficients of adsorbed and dissolved molecules, D_h and D_d, indicated with R in Equation (163), is the measure of the partial immobilization of the Langmuir's type species. Neglecting the downstream partial solute pressure, by means of Equations (162) and (163) one finds the relation for the permeability in the form:

$$P = k_d D\left(1 + \frac{RCH'b}{k_d(1 + bp_A^0)}\right) \tag{164}$$

Extending the idea of incomplete immobilization, Barrer in 1984 investigated the different diffusion pathways allowable in such systems and how they affect the overall transport in glassy polymers [108]. Four diffusion steps were hypothesized to occur and analyzed in detail. For each of these steps, a corresponding diffusive flux is generated:

$$D_{dd} => \quad J_{dd} = -D_{dd}\frac{dC_D}{dx} \text{ flux within the dissolved phase} \tag{165}$$

$$D_{dh} => \quad J_{dh} = -D_{dh}\left[\frac{dC_D}{dx}(1-\theta) + \frac{C_D}{CH'}\frac{dC_H}{dx}\right] \text{ flux from the dissolved to theadsorbed phase} \tag{166}$$

$$D_{hd} => \quad J_{hd} = -D_{hd}\frac{dC_H}{dx} \text{ flux from the adsorbed to the dissolved phase} \tag{167}$$

$$D_{hh} => \quad J_{hh} = -D_{hh}\frac{dC_H}{dx} \text{ flux within the adsorbed phase} \tag{168}$$

where θ in Equation (166) is the ratio between the holes adsorbed concentration, C_H and the saturation holes concentration, CH':

$$\theta = \frac{C_H}{CH'} \tag{169}$$

The sum of the four parallel contributions, written as function of dual mode concentration gradients in Equations (165)–(168), gives the solute total flux:

$$J = J_{dd} + J_{hd} + J_{dh} + J_{hh} = -[D_{dd} + D_{hd}(1-\theta)]\frac{dC_D}{dx} - \left[D_{hh} + D_{hd} + D_{dh}\frac{C_D}{CH'}\right]\frac{dC_H}{dx} \tag{170}$$

Barrer also showed that the overall solute diffusivity may be expressed as:

$$D = \left\{D_{dd}(C_D^0 - C_D^L) + (D_{hd} + D_{hh})(C_H^0 - C_H^L) - \frac{D_{dh}}{M}\left[(\theta_0 - \theta_L) + 2ln\frac{(1-\theta_0)}{(1-\theta_L)}\right]\right\}\frac{1}{(C_A^0 - C_A^L)} \tag{171}$$

where D is the overall solute diffusivity, and superscripts 0, L indicate the upstream and downstream side, respectively. If the downstream solute pressure goes to zero, $C_A^L \ll C_A^0$, Equation (171) becomes:

$$D = \left\{D_{dd}C_D^0 + (D_{hd} + D_{hh})C_H^0 - \frac{D_{dh}}{M}[\theta_0 + 2ln(1-\theta_0)]\right\}\frac{1}{C_A^0} \tag{172}$$

Equation (172) allows to write the following expression for the overall solute permeability:

$$P = k_d\left\{D_{dd} + \left[(D_{hd} + D_{hh})KCH' - D_{dh}\right](1-\theta_0) - 2D_{dh}\frac{(1-\theta_0)}{\theta_0}ln(1-\theta_0)\right\} \quad (173)$$

From Equation (173) we can obtain useful information about two limiting cases. For sufficiently low pressure values, the Langmuir saturation ratio, θ, goes to zero indicating that the holes are nearly free. In this case, Equation (173) reduces to:

$$P = k_d\left\{D_{dd} + \left[(D_{hd} + D_{hh})MCH' + D_{dh}\right]\right\} \quad (174)$$

At the opposite, the other limiting case arises for sufficiently high pressure values. In such case the Langmuir sites saturation occurs, θ goes to one and the permeability approaches the following value:

$$P = k_d D_{dd} \quad (175)$$

Note that the latter equation states that in the saturated condition, the permeability is related only to the mobility of the dissolved phase, as in the case of Equation (160).

In Equation (176) below, we reported the ratio between Equations (174) and (175). As we can see, that ratio is always higher than or equal to one, indicating that in these systems, for which the adsorbed species are partly free to move across the film, the permeability decreases as the holes undergo saturation.

$$\beta = k_d\left\{1 + \frac{\left[(D_{hd} + D_{hh})MCH' + D_{dh}\right]}{D_{dd}}\right\} \quad (176)$$

The interpretation provided by Barrer in 1984 represent the most complete description of the possible mechanism of gas transport in gassy polymer systems offered by the dual mode theory. The previous key results, Equations (165)–(168) and (170)–(176), were also confirmed one year later, in 1985, by Fredrickson and Helfand [122], who were not aware of Barrer's published results. Moreover, they demonstrated that the same relations could be derived starting from a microscopic scale, using a lattice matrix structure, and a mathematical treatment based on probability distribution functions. Different authors, investigating the facilitated transport by fixed carriers, explained and modeled the experimental data using the dual mode as a true physical theory. It is easy to see that the Langmuir sorption model, reported in Equation (158), is the equivalent of Equation (24) reported in the mathematical formulation section.

To the best of our knowledge, Okahata et al. [24] were the first ones to test the dual mode model on FSC systems for oxygen/nitrogen separation membranes in 1986. They synthesized a fixed sites carrier membrane by dispersing, in poly-butyl methacrylate, a cobalt-porphyrin-imidazole complex (CoPIm) that reacts reversibly and rapidly with molecular oxygen and the permeability results were modeled with the Paul and Koros interpretation of the dual mode [120]. Yoshikawa et al. [106], in 1988 investigated the selective permeation of carbon dioxide, over nitrogen and oxygen, in poly(4-vinylpyridine-co-acrilonitrile) membranes. They found that, while for the inert gases, O_2 and N_2, the permeability was independent from the upstream pressure, for the carbon dioxide the behavior was quite different. Their data, shown below in Figure 9a, evidence a decrease in permeability as the upstream pressure increases, in agreement with the saturation mechanism explained. Even in this case, the experimental permeability of carbon dioxide was modeled by the partial immobilization model of Paul and Koros, Equation (164) [120].

A facilitation effect was detected by Nishide et al. [25,105] in the oxygen permeation across a polymeric membranes containing cobalt porphyrin as fixed carrier. Also in this case, oxygen permeability was found to depend on the upstream pressure. The experimental measurements were interpreted with the Barrer equation for permeability, Equation (173), [105], and with the Paul and Koros model, reported in Equation (168), in ref. [25], providing satisfactory fitting results (Figure 9b). The permeability equation derived by Paul and Koros was also used successfully by Yoshikawa et al. in 1995 [107] in the facilitated transport of carbon dioxide across pyridine moiety fixed carrier membrane.

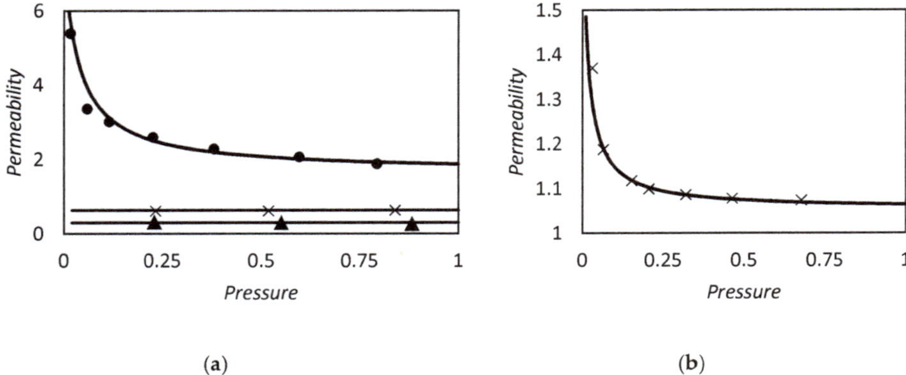

Figure 9. Dual mode model results, permeability as function of upstream pressure. (**a**) Yoshikawa et al. [106]. Circles, crosses and triangles are CO_2, O_2 and N_2 measured, respectively; (**b**) Nishide et al. [105]. Lines are the model predictions. Permeability is in ($1 \times 10^{-9} \times$ cm^3·cm/(cm^2·s·atm)), pressure is in (atm).

3.1.2. Cussler et al., 1989. On the Limits of Facilitated Diffusion

Cussler et al. [100], in 1989, explained the fixed carrier transport mechanism and provided an expression for the solute flux. In particular, they considered the membrane as a series assembly of contiguous lamellae, with one carrier site per lamella. Since the carrier is fixed, it cannot diffuse freely in the membrane, but it is allowed to oscillate around its equilibrium position in the matrix, due to vibrational energy. The same limitation still holds also for the reaction product. Moreover, the authors considered that the free solute does not exist in the polymer phase, so that pure physical diffusion cannot occur. Another hypothesis in the present treatment is that the solute-carrier chemical reaction, Equation (3), takes place only at the membrane surfaces while, inside the membrane, only an exchange mechanism between carrier sites is present. Hence, in some conditions, the complex could hop from one carrier site to another.

Figure 10 reports schematically the description of transport used by the authors. The continuous line circles represent the fixed sites in equilibrium position, while the dashed ones represent the carriers located at the two ends of oscillation. Hence, l indicates the space between contiguous equilibrium positions, while l_0 is the amplitude of the oscillation. If the distance between equilibrium locations is higher than the oscillation width, the carriers cannot get in touch and no solute exchange is possible. In this case, no diffusion flux can be observed. Conversely, if $l_0 \geq l$ two carriers could get in contact and an exchange mechanism can take place. The latter condition is therefore necessary for the solute transport. If that condition is achieved, the equation for the flux is:

$$J_A = \frac{D}{L}\left(C_C^L - C_C^0\right)\left[\left(2 - \frac{l_0}{l}\right) + \left(\frac{l_0}{l} - 1\right)\left(\frac{1}{2} + \frac{1 + cosh(\psi)}{\psi sinh(\psi)}\right)\right]^{-1} \tag{177}$$

$$\psi = \sqrt{\frac{2kC_T(l_0 - l)^2}{l_0 D}} \tag{178}$$

where ψ, defined in Equation (178) above, is the Thiele modulus (or Damköhler number) accounting for the ratio between chemical reaction and diffusion rate. In the previous equation, the constant reaction rate, k, represents the solute exchange rate between two contiguous sites and D is not an intermolecular complex diffusivity, but rather it is the intramolecular diffusion coefficient, i.e., the diffusivity across polymer chains by the proposed jumping mechanism. If the equilibrium condition is considered, Equation (177) becomes:

$$J_A = \frac{D}{L}\left(\frac{C_T l}{l_0}\right)\frac{K_{eq}C_A^0}{(1+K_{eq}C_A^0)}\left[\left(2-\frac{l_0}{l}\right)+\left(\frac{l_0}{l}-1\right)\left(\frac{1}{2}+\frac{1+cosh(\psi)}{\psi sinh(\psi)}\right)\right]^{-1} \quad (179)$$

where the first term in brackets represents the average total carrier concentration in every lamella.

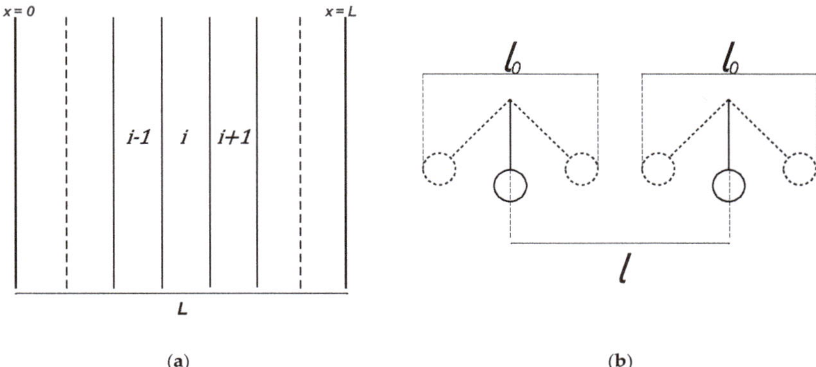

(**a**) (**b**)

Figure 10. Representation of membrane thickness in Cussler et al. [100]. (**a**) Series arrangement of contiguous lamellae. L is the membrane thickness. (**b**) Contiguous fixed sites. l_0 is the oscillation width, l is the distance between equilibrium positions. For $l > l_0$ solute exchange between contiguous sites cannot occur.

The latter equation presented was used to obtain some important results in the two limiting case of facilitated transport.

For the fast reaction case, indeed, the Thiele modulus is large and the flux is given simply by:

$$J_A = \frac{D}{L}\frac{C_T K_{eq}C_A^0}{(1+K_{eq}C_A^0)}\left(\frac{2l/l_0}{3-l_0/l}\right) \quad (180)$$

The same results of Equation (180) could be easy obtained for the mobile carrier case by neglecting the pure physical diffusion mechanism with the only difference of the term in brackets (see the mathematical background section). If we consider $l = l_0$, the difference between the two cases vanishes.

On the other hand, the most interesting case is the one for which the diffusion mechanism is faster than the reaction. In this regime ψ goes to zero, and it is possible to obtain an expression for the apparent diffusion coefficient related to both morphological and reaction parameters. In this case, the equation for the flux becomes:

$$J_A = \frac{[kC_T l^3(l_0-l)/l_0^2]}{L}\frac{C_T K_{eq}C_A^0}{(1+K_{eq}C_A^0)} \quad (181)$$

Equation (181) shows the flux expression for fast diffusion regime. The apparent diffusion coefficient here is the term in square brackets and contains both characteristic lengths involved, l_0, l, as well as the exchange solute rate, k.

As stated, the solute exchange between adjacent sites is possible only if $l_0 \geq l$, and this implies the presence of a percolation threshold, i.e. a carrier concentration limit below which no transport is possible (Figure 11a). In fact, while the oscillating width is strictly correlated to the polymer nature, the distance l is strongly carrier concentration dependent. The higher the concentration, the shorter the distance, and vice-versa. Although the authors did not consider the free solute diffusion, their model is able to capture the transport nature of some of these carrier systems.

As reported in many works about FSC systems [114,123–126], a low carrier concentration limit could exist and the solute flux is observed only above certain concentration values. In Figure 11b, we report the carrier-mediated flux of fructose in plasticized cellulose triacetate membrane using trioctymethilammonium chloride (TOMAC) as carrier, by Riggs and Smith [125]. However, the mechanism proposed by Cussler cannot explain cases in which the solute flux differs from zero, even at low carrier concentrations, as the facilitated transport of oxygen reported by Tsuchida et al. [127] or the transport of organic acids studied by Yoshikawa et al. [128].

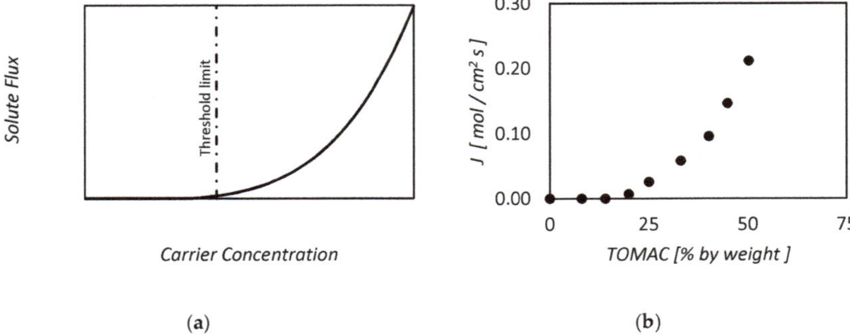

(**a**) (**b**)

Figure 11. Solute flux as function of carrier concentration. (**a**) Qualitative behavior in presence of a threshold limit. For carrier concentration below the threshold, the solute flux does not occur; (**b**) experimentally detected fructose flux in plasticized cellulose triacetate as function of carrier (TOMAC) concentration, Riggs and Smith [125].

3.1.3. Noble, 1990. Analysis of Facilitated Transport with Fixed Site Carrier Membranes

Noble in 1990 presented the first of three papers about the FSC facilitated transport [109]. The other two papers [110,111] were published in 1991 and 1992. The works deeply investigate the nature of facilitation in these systems and provide a mathematical relation between the four diffusion coefficients, previously introduced in the dual mode framework by Barrer [108], and the chemical reaction mechanism between carrier and solute molecules.

In the case of large excess of carrier, which implies that the concentration of free carrier, C_C, is nearly constant and not too different from the total carrier concentration C_T, Noble found that Equation (13) describes the complex mass balance also in the fixed carrier case. The derivation of such equation requires the existence of the four exchange mechanisms introduced by Barrer, [108], conversely to what Cussler did [100].

In Figure 12 the transport paths are shown. The horizontal lines at the top indicate the pure free solute physical diffusion. The double lines indicate the exchange mechanisms between adjacent fixed sites, (horizontal ones) or between a fixed site and the free solute region (vertical ones).

Even if the Equation (13) still holds in this case, the diffusion coefficient appearing, D_{AC}, is now function of the morphology of the matrix and of the mobility between sites. That function was found to be in the form:

$$D_{AC} = l^2\bar{k} \tag{182}$$

where l is the distance between two fixed sites and $\bar{k}$ is a mobility parameter.

For zero downstream pressure, with the assumption of excess carrier and since the mathematical transport problem for solute and reaction product is identical to that in liquid membranes, the solution for the facilitation factor is equivalent to the one of Smith and Quinn [93], already discussed in the previous section and hereafter shown:

$$F = \frac{\left(1 + \frac{\alpha K}{1+K}\right)}{1 + \frac{\alpha K}{1+K}\frac{tanh(\lambda)}{\lambda}}$$

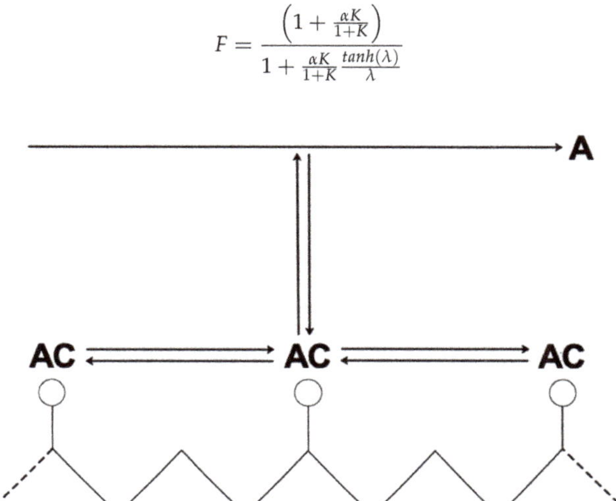

Figure 12. Solute transport pathways in fixed sites carrier systems used by Noble [109].

The parameters reported in Equation (116) have been already explained in previous section; the "only" difference between the present and the mobile carrier case is the physical meaning of the complex diffusivity, here given by Equation (182) but not fully explained. In particular, the relation between the complex diffusivity and the chemical reaction was not given in the present work and presented only later, in the subsequent works of 1991 and 1992.

If the chemical equilibrium is taken into account, $\tanh(\lambda)/\lambda$ goes to zero and the facilitation factor is simply given by:

$$F = \left(1 + \frac{\alpha K}{1+K}\right) \tag{183}$$

This result, valid in the chemical equilibrium regime, is analogous to the one achievable by the dual mode theory and was tested to describe the experimental data of Tsuchida et al. [127] concerning the facilitated transport of molecular oxygen in a polymer with cobalt Schiff complex bases as fixed carrier.

By plotting $E = (F-1)^{-1}$ against feed side gas pressure, it is possible to retrieve the parameters needed to model the system [129–131]. Figure 13 reports the results obtained by using this approach: as shown in Figure 13a, after a certain value of upstream pressure linearity dependences is lost and, after that, the solution given in Equation (183) is no longer valid. However, the use of the latter equation in its range of applicability could be sufficient to retrieve the parameters values required to describe the system in all the range of interest, as shown in Figure 13b. With simple algebra it is possible to demonstrate that the intercept of the plot given in Figure 13a is $(\alpha K)^{-1}$ while the slope is α^{-1}.

In his work in 1990, Noble obtained two key results which he then used in the following studies. They were:

- the mathematical derivation of the mass balance analog of the mobile carrier case, which allows to use, in analytical approximation methods already known, Equation (2.89) while an excess of carrier is considered.
- the functional dependence of the actual complex diffusivity in such systems on morphological and chemical parameters, Equation (182).

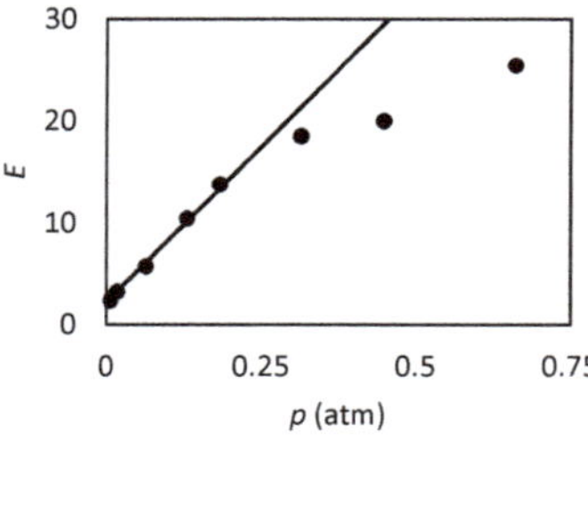

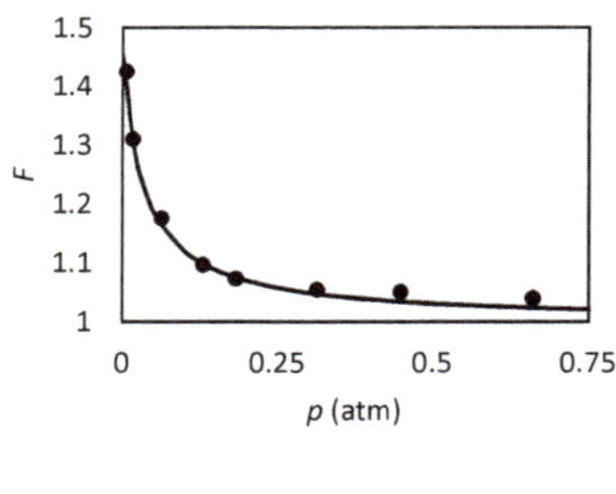

(**a**) (**b**)

Figure 13. Modeling results using Equation (3.22). (**a**) Plot of $E = (F - 1)^{-1}$ as function of upstream pressure. At high pressure, the linearity is lost. (**b**) Facilitation factor calculated by using parameters retrieved in the low pressure range of figure a. Circles are experimental data from Tsuchida et al. [127], lines come from Equation (183).

3.1.4. Noble, 1991, Facilitated Transport Mechanism in Fixed Site Carrier Membranes

In 1991 Noble extended his previous work and, starting from Equation (183), derived a series of equivalences between the Barrer's diffusion coefficients [108] and the equilibrium reaction constant in the excess carrier condition and diffusion limited regime [110]. Moreover, he found that the facilitation could be achieved, theoretically, even if the direct diffusion between carrier sites, D_{hh} in Barrer notation Equation (168), is equal to zero.

To derive the proper set of equivalences, he started from the equation for the total solute flux, Equation (170). The latter can be rewritten in terms of carrier, free solute and complex concentrations as:

$$J = J_{dd} + J_{hd} + J_{dh} + J_{hh} = -\left[D_{dd} + D_{hd}\left(\frac{C_C}{C_T}\right)\right]\frac{dC_A}{dx} - \left[D_{hh} + D_{hd} + D_{dh}\left(\frac{C_A}{C_T}\right)\right]\frac{dC_{AC}}{dx} \tag{184}$$

The above equation ensures that facilitation is allowable even if $D_{hh} = 0$, as in the case of low carrier concentration (i.e., when the sites are too far apart and the hopping mechanism cannot occur). Indeed, if Equation (184) holds, the condition for pure physical diffusion is achieved when the carrier is saturated, i.e., when $C_C \to 0$ and also $\frac{dC_{AC}}{dx} \to 0$. For this reason, thanks to the additional transport pathways considered, no percolation threshold appears a priori even for $D_{hh} = 0$.

In the chemical equilibrium regime, the complex gradient could be related to the free solute gradient by Equation (159). As a result, the solute gradient induces a complex gradient which enhances the transport up to the saturation. The same result may be casted in the facilitation factor terms that, even by neglecting the direct site-site exchange, $D_{hh} = 0$, the facilitation factor is not equal to one, but has the following form:

$$F = 1 + \frac{D_{dh}C_C}{D_{dd}C_T} + \frac{D_{hd}}{D_{dd}}\left(\frac{KC_T}{C_A^0}\right)\frac{C_C}{C_T} \tag{185}$$

Equation (185) again shows that facilitation occurs up to the carrier saturation, i.e., when free carrier sites are no longer present, $C_C = 0$.

The relationship between the actual complex diffusivity and the exchange mechanism between the free diffusion path and the fixed sites region was found by equating Equations (185) and (183) and is given by:

$$D_{AC} = D_{hd} + D_{dh}\left(\frac{C_A^0}{KC_T}\right) \tag{186}$$

The above equation is a crucial point in the analysis provided by Noble in 1991. He demonstrated in a rigorous way, even if in the excess carrier case, that the effective complex diffusivity depends from the chemical reaction, by dimensionless chemical equilibrium K, and from the morphological structure of the matrix, strictly affected by the total carrier concentration, C_T. Note that, since K is the dimensionless reaction equilibrium constant, $K_{eq}C_A^0$, the complex diffusivity, D_{AC}, does not depend on the upstream concentration.

Moreover, also the two diffusion coefficients D_{hd} and D_{dh} could be correlated to the chemical reaction, responsible of the complex formation and consumption.

For the low pressure range, Noble found that the ratio between the two diffusion coefficients is related to dimensionless constant and to the total carrier concentration through a simple mathematical relationship which correlates the two phenomena:

$$K = \frac{D_{dh}C_A^0}{D_{hd}C_T} \tag{187}$$

In chemical equilibrium regime, therefore, the exchange diffusion mechanism acting between fixed sites and free diffusional path is related to the chemical equilibrium constant. In other words, the higher the forward kinetic constant, k_f, the higher the exchange of solute from the physical diffusion region to the fixed site.

Equation (187) can be rearranged and inserted in Equation (186) to provide the final relation, Equation (188), that allows, together with Equation (3.26), to determine D_{dh} and D_{hd} once D_{AC} is known.

$$D_{AC} = 2D_{hd} \tag{188}$$

The main result of the work of 1991 is the demonstration that, theoretically, facilitation can occur even if $D_{hh} = 0$, so that the direct jumping mechanism is not the only one influencing the carrier-mediated transport. The complex diffusivity in such systems is related to chemical reaction parameters and morphological, carrier concentration dependent, parameters by Equations (186), (187) and (188). That solution will be improved by Noble in a subsequent, and more detailed, work [111], hereafter reported, where the final equation that combines all the variables influencing the complex mobility is given.

3.1.5. Noble, 1992, Generalized Microscopic Mechanism of Facilitated Transport in Fixed Site Carrier Membranes

In 1992 [111] Noble provided his general theory for the facilitated transport across FSC membranes. The direct hopping mechanism between fixed sites was included in this final description. He found a general equation for the effective complex diffusivity, D_{AC}, that accounts for the morphological nature of the systems and for the chemical complexing reaction between solute and carrier sites. The dual mode equations by Barrer were used, again, for the total flux, adapted in Equation (184).

As stated above, when two sites are far apart it is not possible for the solute to jump from one site to another. From a macroscopic point of view, considering a homogeneous distribution of carrier sites, this happens for low carrier concentrations. Cussler concluded that in this case no facilitation can occur, [100], while previously Noble had found that the facilitation is still allowable due to the presence of the exchange mechanism related to the complexation reaction [109].

In general, the flux of solute across these systems is function of both free solute and complex gradients across the membrane boundaries by Equation (184). However, as carrier concentration becomes small, in certain zones, Equation (184) reduces simply to:

$$J = J_{dd} = -D_{dd}\frac{dC_A}{dx} \tag{189}$$

because carrier is no longer present locally to react with the solute. However, near the sites, Equation (184) still holds.

In Figure 14 we report the internal structure of the membrane used by Noble in 1992. Similar to Figure 10, the circles are the fixed carriers, while l_1 and l_2 are the two length scales. The first one represents the space of influence of the carrier, in other words the space around a fixed site for which Equation (184) describes the local solute flux. The second, l_2, measures the space between carriers in which the solute transport follows Equation (189).

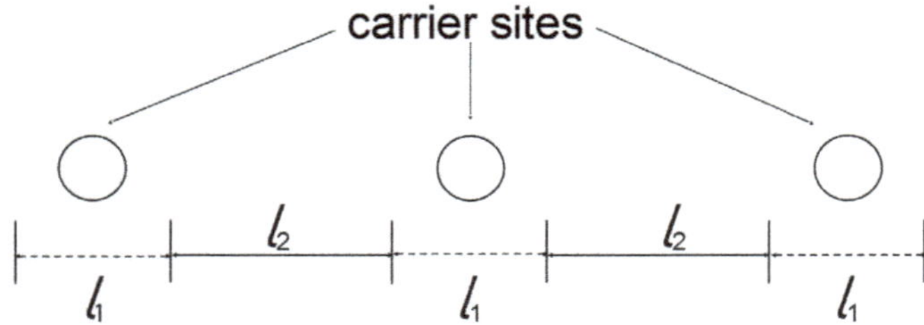

Figure 14. Membrane environment in the present work of Noble. Between sites, in the length space l_2, pure diffusion mechanism exists. Closely to the sites position, l_1, Equation (184) holds.

By taking in account these two lengths, the facilitation factor was derived as:

$$F = \frac{A + \beta B}{D_{dd} f + (A + \beta B)(1 - f)} \tag{190}$$

where f is the length ratio defined as:

$$f = \frac{l_1}{l_1 + l_2} \tag{191}$$

and the other symbols appearing in Equation (190) are defined below:

$$\beta = \frac{K_{eq} C_T}{(1 + K_{eq} C_A)^2} \tag{192}$$

$$A = D_{dd} + D_{dh} \frac{C_C}{C_T} \tag{193}$$

$$B = D_{hh} + D_{hd} + D_{dh} \frac{C_A}{C_T} \tag{194}$$

Assuming the chemical equilibrium and considering the case of low upstream pressure, the facilitation factor reduces to:

$$F = 1 + \beta \frac{D_{AC}}{D_{dd}} \tag{195}$$

which can be equated to Equation (190), to obtain the complex diffusion coefficient.

$$D_{AC} = \frac{1}{\beta} \left[\frac{A + \beta B - D_{dd}}{1 + \left(\frac{A + \beta B}{D_{dd}} \right) \left(\frac{1 - f}{f} \right)} \right] \tag{196}$$

The resulting equation, Equation (196), shows how the actual diffusion coefficient of the complexed species is affected by morphological and chemical reaction effects.

From the above results, it is possible to analyze two opposite cases. First, if the sites are too distant for a direct hopping mechanism, the diffusion coefficient related to that, D_{hh}, is obviously zero and the complex diffusivity can be rewritten in the following form:

$$D_{AC} = D_{dd} \left(\frac{f}{1 - f} \right) \left(\frac{R}{D_{dd} + \beta R} \right) \tag{197}$$

where R is a parameter containing the reaction contribution to the transport and defined as:

$$R = D_{hd} + D_{dh}\left(\frac{C_A}{C_T} + \frac{C_T - C_{AC}}{K_{eq}C_T{}^2}\right) \tag{198}$$

On the contrary, if the sites are adjacent, the direct jump of solute from one carrier site to the next one is possible and $D_{hh} \neq 0$. This happens if the length l_2 approaches zero; the complex diffusivity coefficients is then given by:

$$D_{AC} = D_{hh} + R \tag{199}$$

As we can see, the pure physical diffusion in this case does not influence the complex transport as in Equation (199), D_{dd} does not appear.

These two latter results, Equations (197)–(199), complete the description of Noble about the facilitation in FSC membranes that started three years earlier, in 1990 [109]. The analysis provided by the author is based on some approximations and simplifications, excess of carrier, chemical equilibrium, dilute solute concentration, but still provides an interesting picture of transport in these systems.

In conclusion, in the Generalized Microscopic Mechanism of Facilitated Transport in Fixed Site Carrier Membranes, the general mathematical formulation of the complex diffusion coefficient was derived in terms of morphological and reaction parameters. It was demonstrated theoretically that facilitation can occur even if the carrier sites are too far and the solute cannot hop directly from one to another.

3.1.6. Kang et al., 1995. Analysis of Facilitated Transport in Solid Membranes with Fixed Site Carriers

An interesting theoretical model for the interpretation of facilitated transport in FSC membranes was given in 1996 by Kang et al. [112,113]. Their work, divided in two papers, studies the permeability in these systems by means of analogies with resistor–capacitor circuits (RC), in parallel arrangement, in alternating voltage regime. In the first paper they reported the single circuit case, [112], while in the second one, [113], the solution was extended to the case of a series RC circuit arrangement. Here we present the general solution given for a series circuits, since the single circuit case is a particular solution of that.

In both RC circuits and in FSC membranes, the transport of particles occurs due to a driving force imposed at the boundaries of the system. For the electrical circuit case, the particles are electrons and the driving force is the voltage gradient; for the FSC case, the flux of solute molecules is related to a concentration gradient. Moreover, in the latter case the solute flows through different pathways, as already mentioned by Noble [111]. Even if a parallel RC circuit is considered, the electrons flow separately in two distinct paths and, for the Kirchoff's rules, the total electron flux is equal to the sum of the two single contributions, Equation (200)

$$j_e = j_r + j_c \tag{200}$$

where j_e is the total electrons flux and j_r, j_c are the resistor and capacitor contributions respectively.

Furthermore, as in the case of carrier saturation, for which the facilitation effect disappears, when the charge on the condenser element reaches its maximum value, also the second term in the right hand side of Equation (200) vanishes, leaving only the resistance contribution.

With the above considerations, the authors concluded that the pure physical diffusion could be compared to the resistor element and that the fixed sites behave similarly to the condenser in the RC parallel configuration. In Figure 15, below, we report the schematic pictures of a general RC circuit and a FSC systems. The general relationship between voltage driving force and resistor electron flux is expressed by the Ohm's law:

$$j_e = \sigma\frac{\Delta V}{L} \tag{201}$$

where σ is the conductivity, ΔV is the voltage difference, L is the length of the system.

By comparing the previous equation with the Fick's law written in terms of permeability, for zero downstream pressure, the following analogies could be revealed:

As shown in Table 1, the complex concentration is the one in chemical equilibrium with carrier and solute.

Table 1. Analogies between electrons transport and mass transport in resistor–capacitor circuits (RC) parallel circuit and facilitated transport systems with fixed sites carrier [112].

	RC circuit	Facilitated Systems
Flux	$j_e = \sigma \frac{\Delta V}{L}$	$J = P \frac{p^0}{L}$
Driving Force	$\frac{\Delta V}{L}$	$\frac{p^0}{L}$
Proportionality	σ	P
Capacitor Effect	$q = CV$	$C_{AC} = \frac{C_T C_A K_{eq}}{1+C_A K_{eq}}$

The alternating voltage regime of RC circuit is then replaced by a fluctuating local concentration. That fluctuation is attributed to the presence of chemical equilibrium reaction: the forward reaction lowers the solute concentration, while the inverse reaction has the opposite effect.

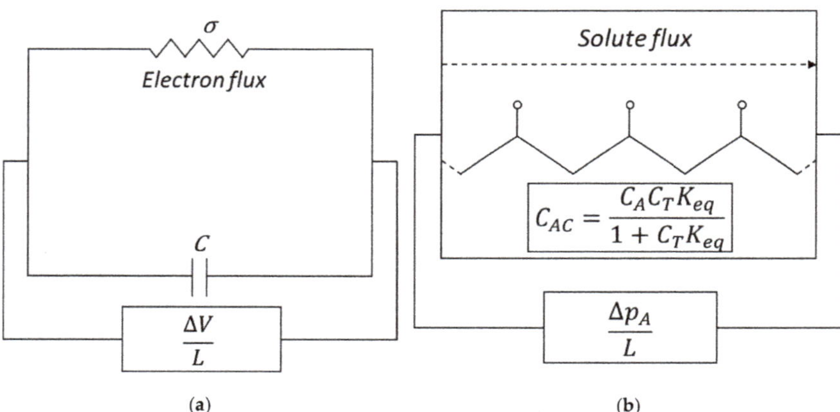

Figure 15. Analogies between RC parallel circuit and fixed sites carrier systems. (**a**) Parallel configuration of single RC circuit; (**b**) fixed sites carrier systems. Analogies among driving forces and capacitance effects according to Table 1.

As a result, the solute concentration value is the sum of two different terms: the first equals the equilibrium value and the other is expressed as sinusoidal function of time, related to the direct and inverse reaction rate. If we consider a system composed by a series of *n-RC* parallel circuits, the following solution for the electrical conductivity can be derived:

$$\frac{\sigma_{RC}}{\sigma} = 1 + \alpha_V \sqrt{n^2 + \left(\frac{C\omega L}{A_s \sigma}\right)^2} \tag{202}$$

where α_V is the extent of voltage fluctuation, ω is the frequency, A_s is the flux surface, C is the capacitance, σ_{RC} is the total conductivity, σ is the resistance conductivity and n is the number of circuits composing the series.

By using the proper substitution, the solution for the facilitation factor, expressed as ratio between permeability of facilitated systems divided by the pure physical diffusion contribution, is reported in Equation (203)

$$F = 1 + \alpha_p \sqrt{n^2 + [2\pi\psi\gamma ln(1+K)]^2} \tag{203}$$

where:

$$\alpha_p = \frac{C_A^d}{C_A^0}$$

$$\psi = \frac{k_r L^2}{D_A}$$

$$\gamma = \frac{C_T}{C_A^0}$$

$$K = K_{eq} C_A^0$$

The parameter n in Equation (203) is the number of RC circuits in the series. By setting $n = 1$ the single case is obtained [112].

The extent of concentration fluctuation, α_p, is essentially a fitting parameter since C_A^d represents the perturbation amplitude, that does not have a clear physical meaning. Moreover, the concentration fluctuations, if existing, probably have a more complicated mathematical form than the one used by the authors as sinusoidal function.

Both the RC models [112,113] were tested successfully against the experimental data measured by Ohyanagi et al. [26] about the oxygen facilitated transport, Figure 16. As one can expect, the series RC circuits model provided the best agreement due to the presence of one additional parameter, the number of circuits, n, that, for the membranes case, represents the number of sublayers in which the thickness is divided, and can be calculated from the following:

$$n = N_{AV} C_T (\pi r_s L)$$

where N_{AV} is the Avogadro number and r_s is the kinetic radius of solute.

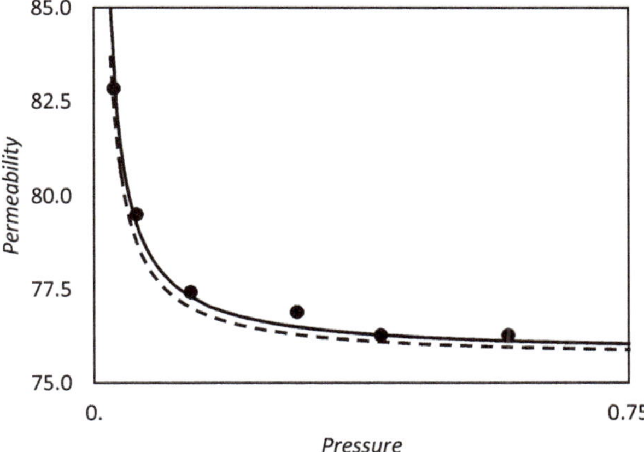

Figure 16. Oxygen permeability in a poly-dimethyl siloxane (PDMS) membrane with metallo-porphyrin fixed carrier. Circles are experimental data from Ohynagi et al. [26], $n = 228$ and $n = 1$ for continuous and broken line, respectively. Permeability is in $(1 \times 10^{-9} \times cm^3 \cdot cm/(cm^2 \cdot s \cdot atm))$, pressure is in (atm).

3.1.7. Zarca et al., 2017. A Practical Approach to Fixed-Site Carrier Facilitated Transport Modeling for the Separation of Propylene/Propane Mixtures Through Silver-Containing Polymeric Membranes

Recently, in 2017, Zarca et al. [114] modeled the facilitated transport of propylene in silver salts, $AgBF_4$ embedded in polymeric membranes (PVDF-HFP) with an analytical approach. Furthermore, that approach was extended to consider also the presence of mobile carriers inside the membranes [18] (see next paragraph).

The basic idea of the model proposed by Zarca et al. is that the free solute follows a pure physical diffusion mechanism while the complex diffusion is somehow related to the chemical reaction, similar to the dual mode based transport model of Noble [111]. Some differences arise when comparing the approach of Zarca et al. to the one of Noble. Noble considers four diffusion coefficients, while in this approach there are only two ones: the first one accounts for the pure physical diffusion mechanism, D_{dd} in Noble's notation, and the second one incorporates all the effects of the chemical reaction. The exchange between different zones of the membrane has been not considered explicitly, and a hopping mechanism was hypothesized.

The expression for the solute flux used by the authors is given hereafter, Equation (204):

$$J_A = -D_{dd}\frac{dC_A}{dx} - A\frac{dC_A}{dx} \tag{204}$$

or in integrated form as:

$$J_A = D_{dd}\frac{(C_A^0 - C_A^L)}{L} + A\frac{(C_A^0 - C_A^L)}{L} \tag{205}$$

where A is an effective diffusivity. Note that in both the above equations, the flux is given as function of the pure solute concentration gradient only. Considering the concentration on the boundaries in equilibrium with the gas phase, by definition of permeability, Equation (205) becomes:

$$J_A = P\frac{(p_A^0 - p_A^L)}{L} + K_H\frac{(p_A^0 - p_A^L)}{L} \tag{206}$$

The effective permeability, or hopping constant as the authors called it, K_H is the term which contains all the reaction effects. In the definition of this key parameter, Equation (207), an Arrhenius type of law was included to explicitly consider the temperature dependence of the facilitated transport mechanism:

$$K_H = \alpha\left(\frac{C_T}{1 + K_{eq}C_A^0}\right)^{\beta} e^{\frac{E_A}{R}\left(\frac{1}{293} - \frac{1}{T}\right)} \tag{207}$$

The first term in brackets is the free carrier concentration in chemical equilibrium with the upstream side concentration of free solute, C_A^0. In the original paper, the equilibrium constant was given in heterogeneous form accounting both for concentration (carrier and complex) and pressure (solute in gas phase). Here we have converted that expression in homogeneous terms and used concentrations only. However, the conversion between the two is straightforward.

The model contains four parameters: K_{eq} and E_A (activation energy) which have a clear physical meaning, and α and β that are pure fitting parameters. In particular, β is a mathematical correction that allows the model to describe the percolation threshold limit. It was shown that the proposed method is in good agreement with the experimental data for facilitated transport of propylene in polymeric membranes containing silver ions, reported by the same authors, Figure 17. Its mathematical simplicity, coupled with the capability of fitting experimental data, in a larger range of experimental conditions than the one of Noble [111] based on the excess carrier hypothesis, makes this model a promising one for the facilitated transport systems.

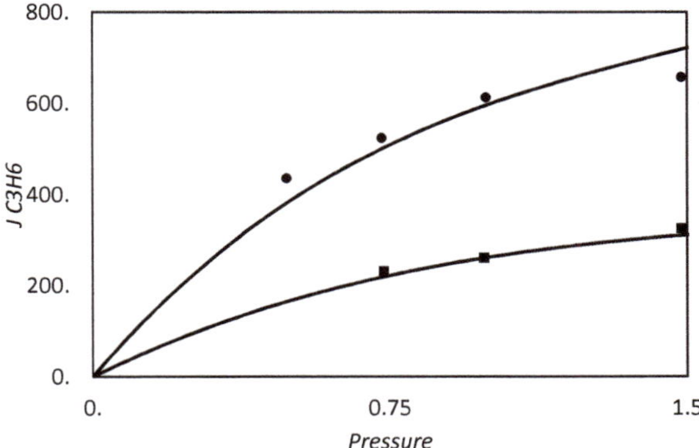

Figure 17. Ethylene flux as function of upstream pressure in membranes of poly-vinyldene fluoride (PVDF) containing $AgBF_4$ as fixed carrier at two different temperatures. Symbols are experimental data from Zarca et al., lines are model results. From the top, T = 303.15 and 293.15 K. Flux is expressed in (1×10^{-6} mol/cm^2·s), pressure is in (atm).

3.1.8. Zarca et al., 2017. Generalized Predictive Modeling for Facilitated Transport Membranes Accounting For Fixed and Mobile Carriers

As extension of the previous model, the authors included a third term in the flux expression to consider the case of hybrid fixed-mobile carriers systems [18].

Object of their study was the facilitated separation between propylene and propane but, in this case, an ionic liquid, $BMImBF_4$, was also incorporated in the poly-vinyldene fluoride-co-hexafluoropropylene (PVDF-HFP) polymeric membranes, together with a silver salt, $AgBF_4$. The resulting transport mechanism is a combination of the two effects of mobile and fixed carrier. Silver ions bounded to the polymer backbone act as fixed carriers while, the unbounded ones, due to the presence of liquid ion, are free to diffuse in the system, acting as mobile carrier. For these hybrid systems the total solute flux has the form:

$$J_A = P\frac{(p_A^0 - p_A^L)}{L} + K_H\frac{(p_A^0 - p_A^L)}{L} + P_{comp}\frac{(p_A^0 - p_A^L)}{L} \tag{208}$$

where K_H is the enhanced permeability related to fixed sites, given by Equation (207) and P_{comp} is the additive permeation term accounting for the presence of mobile carrier facilitation. Since in mobile facilitated transport the percolation limit does not exist, the parameter accounting for that in Equation (207), β, is set equal to one.

Following a kind of solution–diffusion model, P_{comp} is calculated as the product of the chemical induced solubility, by means of chemical reaction in liquid phase with free silver ions, and the diffusion coefficient of the solute-complex in the ionic liquid.

The solubility induced by the chemical reaction has been derived from the chemical equilibrium as the concentration of the complexed species divided by the upstream partial pressure of solute A. To do that, the authors considered that the free solute concentration in the liquid phase could be explained by the Henry's law using the upstream partial pressure, so that the equilibrium complex concentration is given by:

$$C_{AC} = \frac{K'_{eq}C_T H p_A^0}{1 + K'_{eq}H p_A^0} \tag{209}$$

where H is the Henry's constant and K'_{eq} is the chemical equilibrium constant in the liquid environment. The solubility coefficient, S_{AC}, is obtained by dividing Equation (209) for the upstream partial pressure and is shown in Equation (210).

$$S_{AC} = \frac{K'_{eq} C_T H}{1 + K'_{eq} H p_A^0} \tag{210}$$

In the end, the permeability through the mobile carrier mechanism is:

$$P_{comp} = \frac{K'_{eq} C_T H}{1 + K'_{eq} H p_A^0} D_{comp} \tag{211}$$

where D_{comp} indicates the diffusion coefficient of the complex species in the ionic liquid environment. Arrhenius–like functions were used to describe the thermal dependence of the equilibrium constant, K'_{eq}, of the diffusion coefficient and of the Henry's constant, so that the present model could in principle describe facilitation at different temperatures.

The only pure fitting parameter is α, since β is set equal to one a priori in this case. The experimental data for the system under study were modeled with satisfactory results, as we can see in Figure 18a.

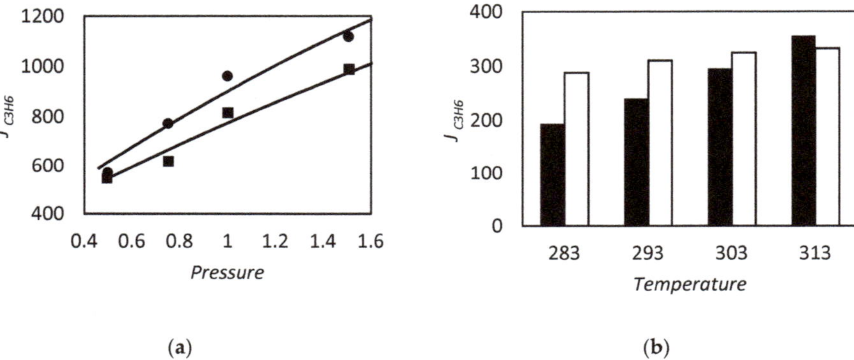

(a) (b)

Figure 18. Ethylene flux in polymeric membranes (PVDF-HFP)/BMImBF$_4$/AgBF$_4$. (**a**) Influence of ethylene pressure at two different temperature. Symbols are experimental data, lines are model calculations from Zarca et al. [18]. From the top, T = 303.15 K, 293.15 K; (**b**) Prediction of fixed site (black) and mobile carrier (white) contribution to total flux as function of temperature. Flux is expressed in (1×10^{-6} mol/cm^2·s), pressure is in (atm), temperature is in (K).

Furthermore, the temperature dependence relations and the co-presence of different terms for the two distinct facilitation mechanisms, allow the model to predict the experimental conditions for which one effect overcomes the other, as shown in Figure 18b, giving interesting and practical information.

Even if containing fitting parameters, the method proposed by Zarca et al. seems to be a useful calculation tool for hybrid facilitated carrier membranes performance.

4. Conclusions

In this work, different mathematical approaches developed to tackle the mass transfer problem in facilitated transport membranes are reviewed, and their range of applicability and accuracy analyzed for both mobile and fixed carrier systems.

In particular, the analysis is carried out for a simplified case where only one reaction occurs between the carrier and the target permeants. Even for this simplified system, there is no general analytical solution, so that most modeling approaches are based on simplified solutions approximating the real membrane behavior.

Two opposite transport regimes define the limiting cases of such systems: the pure physical diffusion of unreacted solute and the chemical equilibrium regime. Both scenarios are oversimplifications of the problem and lead to unreliable quantitative results in most operative conditions. Nevertheless, such conditions can be used as reference states to develop approximate methods for the solution of the facilitated transport problem.

In this direction, for the mobile carrier case, Goddard et al. [84] divided the membrane thickness in three sublayers, assuming the chemical equilibrium in the middle while, in the adjacent boundary zones, the same order of magnitude was considered for the diffusion and reaction mechanisms.

Smith et al. [88] studied the transport in 'thin' or 'thick' films considering that, near the boundaries, the reaction should not impact the solute transport, compared to the inner core while Kreuzer and Hoofd derived their solution [86,87], considering that the species concentrations deviate from the equilibrium ones by a position departure function which goes quickly to zero as the distance from boundaries increase.

The same approach was also used by Teramoto [74] and Morales-Cabrera et al. [98], and led to the conclusion that for Damköhler number $\gg 1$, the internal equilibrium core is a reasonable assumption. Moreover, the last two methods are also able to predict the presence of an optimal range of equilibrium constants that maximize the facilitation effect when parameters such as species diffusivities, solute partial pressure or concentration change.

In conclusion, among the models reported in this work for the mobile carrier case, the ones from Teramoto [74] and from Morales-Cabrera et al. [98], seem the more flexible ones as they can be applied in a wide range of operative conditions, spacing from the pure physical to chemical equilibrium regime, without the assumption of equal diffusivity of carrier and reaction product. However, they consist in a system of equations that need to be solved by using a numerical technique and it can be not straightforward.

In cases for which the equality between carrier and complex diffusivities is reasonable, and if the excess of carrier concentration can be assumed, the simple models provided by Smith and Quinn [93], Noble et al. [72] and Jeema and Noble [96], seem to be the best choice due to their descriptive ability coupled with a very simple mathematical form.

In the fixed carrier systems, the only species that can freely diffuse across the film are, obviously, the solute of interest and the other dissolved species, since both the carrier and the complex (reaction product) are immobilized in the matrix and can only vibrate around their equilibrium position. It follows that the chemical reaction does not solely enhance the overall solute solubility in the systems, but acts as a catalyst in the diffusion process.

Noble [111] formalized this effect by considering all the possible diffusion paths accessible by the solute in the solid matrix, starting from the generalized dual mode theory of Barrer [108]. He concluded that, in principle, a facilitation may be detected even if the carrier sites are too far one from another and a pure hopping mechanism is not allowed, contrarily for example to what foreseen by Cussler et al. in their modeling approach [100]. Another important result from Noble's work was that fixed carrier facilitated transport systems can be described by a simple modeling approach, mathematically equivalent to that developed by Smith and Quinn for mobile carrier systems [93].

Other interesting approaches in the analysis of fixed carrier systems comes from the analogy with RC parallel circuit model provided by Kang et al. [112,113] and from the description recently proposed by Zarca et al. for pure fixed sites, [114] and for the hybrid cases (mobile + fixed carriers), [18]. The resulting models are based on the use of some fitting parameters but, nevertheless, seem to be a very useful tool to indagate these peculiar systems.

This work shows that there are many possible tools that can be used to describe and understand facilitated transport membrane and can be seen as a valid support in the development of facilitated transport membranes and in the correct exploitation of the already existing ones.

However, it is worth to mentioning that for both the mobile and fixed carrier systems one of the main issue is the knowledge of the physical and chemical parameters needed to quantitatively describe the transport. In many cases, the experimental determination of these latter is not so easy, in particular for the fixed carrier cases. To overcome this point, the use of advanced simulations by the molecular dynamics technique and/or the use of advanced thermodynamic models, or equation of state, that can describe complex multicomponent systems starting from the sole knowledge of the pure components behavior, could be a way to open new windows in this modeling field.

Supplementary Materials: The following are available online at http://www.mdpi.com/2077-0375/9/2/26/s1, S1: Supplementary information reporting additional details on the derivation of the following models: Goddard, Shultz and Bassett, 1969, On membrane diffusion with near equilibrium reaction. Kreuzer and Hoofd, 1970, Facilitated diffusion of oxygen in the presence of hemoglobin. Yung and Probstein, 1973, Similarity considerations in facilitated transport. Smith, Meldon, Colton, 1973, An analysis of carrier facilitated transport. Basaran et al., 1989, Facilitated Transport with Unequal Carrier and Complex Diffusivities. Morales-Cabrera et al., 2002, Approximate Method for The Solution of Facilitated Transport Problems in Liquid Membranes.

Author Contributions: In the present review, most of the work related to bibliographyc search and analysis of the models was carried out by R.R., M.G.D.A. and M.G.B. were mainly involved in structuring the review and in the supervision of R.R. work, in order to ensure the final review was complete and rigorous.

Funding: This work has been performed in the framework of the European Project H2020 NANOMEMC^2 "NanoMaterials Enhanced Membranes for Carbon Capture", funded by the Innovation and Networks Executive Agency (INEA) Grant Agreement Number: 727734.

Conflicts of Interest: The authors declare no conflict of interest.

References

1. Schultz, J.S.; Goddard, J.D.; Suchdeo, S.R. Facilitated transport via carrier-mediated diffusion in membranes: Part I. Mechanistic aspects, experimental systems and characteristic regimes. *AIChE J.* **1974**, *20*, 417–445. [CrossRef]
2. Goddard, J.D.; Schultz, J.S.; Suchdeo, S.R. Facilitated transport via carrier—mediated diffusion in membranes: Part II. Mathematical aspects and analyses. *AIChE J.* **1974**, *20*, 625–645. [CrossRef]
3. Osterhout, W.J.V. Some models of protoplasmic surfaces. *Cold Spring Harb. Symp. Quant. Biol.* **1940**, *8*, 51–62. [CrossRef]
4. Scholander, P.F. Oxygen Transport through Hemoglobin Solutions. *Science* **1960**, *131*, 585–590. [CrossRef] [PubMed]
5. Wittemberg, J.B. The Molecular Mechanism of Hemoglobin-Facilitated Oxygen Diffusion. *J. Biol. Chem.* **1966**, *241*, 104–114.
6. Noble, D.R.; Koval, A.C.; Pellegrino, J. Facilitated Transport Membrane Systems. *Chem. Eng. Prog.* **1989**, *85*, 58–70.
7. Kemena, L.L.; Noble, R.D.; Kemp, N.J. Optimal regimes of facilitated transport. *J. Memb. Sci.* **1983**, *15*, 259–274. [CrossRef]
8. Basaran, O.A.; Burban, P.M.; Auvil, S.R. Facilitated transport with unequal carrier and complex diffusivities. *Ind. Eng. Chem. Res.* **1989**, *28*, 108–119. [CrossRef]
9. Way, J.D.; Noble, R.D.; Flynn, T.M.; Sloan, E.D. Liquid membrane transport: A survey. *J. Memb. Sci.* **1982**, *12*, 239–259. [CrossRef]
10. Basu, A.; Akhtar, J.; Rahman, M.H.; Islam, M.R. A Review of Separation of Gases Using Membrane Systems. *Pet. Sci. Technol.* **2004**, *22*, 1343–1368. [CrossRef]
11. Sholl, D.S.; Lively, R.P. Seven chemical separations to change the world. *Nature* **2016**, *532*, 435–437. [CrossRef] [PubMed]
12. Strathmann, H. Membrane separation processes: Current relevance and future opportunities. *AIChE J.* **2001**, *47*, 1077–1087. [CrossRef]
13. Teramoto, M.; Matsuyama, H.; Yamashiro, T.; Katayama, Y. Separation of ethylene from ethane by supported liquid membranes containing silver nitrate as a carrier. *J. Chem. Eng. Jpn.* **1986**, *19*, 419–424. [CrossRef]
14. Ho, W.S.; Dalrymple, D.C. Facilitated transport of olefins in Ag+-containing polymer membranes. *J. Memb. Sci.* **1994**, *91*, 13–25. [CrossRef]
15. Funke, H.H.; Noble, R.D.; Koval, C.A. Separation of gaseous olefin isomers using facilitated transport membranes. *J. Memb. Sci.* **1993**, *82*, 229–236. [CrossRef]

16. Faiz, R.; Li, K. Olefin/paraffin separation using membrane based facilitated transport/chemical absorption techniques. *Chem. Eng. Sci.* **2012**, *73*, 261–284. [CrossRef]
17. Azizi, S.; Kaghazchi, T.; Kargari, A. Propylene/propane separation using N-methyl pyrrolidone/$AgNO_3$ supported liquid membrane. *J. Taiwan Inst. Chem. Eng.* **2015**, *57*, 1–8. [CrossRef]
18. Zarca, R.; Ortiz, A.; Gorri, D.; Ortiz, I. Generalized predictive modeling for facilitated transport membranes accounting for fixed and mobile carriers. *J. Memb. Sci.* **2017**, *542*, 168–176. [CrossRef]
19. Jung, J.P.; Park, C.H.; Lee, J.H.; Park, J.T.; Kim, J.-H.; Kim, J.H. Facilitated olefin transport through membranes consisting of partially polarized silver nanoparticles and PEMA-g-PPG graft copolymer. *J. Memb. Sci.* **2018**, *548*, 149–156. [CrossRef]
20. Weston, M.H.; Colón, Y.J.; Bae, Y.-S.; Garibay, S.J.; Snurr, R.Q.; Farha, O.K.; Hupp, J.T.; Nguyen, S.T. High propylene/propane adsorption selectivity in a copper(catecholate)-decorated porous organic polymer. *J. Mater. Chem. A* **2014**, *2*, 299–302. [CrossRef]
21. Choi, H.; Lee, J.H.; Kim, Y.R.; Song, D.; Kang, S.W.; Lee, S.S.; Kang, Y.S. Tetrathiafulvalene as an electron acceptor for positive charge induction on the surface of silver nanoparticles for facilitated olefin transport. *Chem. Commun.* **2014**, *50*, 3194. [CrossRef] [PubMed]
22. Chae, I.S.; Kang, S.W.; Park, J.Y.; Lee, Y.-G.; Lee, J.H.; Won, J.; Kang, Y.S. Surface Energy-Level Tuning of Silver Nanoparticles for Facilitated Olefin Transport. *Angew. Chem. Int. Ed.* **2011**, *50*, 2982–2985. [CrossRef] [PubMed]
23. Kang, Y.S.; Kang, S.W.; Kim, H.; Kim, J.H.; Won, J.; Kim, C.K.; Char, K. Interaction with Olefins of the Partially Polarized Surface of Silver Nanoparticles Activated byp-Benzoquinone and Its Implications for Facilitated Olefin Transport. *Adv. Mater.* **2007**, *19*, 475–479. [CrossRef]
24. Okahata, Y.; Noguchi, H.; Seki, T. Highly selective transport of molecular oxygen in a polymer containing a cobalt porphyrin complex as a fixed carrier. *Macromolecules* **1986**, *19*, 494–496. [CrossRef]
25. Nishide, H.; Ohyanagi, M.; Okada, O.; Tsuchida, E. Dual-mode transport of molecular oxygen in a membrane containing a cobalt porphyrin complex as a fixed carrier. *Macromolecules* **1987**, *20*, 417–422. [CrossRef]
26. Ohyanagi, M.; Nishide, H.; Suenaga, K.; Tsuchida, E. Effect of Polymer Matrix and Metal Species on Facilitated Oxygen Transport in Metalloporphyrin (Oxygen Carrier) Fixed Membranes. *Macromolecules* **1988**, *21*, 1590–1594. [CrossRef]
27. Chen, S.H.; Lai, J.Y. Polycarbonate/(N,N′-dialicylidene ethylene diamine) cobalt(II) complex membrane for gas separation. *J. Appl. Polym. Sci.* **1996**, *59*, 1129–1135. [CrossRef]
28. Shentu, B.; Nishide, H. Facilitated Oxygen Transport Membranes of Picket-fence Cobaltporphyrin Complexed with Various Polymer Matrixes. *Ind. Eng. Chem. Res.* **2003**, *42*. [CrossRef]
29. Preethi, N.; Shinohara, H.; Nishide, H. Reversible oxygen-binding and facilitated oxygen transport in membranes of polyvinylimidazole complexed with cobalt-phthalocyanine. *React. Funct. Polym.* **2006**, *66*, 851–855. [CrossRef]
30. Li, H.; Choi, W.; Ingole, P.G.; Lee, H.K.; Baek, I.H. Oxygen separation membrane based on facilitated transport using cobalt tetraphenylporphyrin-coated hollow fiber composites. *Fuel* **2016**, *185*, 133–141. [CrossRef]
31. Powell, C.E.; Qiao, G.G. Polymeric CO_2/N_2 gas separation membranes for the capture of carbon dioxide from power plant flue gases. *J. Memb. Sci.* **2006**, *279*, 1–49. [CrossRef]
32. Adewole, J.K.; Ahmad, A.L.; Ismail, S.; Leo, C.P. Current challenges in membrane separation of CO_2 from natural gas: A review. *Int. J. Greenh. Gas Control* **2013**, *17*, 46–65. [CrossRef]
33. Norahim, N.; Yaisanga, P.; Faungnawakij, K.; Charinpanitkul, T.; Klaysom, C. Recent Membrane Developments for CO_2 Separation and Capture. *Chem. Eng. Technol.* **2018**, *41*, 211–223. [CrossRef]
34. *Summary for Policymakers of IPCC Special Report on Global Warming of 1.5 °C Approved by Governments*; IPCC: Geneva, Switzerland, 2018.
35. Guha, A.K.; Majumdar, S.; Sirkar, K.K. Facilitated Transport of CO_2 through an Immobilized Liquid Membrane of Aqueous Diethanolamine. *Ind. Eng. Chem. Res.* **1990**, *29*, 2093–2100. [CrossRef]
36. Davis, R.A.; Sandall, O.C. CO_2/CH_4 separation by facilitated transport in amine-polyethylene glycol mixtures. *AIChE J.* **1993**, *39*, 1135–1145. [CrossRef]
37. Teramoto, M.; Nakai, K.; Huang, Q.; Watari, T.; Matsuyama, H. Facilitated Transport of Carbon Dioxide through Supported Liquid Membranes of Aqueous Amine Solutions. *Ind. Eng. Chem. Res.* **1996**, *35*, 538–545. [CrossRef]

38. Matsuyama, H.; Terada, A.; Nakagawara, T.; Kitamura, Y.; Teramoto, M. Facilitated transport of CO_2 through polyethylenimine/poly(vinyl alcohol) blend membrane. *J. Memb. Sci.* **1999**, *163*, 221–227. [CrossRef]
39. Huang, J.; Zou, J.; Ho, W.S.W. Carbon Dioxide Capture Using a CO_2-Selective Facilitated Transport Membrane. *Ind. Eng. Chem. Res.* **2008**, *47*, 1261–1267. [CrossRef]
40. Zou, J.; Ho, W.S.W. CO_2-selective polymeric membranes containing amines in crosslinked poly(vinyl alcohol). *J. Memb. Sci.* **2006**, *286*, 310–321. [CrossRef]
41. Deng, L.; Kim, T.-J.; Hägg, M.-B. Facilitated transport of CO_2 in novel PVAm/PVA blend membrane. *J. Memb. Sci.* **2009**, *340*, 154–163. [CrossRef]
42. Sandru, M.; Kim, T.-J.; Capala, W.; Huijbers, M.; Hägg, M.-B. Pilot scale testing of polymeric membranes for CO_2 capture from coal fired power plants Selection and/or peer-review under responsibility of GHGT. *Energy Procedia* **2013**, *37*, 6473–6480. [CrossRef]
43. He, X.; Lindbråthen, A.; Kim, T.-J.; Hägg, M.-B. Pilot testing on fixed-site-carrier membranes for CO_2 capture from flue gas. *Int. J. Greenh. Gas Control* **2017**, *64*, 323–332. [CrossRef]
44. Ansaloni, L.; Salas-Gay, J.; Ligi, S.; Baschetti, M.G. Nanocellulose-based membranes for CO_2 capture. *J. Memb. Sci.* **2017**, *522*, 216–225. [CrossRef]
45. Ansaloni, L.; Zhao, Y.; Jung, B.T.; Ramasubramanian, K.; Baschetti, M.G.; Ho, W.S.W. Facilitated transport membranes containing amino-functionalized multi-walled carbon nanotubes for high-pressure CO_2 separations. *J. Memb. Sci.* **2015**, *490*, 18–28. [CrossRef]
46. Fernández-Barquín, A.; Rea, R.; Venturi, D.; Giacinti-Baschetti, M.; De Angelis, M.G.; Casado-Coterillo, C.; Irabien, Á. Effect of relative humidity on the gas transport properties of zeolite A/PTMSP mixed matrix membranes. *RSC Adv.* **2018**, *8*, 3536–3546. [CrossRef]
47. Otvagina, K.V.; Mochalova, A.E.; Sazanova, T.S.; Petukhov, A.N.; Moskvichev, A.A.; Vorotyntsev, A.V.; Afonso, C.A.M.; Vorotyntsev, I.V. Preparation and Characterization of Facilitated Transport Membranes Composed of Chitosan-Styrene and Chitosan-Acrylonitrile Copolymers Modified by Methylimidazolium Based Ionic Liquids for CO_2 Separation from CH_4 and N_2. *Membranes* **2016**, *6*, 31. [CrossRef] [PubMed]
48. He, X.; Kim, T.-J.; Hägg, M.-B. Hybrid fixed-site-carrier membranes for CO_2 removal from high pressure natural gas: Membrane optimization and process condition investigation. *J. Memb. Sci.* **2014**, *470*, 266–274. [CrossRef]
49. Hanioka, S.; Maruyama, T.; Sotani, T.; Teramoto, M.; Matsuyama, H.; Nakashima, K.; Hanaki, M.; Kubota, F.; Goto, M. CO_2 separation facilitated by task-specific ionic liquids using a supported liquid membrane. *J. Memb. Sci.* **2008**, *314*, 1–4. [CrossRef]
50. Liu, Z.; Liu, C.; Li, L.; Qin, W.; Xu, A. CO_2 separation by supported ionic liquid membranes and prediction of separation performance. *Int. J. Greenh. Gas Control* **2016**, *53*, 79–84. [CrossRef]
51. Shin, J.Y.; Yamada, S.A.; Fayer, M.D. Carbon Dioxide in a Supported Ionic Liquid Membrane: Structural and Rotational Dynamics Measured with 2D IR and Pump–Probe Experiments. *J. Am. Chem. Soc.* **2017**, *139*, 11222–11232. [CrossRef]
52. Olivieri, L.; Aboukeila, H.; Giacinti Baschetti, M.; Pizzi, D.; Merlo, L.; Sarti, G.C. Humid permeation of CO_2 and hydrocarbons in Aquivion® perfluorosulfonic acid ionomer membranes, experimental and modeling. *J. Memb. Sci.* **2017**, *542*, 367–377. [CrossRef]
53. Kutchai, H.; Jacquez, J.A.; Mather, F.J. Nonequilibrium-Facilitated Oxygen Transport in Hemoglobin Solution. *Biophys. J.* **1970**, *10*, 38–54. [CrossRef]
54. Ward, W.J. Analytical and experimental studies of facilitated transport. *AIChE J.* **1970**, *16*, 405–410. [CrossRef]
55. Collatz, L. *The Numerical Treatment of Differential Equations*; Springer: Berlin, Germany, 1960; ISBN 3642884342.
56. Kirkköprü-Dindi, A.; Noble, R.D. Optimal regimes of facilitated transport for multiple site carriers. *J. Memb. Sci.* **1989**, *42*, 13–25. [CrossRef]
57. Carey, G.F.; Finlayson, B.A. Orthogonal collocation on finite elements. *Chem. Eng. Sci.* **1975**, *30*, 587–596. [CrossRef]
58. Jain, R.; Schultz, J. A Numerical technique for solving carrier-mediated transport problems. *J. Memb. Sci.* **1982**, *11*, 79–106. [CrossRef]
59. Barbero, A.J.; Manzanares, J.A.; Mafé, S. A Computational Study of Facilitated Diffusion Using The Boundary Element Method. *J. Non-Equilib. Thermodyn.* **1995**, *20*, 332–341. [CrossRef]
60. Ramachandran, P.A. Application of the boundary element method to non-linear diffusion with reaction problems. *Int. J. Numer. Methods Eng.* **1990**, *29*, 1021–1031. [CrossRef]

61. Ebadi Amooghin, A.; Moftakhari Sharifzadeh, M.M.; Zamani Pedram, M. Rigorous modeling of gas permeation behavior in facilitated transport membranes (FTMs); evaluation of carrier saturation effects and double-reaction mechanism. *Greenh. Gases Sci. Technol.* **2017**, *8*, 429–443. [CrossRef]
62. Graaf, J.M.D. A system of reaction-diffusion equations: Convergence towards the stationary solution. *Nonlinear Anal. Theory Methods Appl.* **1990**, *15*, 253–267. [CrossRef]
63. Feng, W. Coupled system of reaction-diffusion equations and applications in carrier facilitated diffusion. *Nonlinear Anal. Theory Methods Appl.* **1991**, *17*, 285–311. [CrossRef]
64. Desvillettes, L.; Fellner, K.; Pierre, M.; Vovelle, J.; Desvillettes, L.; Fellner, K.; Pierre, M.; Vovelle, J. Global Existence for Quadratic Systems of Reaction-Diffusion. *Adv. Nonlinear Stud.* **2007**, *7*, 491–511. [CrossRef]
65. Britton, N.F. *Reaction-Diffusion Equations and Their Application to Biology*; Academic Press: London, UK; Orlando, FL, USA, 1986.
66. Caputo, M.C.; Vasseur, A. Global Regularity of Solutions to Systems of Reaction–Diffusion with Sub-Quadratic Growth in Any Dimension. *Commun. Partial Differ. Equ.* **2009**, *34*, 1228–1250. [CrossRef]
67. Lo, W.-C.; Chen, L.; Wang, M.; Nie, Q. A Robust and Efficient Method for Steady State Patterns in Reaction-Diffusion Systems. *J. Comput. Phys.* **2012**, *231*, 5062–5077. [CrossRef]
68. Friedlander, S.K.; Keller, K.H. Mass transfer in reacting systems near equilibrium. *Chem. Eng. Sci.* **1965**, *20*, 121–129. [CrossRef]
69. Keller, K.H.; Friedlander, S.K. The Steady-State Transport of Oxygen through Hemoglobin Solutions. *J. Gen. Physiol.* **1966**, *49*, 663–679. [CrossRef]
70. Wyman, J. Facilitated Diffusion and the Possible Role of Myoglobin as a Transport Mechanism. *J. Biol. Chem.* **1966**, *241*, 115–121.
71. La Force, R.C.; Fatt, I. Steady-state diffusion of oxygen through whole blood. *Trans. Faraday Soc.* **1962**, *58*, 1451. [CrossRef]
72. Noble, R.D.; Way, J.D.; Powers, L.A. Effect of external mass-transfer resistance on facilitated transport. *Ind. Eng. Chem. Fundam.* **1986**, *25*, 450–452. [CrossRef]
73. Olander, D.R. Simultaneous mass transfer and equilibrium chemical reaction. *AIChE J.* **1960**, *6*, 233–239. [CrossRef]
74. Teramoto, M. Approximate Solution of Facilitation Factors in Facilitated Transport. *Ind. Eng. Chem. Res.* **1994**, *33*, 2161–2167. [CrossRef]
75. Fatt, I.; La Force, R.C. Theory of Oxygen Transport through Hemoglobin Solutions. *Science* **1961**, *133*, 1919–1921. [CrossRef]
76. Zhao, Y.; Ho, W.S.W. CO_2 -Selective Membranes Containing Sterically Hindered Amines for CO_2 /H_2 Separation. *Ind. Eng. Chem. Res.* **2013**, *52*, 8774–8782. [CrossRef]
77. Tong, Z.; Ho, W.S.W. Facilitated transport membranes for CO_2 separation and capture. *Sep. Sci. Technol.* **2017**, *52*, 156–167. [CrossRef]
78. Tee, Y.-H.; Zou, J.; Ho, W.S.W. CO_2-Selective Membranes Containing Dimethylglycine Mobile Carriers and Polyethylenimine Fixed Carrier. *J. Chin. Inst. Chem. Eng.* **2006**, *37*, 37–47. [CrossRef]
79. Liao, J.; Wang, Z.; Gao, C.; Li, S.; Qiao, Z.; Wang, M.; Zhao, S.; Xie, X.; Wang, J.; Wang, S. Fabrication of high-performance facilitated transport membranes for CO_2 separation. *Chem. Sci.* **2014**, *5*, 2843–2849. [CrossRef]
80. Grünauer, J.; Shishatskiy, S.; Abetz, C.; Abetz, V.; Filiz, V. Ionic liquids supported by isoporous membranes for CO_2/N_2 gas separation applications. *J. Memb. Sci.* **2015**, *494*, 224–233. [CrossRef]
81. Halder, K.; Khan, M.M.; Grünauer, J.; Shishatskiy, S.; Abetz, C.; Filiz, V.; Abetz, V. Blend membranes of ionic liquid and polymers of intrinsic microporosity with improved gas separation characteristics. *J. Memb. Sci.* **2017**, *539*, 368–382. [CrossRef]
82. Zhao, Y.; Winston Ho, W.S. Steric hindrance effect on amine demonstrated in solid polymer membranes for CO_2 transport. *J. Memb. Sci.* **2012**, *415–416*, 132–138. [CrossRef]
83. Blumenthal, R.; Katchalsky, A. The effect of the carrier association-dissociation rate on membrane permeation. *Biochim. Biophys. Acta* **1969**, *173*, 357–369. [CrossRef]
84. Goddard, J.D.; Schultz, J.S.; Bassett, R.J. On membrane diffusion with near-equilibrium reaction. *Chem. Eng. Sci.* **1970**, *25*, 665–683. [CrossRef]
85. Van Dyke, M. *Perturbation Methods in Fluid Mechanics*; Academic Press: New York, NY, USA, 1964; ISBN 10: 0127130500.

86. Kreuzer, F.; Hoofd, L.J.C. Facilitated diffusion of oxygen in the presence of hemoglobin. *Respir. Physiol.* **1970**, *8*, 280–302. [CrossRef]
87. Kreuzer, F.; Hoofd, L.J.C. Factors influencing facilitated diffusion of oxygen in the presence of hemoglobin and myoglobin. *Respir. Physiol.* **1972**, *15*, 104–124. [CrossRef]
88. Smith, K.A.; Meldon, J.H.; Colton, C.K. An analysis of carrier-facilitated transport. *AIChE J.* **1973**, *19*, 102–111. [CrossRef]
89. Hoofd, L.; Kreuzer, F. A new mathematical approach for solving carrier-facilitated steady-state diffusion problems. *J. Math. Biol.* **1979**, *8*, 1–13. [CrossRef]
90. Yung, D.; Probstein, R.F. Similarity considerations in facilitated transport. *J. Phys. Chem.* **1973**, *77*, 2201–2205. [CrossRef]
91. Airy, G.B. ON the Intensity of Light in the neighbourhood of a Caustic. *Trans. Camb. Philos. Soc.* **1838**, *6*, 379–402.
92. Bdzil, J.; Carlier, C.C.; Frisch, H.L.; Ward, W.J.; Breiter, M.W. Analysis of potential difference in electrically induced carrier transport systems. *J. Phys. Chem.* **1973**, *77*, 846–850. [CrossRef]
93. Smith, D.R.; Quinn, J.A. The prediction of facilitation factors for reaction augmented membrane transport. *AIChE J.* **1979**, *25*, 197–200. [CrossRef]
94. Donaldson, T.L.; Quinn, J.A. Carbon dioxide transport through enzymatically active synthetic membranes. *Chem. Eng. Sci.* **1975**, *30*, 103–115. [CrossRef]
95. Folkner, C.A.; Noble, R.D. Transient response of facilitated transport membranes. *J. Memb. Sci.* **1983**, *12*, 289–301. [CrossRef]
96. Jemaa, N.; Noble, R.D. Improved analytical prediction of facilitation factors in facilitated transport. *J. Memb. Sci.* **1992**, *70*, 289–293. [CrossRef]
97. Van Krevelen, D.W.; Hoftijzer, P.J. Kinetics of gas-liquid reactions part I. General theory. *Recl. Trav. Chim. Pays-Bas* **2010**, *67*, 563–586. [CrossRef]
98. Morales-Cabrera, M.A.; Pérez-Cisneros, E.S.; Ochoa-Tapia, J.A. Approximate Method for the Solution of Facilitated Transport Problems in Liquid Membranes. *Ind. Eng. Chem. Res.* **2002**, *41*, 4626–4631. [CrossRef]
99. Morales-Cabrera, M.A. Factores de mejora en membranas liquidas. Master's Thesis, Universidad Autonoma Metropolitana, Mexico City, México, 2000.
100. Cussler, E.L.; Aris, R.; Bhown, A. On the limits of facilitated diffusion. *J. Memb. Sci.* **1989**, *43*, 149–164. [CrossRef]
101. Venturi, D.; Grupkovic, D.; Sisti, L.; Baschetti, M.G. Effect of humidity and nanocellulose content on Polyvinylamine-nanocellulose hybrid membranes for CO_2 capture. *J. Memb. Sci.* **2018**, *548*, 263–274. [CrossRef]
102. Rafiq, S.; Deng, L.; Hägg, M.-B. Role of Facilitated Transport Membranes and Composite Membranes for Efficient CO_2 Capture—A Review. *ChemBioEng Rev.* **2016**, *3*, 68–85. [CrossRef]
103. Langmuir, I. The Adsorption of Gases on Plane Surfaces of Glass, Mica and Platinum. *J. Am. Chem. Soc.* **1918**, *40*, 1361–1403. [CrossRef]
104. Masel, R.I. *Principles of Adsorption and Reaction on Solid Surfaces*; Wiley: Hoboken, NJ, USA, 1996; ISBN 0471303925.
105. Nishide, H.; Kawakami, H.; Suzuki, T.; Azechi, Y.; Soejima, Y.; Tsuchida, E. Effect of polymer matrix on the oxygen diffusion via a cobalt porphyrin fixed in a membrane. *Macromolecules* **1991**, *24*, 6306–6309. [CrossRef]
106. Yoshikawa, M.; Ezaki, T.; Sanui, K.; Ogata, N. Selective permeation of carbon dioxide through synthetic polymer membranes having pyridine moiety as a fixed carrier. *J. Appl. Polym. Sci.* **1988**, *35*, 145–154. [CrossRef]
107. Yoshikawa, M.; Fujimoto, K.; Kinugawa, H.; Kitao, T.; Kamiya, Y.; Ogata, N. Specialty polymeric membranes. V. Selective permeation of carbon dioxide through synthetic polymeric membranes having 2-(N,N-dimethyl)aminoethoxycarbonyl moiety. *J. Appl. Polym. Sci.* **1995**, *58*, 1771–1778. [CrossRef]
108. Barrer, R.M. Diffusivities in Glassy Polymers for the Dual Mode Sorption Model. *J. Memb. Sci.* **1984**, *18*, 25–35. [CrossRef]
109. Noble, R.D. Analysis of facilitated transport with fixed site carrier membranes. *J. Memb. Sci.* **1990**, *50*, 207–214. [CrossRef]
110. Noble, R.D. Facilitated transport mechanism in fixed site carrier membranes. *J. Memb. Sci.* **1991**, *60*, 297–306. [CrossRef]

111. Noble, R.D. Generalized microscopic mechanism of facilitated transport in fixed site carrier membranes. *J. Memb. Sci.* **1992**, *75*, 121–129. [CrossRef]
112. Kang, Y.S.; Hong, J.-M.; Jang, J.; Kim, U.Y. Analysis of facilitated transport in solid membranes with fixed site carriers 1. Single RC circuit model. *J. Memb. Sci.* **1996**, *109*, 149–157. [CrossRef]
113. Hong, J.-M.; Kang, Y.S.; Jang, J.; Kim, U.Y. Analysis of facilitated transport in polymeric membrane with fixed site carrier 2. Series RC circuit model. *J. Memb. Sci.* **1996**, *109*, 159–163. [CrossRef]
114. Zarca, R.; Ortiz, A.; Gorri, D.; Ortiz, I. A practical approach to fixed-site-carrier facilitated transport modeling for the separation of propylene/propane mixtures through silver-containing polymeric membranes. *Sep. Purif. Technol.* **2017**, *180*, 82–89. [CrossRef]
115. Henry, W. Experiments on the Quantity of Gases Absorbed by Water, at Different Temperatures, and under Different Pressures. *Philos. Trans. R. Soc. Lond.* **1803**, *93*, 29–274. [CrossRef]
116. Meares, P. The Diffusion of Gases Through Polyvinyl Acetate. *J. Am. Chem. Soc.* **1954**, *76*, 3415–3422. [CrossRef]
117. Barrer, R.M.; Barrie, J.A.; Slater, J. Sorption and diffusion in ethyl cellulose. Part III. Comparison between ethyl cellulose and rubber. *J. Polym. Sci.* **1958**, *27*, 177–197. [CrossRef]
118. Petropoulos, J.H. Quantitative analysis of gaseous diffusion in glassy polymers. *J. Polym. Sci. Part A* **1970**, *8*, 1797–1801. [CrossRef]
119. Michaels, A.S.; Vieth, W.R.; Barrie, J.A. Diffusion of Gases in Polyethylene Terephthalate. *J. Appl. Phys.* **1963**, *34*, 13–20. [CrossRef]
120. Paul, D.R.; Koros, W.J. Effect of partially immobilizing sorption on permeability and the diffusion time lag. *J. Polym. Sci. Polym. Phys. Ed.* **1976**, *14*, 675–685. [CrossRef]
121. Koros, W.J.; Paul, D.R.; Rocha, A.A. Carbon dioxide sorption and transport in polycarbonate. *J. Polym. Sci. Polym. Phys. Ed.* **1976**, *14*, 687–702. [CrossRef]
122. Fredrickson, G.H.; Helfand, E. Dual-mode transport of penetrants in glassy polymers. *Macromolecules* **1985**, *18*, 2201–2207. [CrossRef]
123. Ogata, N.; Sanui, K.; Fujimura, H. Active transport membrane for chlorine ion. *J. Appl. Polym. Sci.* **1980**, *25*, 1419–1425. [CrossRef]
124. Przewoźna, M.; Gajewski, P.; Michalak, N.; Bogacki, M.B.; Skrzypczak, A. Determination of the Percolation Threshold for the Oxalic, Tartaric, and Lactic Acids Transport through Polymer Inclusion Membranes with 1-Alkylimidazoles as a Carrier. *Sep. Sci. Technol.* **2014**, *49*, 1745–1755. [CrossRef]
125. Riggs, J.A.; Smith, B.D. Facilitated Transport of Small Carbohydrates through Plasticized Cellulose Triacetate Membranes. Evidence for Fixed-Site Jumping Transport Mechanism. *J. Am. Chem. Soc.* **1997**, *119*, 2765–2766. [CrossRef]
126. Smith, B.D.; Gardiner, S.J.; Munro, T.A.; Paugam, M.-F.; Riggs, J.A. Facilitated Transport of Carbohydrates, Catecholamines, and Amino Acids Through Liquid and Plasticized Organic Membranes. *J. Incl. Phenom. Mol. Recognit. Chem.* **1998**, *32*, 121–131. [CrossRef]
127. Tsuchida, E.; Nishide, H.; Ohyanagi, M.; Kawakami, H. Facilitated transport of molecular oxygen in the membranes of polymer-coordinated cobalt Schiff base complexes. *Macromolecules* **1987**, *20*, 1907–1912. [CrossRef]
128. Yoshikawa, M.; Shudo, S.; Sanui, K.; Ogata, N. Active transport of organic acids through poly(4-vinylpyridine-co-acrylonitrile) membranes. *J. Memb. Sci.* **1986**, *26*, 51–61. [CrossRef]
129. Noble, R.D. Relationship of system properties to performance in facilitated transport systems. *Gas Sep. Purif.* **1988**, *2*, 16–19. [CrossRef]
130. Way, J.D.; Noble, R.D. Hydrogen sulfide facilitated transport in perfluorosulfonic acid membranes. In *Liquid Membranes*; American Chemical Society: Washington, DC, USA, 1987; Chapter 9; pp. 123–137.
131. Way, J.D.; Noble, R.D.; Reed, D.L.; Ginley, G.M.; Jarr, L.A. Facilitated transport of CO_2 in ion exchange membranes. *AIChE J.* **1987**, *33*, 480–487. [CrossRef]

Review

Molecular Modeling Investigations of Sorption and Diffusion of Small Molecules in Glassy Polymers

Niki Vergadou [1,*] and Doros N. Theodorou [2]

[1] Molecular Thermodynamics and Modelling of Materials Laboratory, Institute of Nanoscience and Nanotechnology, National Center for Scientific Research Demokritos, Aghia Paraskevi Attikis, GR-15310 Athens, Greece
[2] School of Chemical Engineering, National Technical University of Athens, GR 15780 Athens, Greece
* Correspondence: n.vergadou@inn.demokritos.gr; Tel.: +30-210-650-3960

Received: 12 June 2019; Accepted: 23 July 2019; Published: 8 August 2019

Abstract: With a wide range of applications, from energy and environmental engineering, such as in gas separations and water purification, to biomedical engineering and packaging, glassy polymeric materials remain in the core of novel membrane and state-of the art barrier technologies. This review focuses on molecular simulation methodologies implemented for the study of sorption and diffusion of small molecules in dense glassy polymeric systems. Basic concepts are introduced and systematic methods for the generation of realistic polymer configurations are briefly presented. Challenges related to the long length and time scale phenomena that govern the permeation process in the glassy polymer matrix are described and molecular simulation approaches developed to address the multiscale problem at hand are discussed.

Keywords: polymers; diffusion; transition state theory; sorption; permeability; penetrant; separations; kinetic Monte Carlo; coarse-graining; multiscale modeling

---

1. Introduction

Molecular modeling and simulation play an increasingly important role in the design of products and processes for addressing the grand challenges faced by contemporary society in domains such as health, energy, food, clean water, and the protection of the environment. This is attested by the dramatic increase in the fraction of research papers in science and engineering that use molecular modeling [1]; by the enhanced role of molecular modeling activities in industry [2]; and by economic analyses of the impact of molecular modeling [3]. Very visible and well-funded national initiatives in the area of materials include modeling and simulation as one of their top priorities (see, for example, Materials Genome Initiative in the United States [4]) or are entirely focused on molecular and multiscale modeling (see, for example, OCTA Project in Japan [5] and European Materials Modeling Council in the European Union [6]).

Molecular simulations of sorption and diffusion in membrane materials have been performed since the late 1980s. This is understandable, given that membrane separations constitute a very dynamic and rapidly growing field of modern technology. Early molecular dynamics simulations were instrumental in elucidating atomic-level mechanisms, e.g., of elementary jumps of gaseous penetrants executed between clusters of accessible volume in a glassy polymer. In the 1990s the basic methodology for predicting sorption isotherms and diffusion coefficients from atomistic simulations was established and validated on well-defined model systems. This methodology is exposed in several past reviews of permeation phenomena in amorphous polymers [7–9]. Contemporary simulation efforts have expanded to address permeation phenomena in more sophisticated materials of complex chemical constitution, nanostructured and hybrid organic-inorganic systems. As the application of conventional molecular simulation algorithms is often too computationally demanding for such systems, there is increasing interest in multiscale modeling methods based on systematic coarse-graining of the molecular representation. For some categories of new membrane materials whose synthesis has taken off only recently, such as Metal Organic Frameworks, one sees simulation work going hand-in-hand with, and even guiding, experimental efforts. High throughput computational screening for the identification of new nanoporous material structures that best satisfy the requirements of specific gas separation, sequestration, or storage applications has emerged [10]. Machine learning techniques are starting to be used both in the development of more refined quantum mechanics-based classical force fields (or coarse grained ones based on the atomistic representation) for conducting simulations and in mining the copious data generated by molecular simulations for materials design.

In this article we present a brief review of molecular modeling and multiscale simulations (Figure 1) of sorption and diffusion of small molecules in glassy polymers, a broad category of materials which continue being a workhorse of selective membrane separation technology. The treatment is concise and certainly not intended to be all-encompassing. Emphasis is laid on work conducted in the last couple of decades. An extensive reference list is provided, to which the interested reader can resort for details.

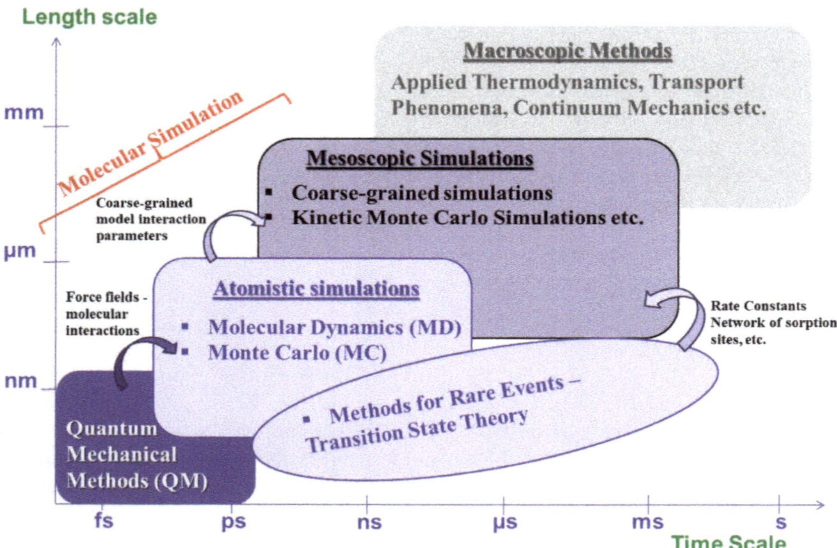

Figure 1. Molecular simulation methods at multiple length- and time scales. Hierarchical multiscale simulations utilize information extracted from the previous scale as input for conducting molecular simulations at longer length and time scales.

The structure of the article is as follows: in Section 2 we present an introduction to basic concepts and discuss equilibrium and non-equilibrium molecular dynamics and Monte Carlo techniques for simulating polymers and sorption and diffusion of small molecules therein. Methodological aspects of the simulations are addressed very briefly—they have been presented extensively in earlier reviews. Emphasis is laid on multiscale methods that incorporate infrequent event analysis of elementary diffusive jumps and on coarse-graining/reverse mapping for the generation of realistic model configurations in which to study permeation phenomena. Both these aspects are necessary for predicting the separation performance of dense glassy polymer matrices, especially if their chemical constitution is complex. Section 3 discusses coarse-grained approaches to the simulation of sorption and diffusion phenomena in polymers per se. The approaches discussed have been designed to address long length and time scale phenomena by cutting down on the computational cost relative to full atomistic analyses; they include methodologies that incorporate mesoscopic kinetic Monte Carlo techniques that utilize information from atomistic simulations (Figure 1) and coarse-grained molecular dynamics simulations accompanied by appropriate time mapping. Some mechanistic aspects of sorption and transport revealed by recent simulation work are presented in Section 4. Emphasis in this Section is laid on the characteristics of the penetrant's hopping diffusional mechanism, on swelling and plasticization of glassy polymer matrices by compressed CO_2 and on mixed gas sorption and diffusion phenomena in polymer glasses. Finally, Section 5 briefly pinpoints challenges that still have to be met, especially in conjunction with new polymer-based membrane materials, and reflects on the future outlook of molecular simulations of sorption and diffusion phenomena in dense amorphous polymeric membranes.

2. Background and Methodology

During the membrane separation process the penetrant molecules are transferred from an upstream fluid phase to the polymer membrane where they dissolve on the membrane surface (sorption phase) and then diffuse through the polymer matrix, exiting eventually at the downstream fluid phase (desorption). The permeation process through dense polymer membranes is described by the widely known solution-diffusion mechanism, considering a thermodynamic process (sorption) and a kinetic process (mass transport). Mass transport phenomena are governed by random molecular motion and permeation is driven by the presence of chemical potential gradients. In practical formulations the gradient of concentration is used instead of the gradient of chemical potential. In a dense polymeric membrane, permeation of a single component is often described locally by the empirical Fick's laws [11]:

$$J = -D\nabla C \tag{1}$$

$$\frac{\partial C}{\partial t} = D\nabla^2 C \tag{2}$$

where C is the concentration of the diffusing component, D is the diffusion coefficient, and t is the time. The concentration profile inside the membrane is considered to reach a steady state after a certain time. Assuming that D is concentration-independent, which is typically the case at low concentrations, Equation (1) can be integrated to [12]:

$$J = D\frac{C_U - C_D}{l} \tag{3}$$

where l is the membrane thickness and C_U and C_D are the upstream and downstream permeant concentration, respectively. Equation (3) can be also expressed in terms of fluid partial vapor pressures or fugacities, p_U and p_D on the two sides of the membrane:

$$J = P\frac{p_U - p_D}{l} \tag{4}$$

in which P denotes the permeability coefficient. At low concentrations in the Henry regime and considering a constant diffusion coefficient over a range of pressures, the permeability of a membrane to a specific kind of penetrant molecules can be expressed as the product of solubility coefficient, S, and diffusivity, D, of the penetrant in the membrane:

$$P = DS \tag{5}$$

where the solubility is considered as the ratio of concentration at a specific pressure to the pressure.

The ideal permselectivity of a membrane material for two penetrant components (i and j) is defined as the ratio of permeabilities for each pair of species in pure:

$$a_{ij} = \frac{P_i}{P_j} \tag{6}$$

The ideal permselectivity can be estimated as the product of diffusivity selectivity and solubility selectivity under the assumption that each penetrant component is not affected by the other:

$$a_{ij} = \frac{D_i}{D_j}\frac{S_i}{S_j} \tag{7}$$

A high performance separation membrane [13] is expected to exhibit high selectivity characteristics combined, simultaneously, with satisfactory, or even, ideally, high permeability to one of the penetrant components. The selectivity performance of membrane materials for gas separations is limited by an upper bound. This empirical limit originates from the fact than an increase in permeability of the more permeable component is accompanied by a decrease in the separation ability. If P_i is the permeability of the more permeable gas, the upper bound [13,14] is determined as a line in the log-log plot of the relationship $P_i = k\left(\frac{P_i}{P_j}\right)^n$, where n is the slope of the line, above which no data exist. Data near or on the upper bound correspond almost exclusively to glassy polymers and this is attributed primarily to the higher solubility coefficients in glasses compared to rubbery polymers due to the increase of free volume below the glass transition temperature (T_g), and in some cases partially to a better diffusion selectivity of glassy polymers in comparison to the rubbery ones. In Figure 2, data on selectivity for O_2/N_2 separation are plotted as a function of O_2 permeability for rubbery polymers and the upper bounds for both rubbery and glassy polymers [15,16] are shown that exhibit about an order of magnitude difference in favour of the separation performance of glassy polymers.

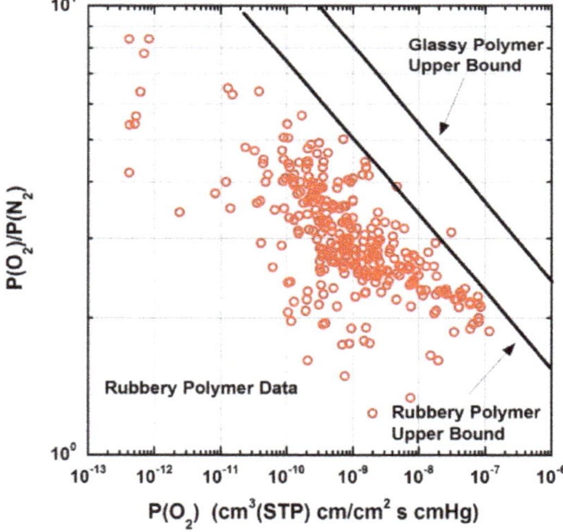

Figure 2. Permselectivity performance of glassy polymers in comparison to rubbery polymers for O_2/N_2 separation. Adapted with permission from [15].

Molecular simulation methods have been developed and are employed to unravel the microscopic mechanisms that are responsible for sorption and transport phenomena in glassy polymeric membranes and to predict their macroscopic permeability and selectivity properties. A compact summary of some representative molecular simulation methods and their applicability is provided in Table 1.

Table 1. Compact summary of representative computational methods and their applicability for the prediction of solubility and diffusivity in glassy polymers.

Method	Application	Advantages	Disadvantages	Refs.
Molecular Dynamics	Numerical integration the system's classical equations of motion	Calculation of thermodynamic, dynamic and transport properties	Not able to correctly sample the dynamics of systems that are characterized by a broad range of time scales or rare events	[17–19]
Monte Carlo	Stochastic method—generation of a Markov chain sequence of configurations	Efficient in sampling long-chain macromolecular systems when coupled with appropriately designed moves ‡	Does not account for the system's time evolution—cannot be used to study the system's dynamics	[18,20,21] ‡[22–28]
Transition State Theory	Infrequent events	Determination of rate constants and penetrant jump pathways	Multidimensional TST that accounts for polymer cooperative motion is computationally intensive	[8,9,29–34]
Transition Path Sampling	Infrequent events	Determination of realistic pathways at finite temperatures	Dependence on the limited initial transition pathways extracted by MD simulations	[35]
Kinetic Monte Carlo	Mesoscopic simulation of a Poisson process	Calculation of penetrant diffusivity by solving numerically the master equations	Requires information on sorption sites network, rate constants and sorption probabilities determined from the atomistic scale	[8,33,34,36–38]
Coarse-grained MD	Simulation of dynamics at a mesoscopic scale	Simulation of longer time- and length- scales	i. Effective time scales in CG-MD that do not correspond to the real dynamics ii. Loss of important chemical detail that governs the penetrant diffusion mechanism	[39–42]
Widom Test Particle Insertion Method	Sorption	Low concentration sorption of small molecules	Inefficient for dense systems/large solutes	[7,43–45]
Iterative Widom Schemes	Sorption	Determination of sorption isotherms	Inefficient for dense systems/large solutes	[46,47]
Thermodynamic Integration	Sorption	Enhanced efficiency for dense systems	Requires conduction of a series of simulations	[48,49]

2.1. Molecular Dynamics

Molecular dynamics (MD) [17–19,21] is based on the solution of the classical equations of motion of a system in a microscopic representation via numerical integration. During an equilibrium MD simulation the evolution of the multiparticle model representing the system is followed over time, and thermodynamic and dynamical properties are computed as averages over its trajectory. MD is widely used for equilibration and property prediction in computational materials science using atomistic or coarse-grained, fully flexible or constrained model systems and a number of statistical ensembles. MD is demanding in terms of computational time and a lot of effort has been expended in the direction of reducing these demands by implementing for example domain decomposition methods for the parallelization of MD algorithms and by developing multiple time step integration schemes [50,51] that allow the use of longer time steps for the interactions that change at a slower rate. The phase-space points visited along a long MD trajectory constitute a representative sample of the equilibrium ensemble dictated by the macroscopic constraints (e.g., constant number of particles, volume, and energy) under which the trajectory was generated, allowing the estimation of thermodynamic properties as ensemble averages. Despite progress in algorithmic developments, the extremely broad range of time scales that characterize the various modes of motion in macromolecular systems, especially in the glassy state (from tens of femtoseconds for covalent bond vibrations to years for physical aging), renders MD an insufficient method for properly equilibrating and correctly sampling the dynamics of these systems or the transport of small molecules therein.

The non-equilibrium MD (NEMD) methodology has been employed in an effort to accelerate penetrant diffusion. Here external forces are applied that retain the simulated system away from equilibrium and enable the determination of the induced flux that develops in response to the imposed forces. For small external forces, the transport properties under steady-state conditions can be directly extracted from the ratio of the flux to the imposed force, as the system is weakly perturbed, remaining in the linear response regime [52,53]. NEMD has been applied for the study of transport phenomena in amorphous polymers. Müller-Plathe et al. [54] were the first to employ NEMD for the study of the transport properties of gases in polyisobutylene. No substantial computational gain compared to equilibrium MD simulations was detected in their work within the linear response regime. Van der Vegt et al. [55] implemented NEMD for the simulation of penetrants at low concentrations using a harmonic external potential and determining the linear regime by a single simulation. NEMD has also been used for the investigation of the permeation mechanism in microporous polymers under the limitation of very small fluxes observed for small external forces [56]. Recently, Anderson and co-workers [57] employed NEMD to examine gas transport in the interfacial region of three polymeric systems (polyethylene, poly (4-methyl-2-pentyne) and polydimethylsiloxane), incorporating in the prediction of penetrant permeability and diffusivity the effect of the chain oscillation amplitude and of the accessible cavity fraction in the polymer matrices.

2.2. *Monte Carlo*

Monte Carlo (MC) simulation [18,21] of a microscopic model aims at generating a large number of configurations sampling in an asymptotic manner the probability density of a statistical mechanical ensemble under specific macroscopic conditions. During an MC simulation, each state is generated depending only on the one that precedes it, by implementing an attempted elementary move that may be accepted or rejected [20] based on the energetics and according to specific selection criteria; the latter are designed to follow the principle of microscopic reversibility (or detailed balance) in the generation of the Markov chain sequence of configurations. The attempted moves often involve a very small subset of degrees of freedom, implementing translations and rotations, insertions, deletions, or swaps of molecules, or volume fluctuations of the entire model system. MC is a stochastic technique that does not account for the system's time evolution and therefore cannot be used for the study of dynamics. This stochastic character of MC, though, coupled with appropriately designed moves, enables the equilibration and sampling of long-chain macromolecular systems orders of magnitude more efficiently than with MD, opening the way to addressing long length scale phenomena.

For the simulation of polymeric systems sophisticated moves [24,27,58] have been introduced, including configurational bias (CBMC) [59–61], concerted rotation (CONROT) [58,62], reptation and various connectivity altering moves such as end-bridging (EBMC) [22,23] and double-bridging (DB) [26]. During an end-bridging move, a chain is divided in two pieces via the excision of a trimer segment located in the chain's inner part as the chain is "attacked" by the end of another chain that lies in close proximity. One piece of the initial chain is then connected to the attacking end by reconstructing the trimer segment that was originally excised. In similar spirit, a double-bridging move is implemented by excising two trimer segments from two individual neighboring chains that are again reconstructed to bridge the remaining chain parts, forming two new chains of the same length as the original ones, but in a completely different conformation. In Figure 3, end-bridging and double bridging MC moves are illustrated for the case of chains with directionality in their macromolecular architecture, in which only the moves depicted with the blue arrows are allowed to be implemented for the chemical structure to be preserved. Using these moves it is possible to equilibrate at all length scales polymer melts with molar mass distributions comparable to those encountered in membrane applications. Equilibrated melt configurations obtained in this way constitute excellent starting points for generating glassy amorphous polymer configurations through cooling. MC methods are often implemented for the study of sorption in membrane materials (see Section 2.3).

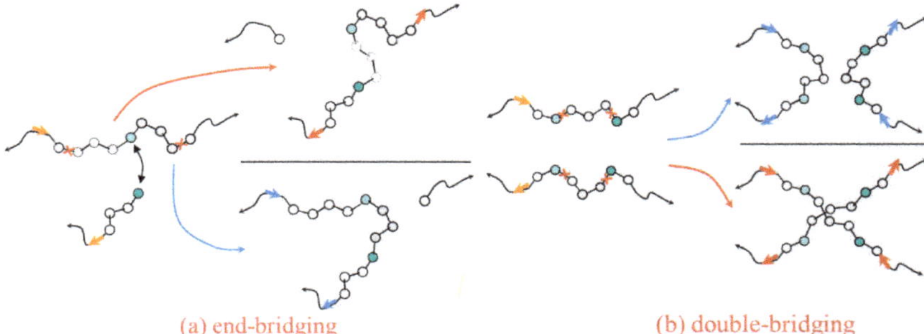

Figure 3. Illustration of the (**a**) end-bridging and (**b**) double-bridging MC moves designed to be implemented for polymeric chains with directionality in their chemical structure. In each case, the moves shown by blue arrows are possible. Intramolecular double-bridging MC moves cannot be performed in the case of directional chains.

2.3. Methods for the Molecular Simulation of Penetrant Sorption

The methods that are used within the context of penetrant sorption mainly involve the implementation of Grand Canonical Monte Carlo (GCMC) and Gibbs Ensemble Monte Carlo (GEMC) [63]. In GCMC the chemical potential, the volume and the temperature are kept constant, while the number of particles changes during the simulation, fluctuating around an equilibrium value. During the GCMC simulation the polymer system is considered to be in contact with a penetrant reservoir at specified temperature and chemical potential, with which it exchanges energy and particles. The trial moves in this case are translations, insertions, and deletions of particles.

GEMC enables the study of phase equilibria in one simulation by simultaneously incorporating two individual simulation boxes between which there is no interface. The attempted moves in this case involve particle translation and rotation, volume change (usually in a manner such that the total volume is preserved) and random particle swaps between the two simulation boxes implemented to simulate phase equilibrium of the two phases under the same temperature, pressure and chemical potential conditions. In systems containing a nonvolatile condensed (e.g., polymer) component at equilibrium with a pure or mixed gas phase, the solubilities of the gaseous permeants can be calculated using MC in a hybrid isothermal-isobaric and grand canonical ensemble [9] that does not require explicit simulation of the gas phase. This is termed the $N_1 f_2 ... f_n PT$ ensemble and keeps fixed the number of molecules of the condensed component, N_1, the fugacities of the gas components, $f_2,...,f_n$, as well as the system temperature and the pressure. Note that only n of the quantities $f_2,...,f_n,P,T$ are independent.

In some cases the acceptance probability of the attempted insertions and deletions in the above MC schemes is very low, such as in the case of low-accessible volume polymer matrices, very strong penetrant/polymer interactions, or bulky penetrant molecules. Semigrand canonical ensemble MC [22,64,65] can be applied in this type of binary or multicomponent mixtures, during which identity exchange trial moves among the various species is attempted. Considering, for example, a binary system consisting of species A and B, MC moves are attempted that involve conversion of one particle of species A to species B or vice versa. The change in free energy (difference of chemical potentials) brought about by such a change in the chemical identity is taken into account in the acceptance of the move.

At low pressures, the Henry's law constants can be calculated using the Widom insertion test particle method [66]. The method involves insertion of a ghost penetrant molecule in the polymer matrix at random positions and orientations and calculation of the interaction energy of the inserted molecule with the matrix. The latter is used subsequently for the determination of the excess chemical potential μ^{ex} of the test molecule. The solubility of the molecule in the polymer matrix can be calculated as:

$$S = \frac{22,400 \text{ cm}^3(\text{STP})}{\text{mol}} \frac{1}{RT} \exp\left(-\frac{\mu^{ex}}{RT}\right) \quad (8)$$

with STP corresponding to a temperature of 273.15 K and a pressure of 101.325 kPa and the solubility obtained in units of $\text{cm}^3(\text{STP})/\left(\text{cm}^3\text{polymer Pa}\right)$. A large number of test particle insertions are implemented in well equilibrated configurations of the polymer systems obtained from MD or MC simulations. The method has been widely applied for the estimation of low concentration sorption of small molecules in polymer systems [7,43–45].

Sorption isotherms can also be extracted utilizing iterative schemes that involve MD simulations of the polymer-penetrant system at constant composition, calculation of the excess chemical potential and gas fugacity and carrying out new MD simulations at a corrected pressure until convergence is reached [46,47].

More elaborate computational techniques are often necessary in cases that involve sorption of large penetrants, high pressures or strong penetrant-polymer interactions, especially at low temperatures. These methods include configurational bias insertions [67], which are used for the bond-by-bond insertion of multi-atom solutes for which Widom test particle insertion is not successful; thermodynamic integration [48,49] during which a small solute is included in a dense system gradually using progressively increasing coupling parameters λ for the solute-matrix interactions and conducting a series of simulations; expanded ensemble techniques [68,69] that also involve a stepwise insertion or deletion of the solute and the implementation of free energy perturbation within a single simulation; excluded-volume map sampling [70–73] and grid search methods [74,75] for efficient sampling in dense systems by blocking regions in which solute insertions are not feasible; the test particle deletion method [76,77] which constitutes an inverse of the Widom method for the calculation of the chemical potential of particles and multisite solutes; extended ensemble MD [78] during which a dynamic coupling parameter is used for the solute-polymer interactions that enhances efficient sampling of the system's phase space; the scission-fusion MC method [67] for the determination of solubility of oligomer solutes in polydisperse polymeric systems of the same chemical constitution as the solute; grand canonical MD [79] that is implemented for vapor-liquid equilibria and sorption in rigid polymer matrices by conducting two individual simulations in different ensembles without the requirement of particle exchanges between the phases; and fast-growth thermodynamic integration [80] where the chemical potential is calculated efficiently from a number of independent thermodynamic integrations for the estimation of sorption of large molecules in dense matrices.

2.4. Molecular Simulation Methods for the Study of Infrequent Events

The wide range of time scales that are involved in the relaxation of various modes of motion in polymeric systems renders their simulation a very complicated task. At temperatures near or below the glass transition temperature (T_g), in particular, the polymer matrix becomes almost static, prohibiting the redistribution of accessible volume. This has profound effects on the diffusion of a small penetrant molecule in a selective glassy polymer. The penetrant spends most of its time executing fast local motions while trapped within pockets of accessible volume, or "sorption sites," within the polymer matrix. Penetrant jumps between sorption sites are infrequent events [8] and cannot be properly sampled by performing even extensively long MD simulations. Transitions from one cavity to another occur rarely and randomly, and generally involve not only translational motion of the penetrant, but also local rearrangements in the polymer matrix that instantaneously open up passages between neighboring pockets of accessible volume.

Transition state theory of infrequent events (TST) [81–83] provides appropriate means for the study of systems that are characterized by time scale separation phenomena and crossing of energy barriers that are much higher than the thermal energy, k_BT. The dynamical evolution of such systems consists of long periods of vibrational motion within a potential energy basin; transitions between different basins (states) are infrequent, corresponding to average times of many vibrational periods. First order rate constants [84] enable the quantification of the average waiting times (which may be very broadly distributed) that elapse between specific transitions occurring and can be calculated invoking TST. The probabilities of occupancy of two states A and B that are separated by a high energy barrier evolve according to a master Equation:

$$\frac{dp_A}{dt} = -k_{A\to B}p + k_{\to}p \tag{9}$$

$$\frac{dp}{dt} = -k_{\to}p + k_{\to}p \tag{10}$$

where p and p are the probabilities of occupancy of states A and B, respectively, and $k_{A\to B}$ the rate constant for a transition from state A to state B.

Implementing TST for the study of a penetrant jump in a polymer matrix involves the determination of transition states between sorption states (Figure 4) as well as of the pathways followed by the system while passing from one state to another [8,9]. In a glassy polymer matrix there is a large number of regions in which a penetrant molecule may reside and there is at least one transition state between any two states. TST enables the calculation of rate constants along individual pathways based on the probability of the system to be on the dividing surface between two states compared to the probability of its initial state, neglecting the existence of any re-crossing events. From a mathematical point of view, a transition state is a first-order saddle point of the potential energy in the polymer-penetrant multidimensional configuration space, i.e., a point where the potential energy gradient equals zero and there is one negative eigenvalue of the Hessian matrix of second derivatives. The jump pathway can be considered as the lowest energy path (around which actual crossing trajectories would fluctuate) that connects two neighboring local minima passing through the transition state and is usually determined by calculating the Intrinsic Reaction Coordinate (IRC) [85]. The general TST expression for the calculation of the associated rate constant is given by:

$$k_{A\to B}^{TST} = \frac{k_B T}{h} e^{\left(-\frac{G^{\ddagger}-G_A}{k_B T}\right)} \tag{11}$$

with $G^{\ddagger}$ and G_A denoting the Gibbs energy of the system lying on the dividing surface and in the initial state, respectively. Using a harmonic approximation for the potential energy and considering the quantum mechanical vibrational partition function, the rate constant can be written as [32]:

$$k_{A\to B}^{TST} = \frac{k_B T}{h} \frac{\prod_{a=1}^{f}\left[1-\exp\left(-\frac{h\nu_a}{k_B T}\right)\right]}{\prod_{a=2}^{f}\left[1-\exp\left(-\frac{h\nu_a^{\ddagger}}{k_B T}\right)\right]} \exp\left(-\frac{V^{\ddagger}-V_A}{k_B T}\right) \tag{12}$$

where ν_a, V_A and $\nu_a^{\ddagger}$ $V^{\ddagger}$ are the vibrational frequencies and potential energy at the initial state and at the transition state, respectively. If re-crossing events are important in the process, then a dynamical correction factor to $k_{A\to B}^{TST}$ has to be calculated and taken into account [82,86,87].

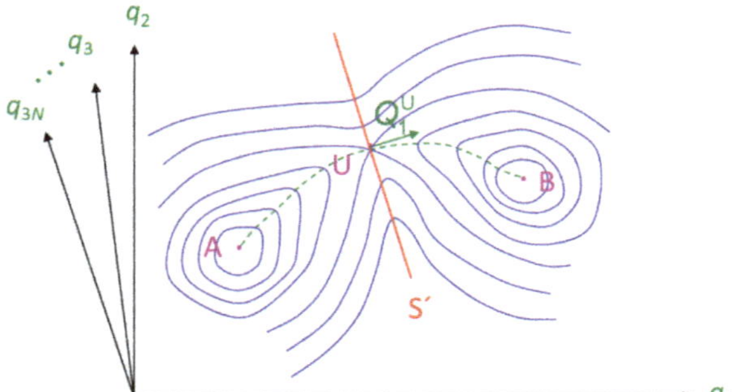

Figure 4. Trajectory, in $3N$ dimensional coordinate space $\mathbf{q}$, that passes from one basin to another, crossing a ($3N$-1) dimensional "dividing surface" separating the two basins. Solid lines correspond to a ($3N$-1)-dimensional hypersurface of constant energy V, Q_1^U is the coordinate along the negative curvature of $V(q)$ (reaction coordinate) and S′ is a ($3N$-1)-dimensional hypersurface normal to the reaction coordinate, constituting the dividing surface between the states A and B.

Gusev et al. [29] were the first to systematically apply TST for the study of diffusivity of spherical probes in rigid minimum energy glassy polymeric matrices generated using molecular mechanics. Identifying the invocation of a static polymer matrix as a crucial drawback of their study, Gusev and Suter [30] implemented the TST method allowing independent harmonic vibrations of the polymer atoms around their equilibrium positions. The elastic motion of the polymer matrix incorporated in their work involved the harmonic motion of polymer atoms at very short times, thus not taking into account any structural relaxation. The amplitude of the elastic motion was determined by an adjustable parameter $< \Delta^2 >$ that directly affected the extracted rate constants. Within the Gusev-Suter approach, TST calculations were conducted in the three-dimensional space of the spherical penetrant. Their method was computationally efficient and in cases that involved very small penetrants able to extract satisfactory results [7,30,88–90].

Greenfield and Theodorou [31,32] developed a multidimensional TST methodology in generalized coordinates, in which polymer degrees of freedom are taken into account for the calculation of transition states and diffusion pathways considering a spherical penetrant. The polymer atoms are in this case allowed to move, enabling the realization of local-chain motions that take place during a penetrant passage from one cavity to another. The inclusion of polymer degrees of freedom in the pathway determination becomes very important as the size of the penetrant increases compared to the typical size of clusters of accessible volume in the glassy polymer matrix. Greenfield used a geometric analysis approach [91] for the determination of the accessible volume and its distribution in the polymer matrix to obtain an initial estimate of sorption sites. Narrow-necking regions connecting regions of accessible volume for small spherical probes serve as the starting points for the calculation of the transition states (Figure 5). The reaction path from one state to the other was calculated using a subset of degrees of freedom that were relevant to the specific transition along the pathway. The method was further extended [33,34,92] and applied for the study of polymers with complex chemical constitution in Cartesian coordinates and in the polymer-penetrant multidimensional space using a fully flexible representation for a CO_2 penetrant.

Rare events can be also studied by implementing transition path sampling methods (TPS) [93,94]. The TPS methodology is used for the calculation of rate constants by sampling dynamical reactive trajectories. Trial trajectories are generated based on a pre-determined transition path, a point of which is randomly selected and the momenta corresponding to it are perturbed. The acceptance probability of the generated trial trajectories can be tuned by the amplitude of the momenta perturbation, with the acceptance increasing for small perturbations.

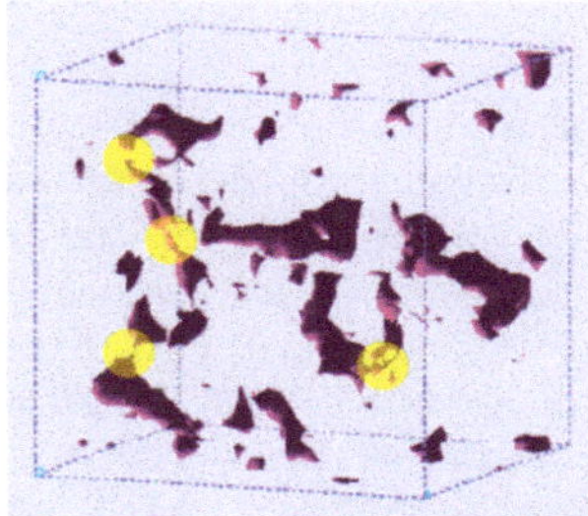

Figure 5. Indicative narrow-necking regions in the accessible volume obtained from geometric analysis, which serve as initial estimates for the transition state search calculations [33]. The simulated polymer is a poly(amide imide) in a cubic box of edge length 28.09 Å with periodic boundary conditions. Atoms are not shown. The dark regions are clusters of accessible volume, as determined with a spherical probe of radius 1.1 Å. Necking regions are identified by probing the same configuration with a smaller probe radius. Some of the necking regions are indicatively highlighted in yellow.

A number of TPS schemes have been developed to address acceptance probability issues [93,95–98]. TPS generally enables the determination of realistic pathways at finite temperatures, but is computationally intensive. It has been implemented for the study of a wide range of infrequent events, such as nucleation [99–102], protein folding and conformational changes [103–106], glassy dynamics [107], reactions [108–110] and penetrant diffusion in non-amorphous materials [111–113]. The use of TPS methods for the study of penetrant diffusion in polymeric systems is scarce and involves the study of diffusion of water in a glassy hydrophilic polymer [35]. TPS implementation in this case directly depends on the limited initial transition pathways extracted by MD simulations for the systems under study.

2.5. Interactions and Generation of Realistic Structures

Accurate predictions of permeability properties in glassy polymeric systems largely rely on the generation of configurations that are able to resemble the real polymer matrix as well as on the reliable representation of the molecular system under study and the mathematical description of its interactions. A particle-based model may be fully atomistic (AA—all atom); united atom (UA), in which hydrogen atoms are considered to form a single interaction site together with the heavy atom to which they are connected; or coarse-grained (CG), consisting of interacting moieties that are derived by lumping together several atoms. The potential energy function in atomistic models is often derived empirically by fitting the structure, volumetric, thermal, and phase equilibrium properties of small molecule analogues, but also from ab initio electronic structure calculations. It usually consists of a bonded and a nonbonded term, that correspond, respectively, to interactions between atoms that are chemically bonded and to interactions between atoms that are not connected with chemical bonds and may belong to the same or to different molecules. Bonded interactions include contributions to the energy from deviations of chemical bond lengths and bond angles from their equilibrium values and also contributions related to dihedral angles and improper torsions. The non-bonded part includes an electrostatic term, related to the interactions between partial atomic charges (and in general between permanent multipoles), and a van der Waals term. The latter comprises pairwise interactions based on dispersion (attractive) forces due to correlations between electron clouds in different atoms; repulsive (excluded volume) interactions resulting from the overlap of electron clouds of individual atoms at very small distances; and, in some instances, polarization contributions that are also attractive and stem from an induced change in the charges of a moiety owing to the field that it experiences from its neighborhood. The force field parameters for bonded interactions, partial charges, and polarization contributions are often determined by quantum mechanical (QM) calculations, while parameters for the dispersion and excluded volume contributions are typically based on empirical methods or experimental data.

In the general case, the local segment packing that is involved in the penetrant sorption and transport mechanisms, is interconnected with the longer length scale characteristics of the polymeric system [114]. Creation of in silico realistic structures of dense amorphous polymers is a hard and complicated task [34,115]. Amorphous glasses, in particular, are far from equilibrium and are trapped in local energy minima that directly depend on their formation history. Incorporating this fact in molecular simulation strategies for the generation of well-defined and realistic configurations is a great scientific challenge that has still not been adequately addressed.

Several methodologies have been attempted for the generation and equilibration of initial structures of polymers in the melt state [28,114,116–121]. Atomistic initial configurations of the macromolecular systems can be rigorously obtained via building the chain in the simulation box bond-by-bond using Flory's rotational isomeric state model [122], while avoiding excluded volume overlaps based on a modified continuous model for local interactions. These initial configurations are then subjected to energy minimization (molecular mechanics) that can be applied in a stagewise manner [123], removing any remaining atomic overlaps. Molecular mechanics is generally fast but not able to result in an efficient statistical mechanical sampling; nevertheless, it provides a convenient initial configuration for the subsequent implementation of systematic simulation methods towards the generation of realistic glassy polymeric structures. The atomistic structures from molecular mechanics can for example be subjected to equilibration in the melt state using the powerful connectivity altering MC moves [22–28] or MD or hybrid MC-MD methods. The extracted equilibrated configurations in the melt state have to be subsequently cooled below T_g [124]. The efficiency of the above scheme is very good for polymer systems of rather simple macromolecular architectures and cannot be easily utilized for polymeric matrices with rigid backbones, strong electrostatic and polar interactions, or bulky side groups.

In the case of polymers with complex chemical constitution, systematic hierarchical approaches are required for the generation of realistic structures which are the starting point for studying the permeability of probe molecules in their bulk. These methods may involve the mapping from the atomistic to a coarse-grained level, equilibration at the coarse-grained level and reverse mapping back to the atomistic representation. The atomistic detail is generally necessary for the study of small-molecule sorption and diffusion phenomena in polymeric matrices. Coarse-graining strategies [115,125–131] involve the substitution of groups of atoms by single interaction sites, thereby reducing the number of system degrees of freedom, while at the same time maintaining the ones that are important for the description of the mechanisms/processes under study. The CG model invoked in each case is thus intimately related to the macromolecular architecture and the scientific problem at hand. Therefore, the mapping to the CG representation is largely empirical, as there is no unique way to coarse-grain. In order to perform simulations at the CG level, an effective potential for the CG model has to be developed based on the corresponding CG degrees of freedom, that is able to reproduce the key characteristics that need to be retained in the CG simulations. The majority of the CG force fields are developed and parameterized by fitting either in a top-down manner, targeting the reproduction of macroscopic properties [132–135], or aiming at the reproduction of microscopic properties of the molecular system based on atomistic simulations and experimental findings, following a bottom up approach [136] or using a combination of both strategies [137,138]. Fitting schemes for the parameterization of CG effective potentials include inverse Boltzmann and iterative Boltzmann inversion (IBI) [139], force matching [140,141], inverse Monte Carlo [142] and relative entropy [143,144] methods. In addition to those, interaction potentials between CG moieties can be also determined directly from atomistic interactions of groups of atoms that are mapped onto the CG moieties [145,146], which may also be anisotropic [129]. Hybrid particle-field methods have been also proposed and implemented for the generation of polymer atomic structures [147].

An ideal CG effective potential would enable the reproduction of target properties and would be also transferable to other thermodynamic state points [137,138,148–151]. The CG model should also allow, in a straightforward manner, reverse mapping [152–161] back to representative atomistic configurations (usually on the basis of geometric criteria) of the well equilibrated structures at the CG level. Special measures must be taken for macromolecular structures with complex chemical constitution to avoid potential overlaps or concatenations [162]. The final atomistic configurations from reverse mapping are then subjected to a final relaxation of the system's local interactions. Figure 6 illustrates a hierarchical backmapping procedure used for the equilibration of polystyrene models at three different representations [161]. The multi-resolution method proposed in this study enables the generation of well-equilibrated high molecular weight polymer configurations that are initially equilibrated at the largest length scale corresponding, in this case, to a soft-blob based representation (Figure 6c). The detailed description of the system is introduced stagewise, first migrating from the soft-blob based to a moderately CG model (Figure 6b) and equilibrating the system locally at this level and subsequently re-inserting the microscopic detail (Figure 6a) and equilibrating again at this stage. The hierarchical backmapping using a sequence of models aims at the efficient generation of realistic long chain polymer melts in a computationally feasible manner, preserving the long-wavelength characteristics of the macromolecular systems under study.

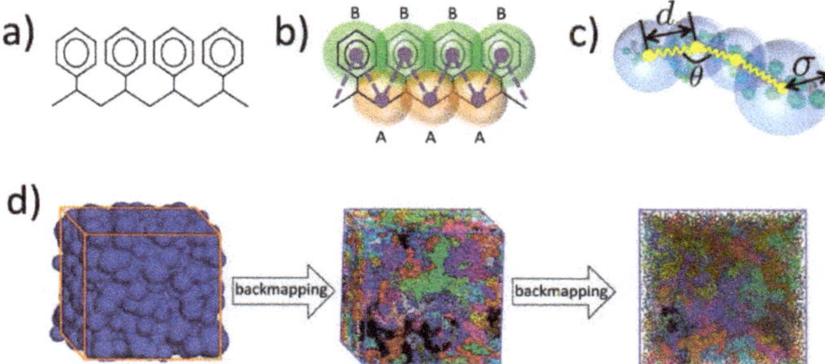

Figure 6. Depiction of a hierarchical backmapping corresponding to (**a**) a united atom representation, (**b**) a moderate CG model consisting of beads of types A and B, and (**c**) a blob-based representation of soft spheres by lumping together a number of beads of representation (**b**). The backmapping procedure is described in (**d**) for the three resolutions. Reproduced with permission from [161].

The amorphous polymer configurations to be used for the prediction of solubility and diffusivity properties have to be validated. Validation of amorphous polymer configurations can be realized via direct comparison with experimental measurements over a range of thermodynamic conditions (temperatures and pressures). Such comparison between predicted and measured values can be conducted for a number of properties, e.g., thermodynamic (density, cohesive energy, solubility parameters), structural (pair distribution functions and structure factors from X-ray and neutron scattering), and dynamical (local dynamics of the chain segments and glass transition). Among the structural properties, of particular relevance to permeability is the accessible volume and how it is distributed and connected; predictions can be compared against Positron Annihilation Lifetime Spectroscopy (PALS). It is generally advantageous to generate a large number of configurations, thereby obtaining a sample of representative "trapped" glassy states, and also to verify the stability of the final glassy configurations [9] prior to using them in subsequent simulations.

3. Coarse-Graining and Multiscale Approaches in Sorption and Diffusivity Prediction

The broad spectra of length- and time-scales present in macromolecular systems necessitate the implementation of elaborate schemes for the prediction of their properties. Apart from the coarse graining methods related to the generation of realistic initial structures discussed in Section 2.5, hierarchical schemes at multiple length and time-scales have been developed and implemented for the study per se of the permeability of small molecules in polymeric systems.

Transition state theory approaches (see Section 2.4) provide a unique means for calculation of the wide range of interstate rate constants characterizing the elementary jumps executed by a penetrant in an amorphous glassy polymer matrix. By coarse graining to a macrostate level, each penetrant jump can be categorized as an intramacrostate or intermacrostate transition, where a "macrostate" is defined as a collection of basins constructed around neighboring potential energy minima separated by energy barriers that are low compared to the thermal energy k_BT. At this level of coarse graining, the diffusivity can be calculated considering a Poisson process of successive uncorrelated penetrant jumps between macrostates. For a transition between two macrostates I and J, the time evolution of the probability that the system is in a macrostate I is calculated as follows:

$$\frac{dp_I}{dt} = -\sum_J k_{I\to J}p_I + \sum_J k_{J\to I}p_J \tag{13}$$

where:

$$k_{I\to J} = \sum_{i\in I}\sum_{j\in J} k_{i\to j}\frac{p_i}{p_I} \tag{14}$$

and $p_I = \sum_{i\in I} p_i$ and $k_{I\to J}$ is the rate constant for the transition from macrostate I to macrostate J. A mesoscopic Kinetic Monte Carlo (KMC) simulation [8,163–165] is implemented for the calculation of the penetrant diffusion coefficients on the extracted (periodic) networks of macrostates and based on the calculated macrostate-to-macrostate rate constants and sorption probabilities [33,34,92]. This multiscale methodology that combines the TST calculated rate constants and sorption site network with mesoscopic simulations at the macrostate coarse-grained level has been very efficient in the prediction of penetrant diffusivity and in the elucidation of the characteristics of the penetrant's diffusional motion. Potential artifacts that may arise from the periodic replication of the formed network can be avoided by generating large irregular networks of macrostates using a reverse Monte Carlo (RMC) approach [36]. Analytical approaches for obtaining the diffusivity from penetrant positions representative of each macrostate and from the rate constants between macrostates, taking advantage of the periodic boundary conditions that characterize the model system in each direction, have also been developed [166].

In an effort to extend the time scale limits of the MD simulations, Neyertz and Brown [37,38] implemented a simplified approach that utilizes MD trajectories of mobile gas penetrants in the polymer matrix to extract information that can be used as input in a kinetic Monte Carlo simulation, referred to as trajectory-extending kinetic Monte Carlo (TEKMC). A sequence of configurations visited during a time interval τ is chosen for analysis from MD trajectories of total length of a few nanoseconds. Based on the penetrant positions, the primary simulation box is divided into subcells (of the same size and shape) and the probability matrix of jumps between subcells is extracted. The method is of low computational cost, requiring, however, proper adjustments of the parameters related to the sampling interval τ and subcell grid size [167]. It has been applied for the study of CO_2 [38] and other small penetrants [37] in fluorinated glassy polymers, leading to predictions in reasonable agreement with experimental measurements. It has also been used for the investigation of gas diffusion in polymer nanocomposites [168] and ion transport in polymer electrolyte melts [169–172]. The method can only be applied provided that a percolating network of subcells is sampled and that cases of non-linked channels are taken into account in the error of the extracted diffusivity. In order that MD be able to provide an interconnected network of subcells as required by this method, MD simulations are restricted to medium to high penetrant concentrations, in which plasticization effects are usually present [38].

Coarse-grained molecular dynamics (CG-MD) simulations have been also investigated for the modeling of transport of large penetrants in amorphous polymer systems in an attempt to reach the normal diffusion regime. The CG models in this case should be rather detailed, preserving a minimum of important information on the chemical structure of the macromolecular system under study which directly affects the diffusional behavior of the penetrants in the polymer matrix. The CG-MD simulations are accompanied by atomistic simulations for the validation of the CG model used. Kremer and co-workers [39,40] attempted a combined experimental and simulation methodology of this type for the study of the transport properties of ethylbenzene in mixture with polystyrene (PS) using CG MD with different CG representations for the PS. Lin et al. [41] incorporated the MARTINI CG force field [133,173,174] to perform CG-MD of a fragrance agent, octanal, in a polymer film consisting of four components. In their simulations, they studied the concentration and temperature dependence as well as the effect of the presence of water on the effective time scale of the CG simulations and on the diffusional behavior of octanal in the polymeric film. Three CG models have been employed by Kumar and co-workers to study the penetrant diffusion mechanism in flexible and rigid polymers for various penetrant sizes and in a wide range of temperatures [42]. In CG-MD simulations, the simulated behavior does not correspond to the time scales of the dynamics of the real system. "Time mapping" procedures are attempted in this case based on atomistic simulation in order to link the time scales involved in the CG simulations with the time scales of the real dynamics. The effective time scale in the CG-MD is strongly dependent on the mapping implementation, with coarser schemes corresponding to larger effective time scales in the CG simulations, and the calculated mechanisms are directly affected by the loss of chemical detail in the local environment of the penetrant molecule.

4. Mechanistic Aspects of Sorption and Transport

The permeation of small penetrants in dense glassy polymeric materials is governed by a number of factors that involve the chemical affinity between the molecule and the polymer as well as characteristics of the glassy polymer matrix. Unlike the mechanism involved in many porous membranes, in which the material participates in the permeation process mostly by the pore structure, in dense amorphous polymeric membranes the polymer is actively engaged in the permeation process.

4.1. Sorption

Sorption of solute molecules in a polymer matrix is directly related to both the amount and distribution of free volume in the amorphous phase and to the interactions of the penetrant with the polymer. In the glassy state, the polymeric system is not in thermodynamic equilibrium. Dense polymer glasses are structurally arrested in configurations that depend on the history of their formation (e.g., cooling rate) and have the ability only to fluctuate within a subspace of configurations neighboring the specific local minimum of potential energy in their almost static structure. Structural relaxation in the direction of equilibrium involves the crossing of very high energy barriers that may take place below T_g in time scales of the order of years, corresponding to physical aging of glassy polymers.

In an equilibrium melt state, the pure polymer as well as the polymer plus penetrant system conforms to a Boltzmann distribution in configuration space. At temperatures near or below the glass transition temperature, the dense amorphous polymer system is no longer in thermodynamic equilibrium. Rather, it is trapped in one among many low energy regions surrounding local minima of the energy, within which it fluctuates thermally. The low energy regions (basins) of the configuration space are separated by high energy barriers that require a lot of time to be overcome. The quenched polymer configurations inherently carry information about the formation history of the glass. Upon cooling, the vitrifying system falls out of equilibrium. The rate of structural relaxation (rate of transitions between different basins in configuration space) decelerates precipitously as the temperature is lowered and becomes much slower than the rate of cooling employed. Redistribution in configuration space cannot keep up with the rate of change in temperature. Thus, the distribution in configuration space becomes arrested, unable to follow the demands of equilibrium, which would dictate passage from higher-energy basins into lower energy ones. The system is trapped within disjoint basins in configuration space, the distribution of these basins being much broader than would be dictated by Boltzmann at the prevailing low temperature. Within each basin however, local reconfiguration is possible; locally, the system can be considered as following a Boltzmann distribution within the confines of the basin. At low penetrant concentrations, the solubility can be studied independently in each disjoint, barrier- surrounded basin and then an average can be taken over individual glassy basins.

As a general rule, glassy polymers have a higher sorption capacity than rubbery polymers, with the glassy sorption isotherms exhibiting a concave curvature in contrast to the rubbery ones that are almost linear. During sorption at high concentrations of highly condensable penetrants such as CO_2, the potential energy barriers that resulted in disjoint subsets of configurations in the dense amorphous glass are altered, enabling a more frequent barrier crossing. This fact leads to volume swelling phenomena that are observed both in glassy and rubbery polymers, due to a penetrant-induced increase in the fractional free volume and to plasticization of the dense amorphous polymer matrix [175–180]. These phenomena directly influence the performance of the material as a separation medium [181], enhancing the permeability and diffusivity as the concentration increases. The presence of the penetrant at high concentrations accelerates the segmental dynamics of the chains, resulting in the case of polymer glasses in the reduction of the glass transition temperature due to polymer softening. Plasticization phenomena are described in terms of a characteristic pressure (known as "plasticization pressure") which is related to a minimum in the gas permeability (Figure 7). Under those circumstances, sorption isotherms of glassy polymers at high pressures may exhibit a rubber type linear behavior, while following a glassy-type curve at lower pressures. Swelling and plasticization behavior of a polymeric system is affected by temperature [182] and this can be considered as an outcome of the interplay of the opposite trends between sorption capacity and chain mobility as a function temperature.

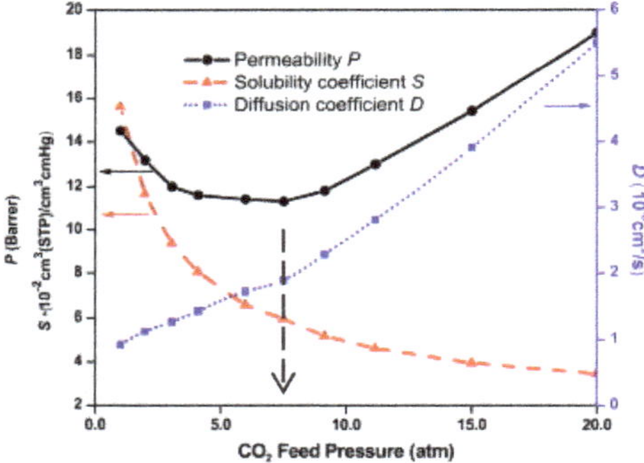

Figure 7. Experimental measurement of CO_2 permeability, solubility and diffusion coefficient as a function of pressure in a polyimide membrane. The black dashed-line arrow is indicative of the characteristic pressure ("plasticization pressure") that corresponds to the minimum of permeability. Reproduced with permission from [183].

Sorption and penetrant-induced plasticization and swelling phenomena have been investigated by molecular simulation of bulk polymeric systems [38,183–189] using either CGMC or implementing iterative multi-stage methods to determine the correct pressure at given penetrant concentration [46,47,190]. Van der Vegt et al. [46,186] have studied sorption of CO_2 in a polyethylene-like membrane and have identified below T_g a hole filling mechanism in combination with a finite positive partial molar volume of CO_2 being responsible for the gas sorption thermodynamics in the glassy polymer. Spyriouni et al. [47] used well equilibrated polystyrene configurations that were obtained implementing a coarse-graining—reverse-mapping scheme [191] that were loaded up with CO_2 in a wide pressure range. They observed a change in the shape of both sorption isotherms and swelling curves with increasing temperature (Figure 8). In this work, a mapping of the accessible volume was conducted to facilitate efficient loading of CO_2. The polymer-solute matrices were subjected to NPT MD simulations during which the solute molecules were swapped between accessible volume cavities to achieve an efficient equilibrium repartitioning. The chemical potential calculations were performed using the Direct Particle Deletion method [47]. They also analyzed the polymer segmental relaxation times in the presence of CO_2 to obtain estimates of the solvent-induced glass transition pressure at various temperatures below the T_g of pure polystyrene, obtaining good agreement with available experimental measurements. Sorption and dilation effects have been modeled in fluorinated polyimides (6FDA-ODA,6FDA-DPX,6FDA-DAM) [183,192,193], in poly (ether sulfone) (PES) and polysulfone [185] polymers and in copolyimide polymers [188] up to high pressures. Performance and plasticization resistance of an emerging class of polymeric membrane materials called 'thermally rearranged polymers' [194–204] that are generated by thermally modifying aromatic polyimide chains, has also been investigated computationally [205–211] in recent years. These phenomena have been also simulated for high free volume polymers of intrinsic microporosity (PIM) [56,189,212,213].

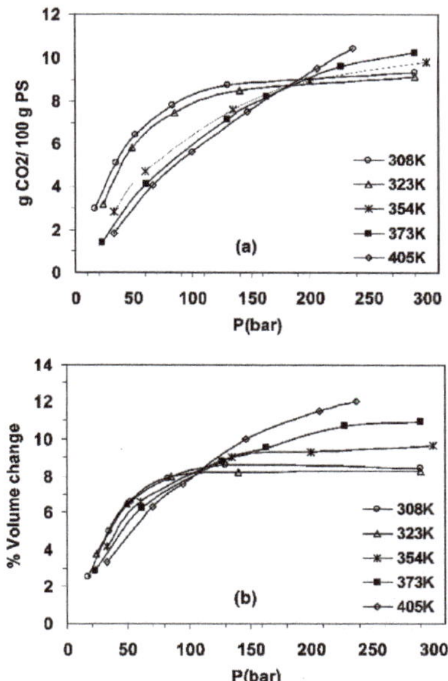

Figure 8. Molecular simulation results on (**a**) isotherms of CO_2 sorption and (**b**) polymer swelling for polystyrene at various temperatures and pressures. Reproduced with permission from [47].

As the penetrant concentrations increase, the volume dilation phenomena are often semi-permanent and, after desorption of the sorbed gas, part of the additional volume that has been created due to swelling during sorption is preserved, a phenomenon that is often referred to as "conditioning" [214]. A hysteretic behavior is thus observed during subsequent sorption cycles, the extent of which depends on the conditions of the experiment and on the amorphous glass thermodynamic history. Plasticization and conditioning effects are in many cases dependent on the thickness of the amorphous polymer membrane [215]. Physical aging generally takes place more rapidly in thin films. [215–217] than in the corresponding thicker ones or the ones of bulk glassy matrices and a different permeability trend is observed during the timescale of the measurements [216]. Physical aging of a glassy polymer matrix is characterized by a decrease in the free volume, leading to a densification [218–220] that affects sorption and permeability properties [221,222] as well as the plasticization behavior of the polymer matrix.

Molecular simulations have been applied incorporating thin membrane models [57,223–230] and explicit gas feed reservoirs that attempt to approximate experimental conditions using MD or NEMD [57]. Such simulations of interfacial phenomena often result in more intense interfacial effects due to the small thickness of the models and the large surface to volume ratio used in simulations, as modeling of dense polymeric films would require the implementation of a very extended surface area corresponding to system sizes that are currently not computationally feasible [223]. In the course of MD of 6FDA-6FpDA fluorinated polyimide membranes [231] using a CO_2 gas reservoir on both sides of the film, a rapid adsorption was initially observed at polymer-gas interfaces that was subsequently followed by a slower uptake. During the sorption process, swelling effects were detected and the additional free volume regions were occupied at short time by the sorbed penetrants. The same models were also used for the study of the effect of structural isomerism in the plasticization of meta-linked and para-linked polyimides, obtaining a plasticization resistance in the meta-linked isomer [228]. In an effort to study the reversibility of polymer membrane dilation, MD simulation has been applied in an explicit film model of 6FDA-6FpDA polyimide with surfaces in contact with a gas reservoir [229]. Two cycles of sorption and post-degassing phases have been applied and metastable states of enhanced solubility after degassing have been identified that depend on the former swelling stages of the material. The two sorption phases were found to differ at short times, as additional void spaces remain after the first cycle of post-degassing relaxation.

Sorption of mixtures of various gases in glassy polymers is common in membrane separations. The mechanism of permeation varies with the concentration of the mixed gas and the competitive character of the sorption among individual species. At low concentrations, each penetrant type may sorb without being affected by the presence of the other components. For an increasing amount of sorbed gases, one species may be more soluble in the polymer matrix leading to plasticization of the membrane, thus modifying the sorption environment of the other components [175,232]. A more complex behavior may arise if strong interactions are present between one of the components and the polymeric membrane material, or among the various penetrants themselves, thus altering the permselectivity performance of the membrane. The mixed gas behavior generally hinges on the chemical identity of the gas pair and its composition along with the pressure conditions and the thickness of the membrane [233]. The actual permselectivity, a^*_{ij}, differs from the ideal one (Equation (6)) [232,234] necessitating the investigation of mixed-gas conditions [235,236] along with the ones of the pure gases for the evaluation of the membrane permselectivity performance.

Mixed-gas simulations [57,223,237] have been recently performed for the study of the actual permeability performance of polymeric membranes. Tanis et al. [237] investigated the permeation properties of pure and mixed gas N_2/CH_4 in two 6FDA-based fluorinated polyimides and in their block copolymer. For the pure CH_4 gas, they extracted a similar methane sorption behavior in the three polyimides, accompanied by a linear-type volume swelling. Small differences were observed in the gas solubility coefficients of each species calculated also for the N_2/CH_4 mixture in a 2:1 composition, indicating that the permselectivity performance is primarily governed by the kinetics of the gas mixture. MD simulation has been conducted for O_2/N_2 simultaneous sorption and separation in a 6FDA-6FpDA polyimide glassy polymer film in the presence of gas reservoirs on each side of the film [223].

High initial pressures were applied and very long simulations of 300ns were employed in order for the system to reach equilibrium both in terms of gas concentration and pressure stabilization. The volume swelling effects for the O_2/N_2 gas mixture were rather limited and the glassy structure was not significantly affected by the air sorption. Figure 9a depicts the average solubility in the film with increasing pressure both including and excluding the probe interactions with the other air molecules, while Figure 9b shows the solubility selectivity with respect to pressure. Both gases were found to occupy the same low energy voids and the calculated solubility selectivity was extracted in good agreement with single gas permeation experimental measurements [238–241].

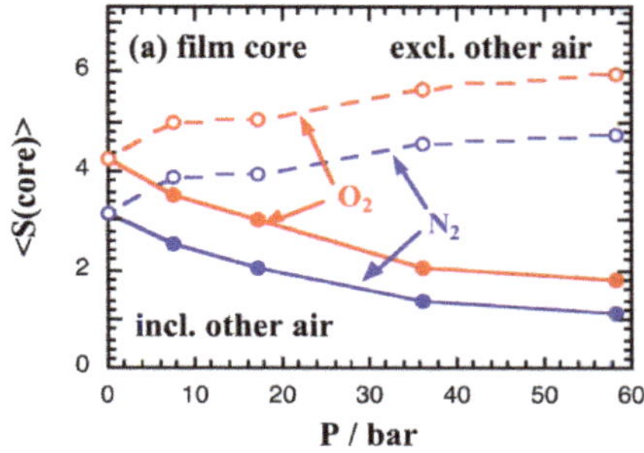

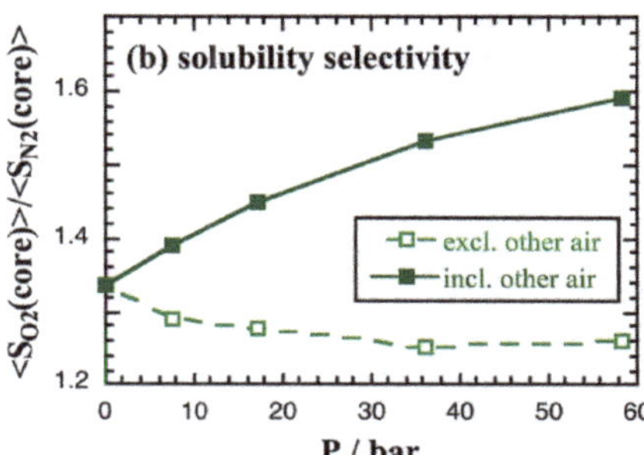

Figure 9. (**a**) Average (dimensionless) solubility as a function of pressure for O_2 and N_2 in a 6FDA-6FpDA polyimide glassy polymer film excluding (dashed lines) or including (solid lines) interactions with the other air molecules. (**b**) Solubility selectivity for O_2/N_2 separation in the same film. Reprinted with permission from [223].

4.2. Diffusion

Diffusion through a dense amorphous polymer matrix is a very slow process at temperatures near or below T_g. In dilute systems, the diffusivity of the penetrant depends on the penetrant size and its interaction with the polymer and also on the size and connectivity of the unoccupied volume in the polymer matrix. Below T_g, there is nearly no redistribution of the unoccupied space and the penetrant spends most of its time trapped in one cavity; a jump from one cavity to another rarely takes place when local fluctuations of the polymer segments open a channel that closes again directly after the jump. Diffusion in the dense glassy amorphous matrix consists of a sequence of infrequent events described by a hopping mechanism [242,243] and is directly related to the rate constants that govern the jumps between the neighboring cavities as well as the distance and connectivity of the pre-existing cavities in the polymer matrix.

The long time scales of structural decorrelation in amorphous polymers (even above T_g) result in the existence of an anomalous diffusion regime at short times which is characterized by $< r^2 > \propto t^n$, $n < 1$, with $< r^2 >$ being the penetrant's mean square displacement and t being the time. The penetrant experiences heterogeneities in its local environment and very long time scales are required to enter the Einstein (or Fickian) regime of normal diffusion. Anomalous diffusion can be associated with the wide range of rate constants that govern the penetrant jumps [244–246] or with the connectivity of the sorption sites [7,36,43], as well as with the penetrant size [70]. TST-based molecular simulation studies have investigated the diffusion mechanism in glassy polymers, revealing important factors that govern the penetrant's diffusional behavior such as the range of jump rate constants [32,33], the size, shape and connectivity of the volume clusters that are accessible to the penetrant molecule [33,91], the energetic and entropic contributions to the jump rate constants [9,32,33,247] and the characteristics of the cooperative motion of the polymer chains during the elementary penetrant jumps [32–34]. In glassy amorphous solids, a spatially non-uniform dynamics is detected and energy and entropy contributions to the rate constants differ from one region of the polymer matrix to the other. A range of rate constants of the order of 10^{-3}–10^3 μs^{-1} were determined for CH_4 in glassy atactic polypropylene [32] with a mean jump length of 5 Å. In the TST study of CO_2 diffusivity in a complex poly (amide imide) system, [-NH-C_6H_4-$C(CF_3)_2$-C_6H_4-NH-CO- $C_6H_4(CH_3)$-$N(CO)_2C_6H_3$-CO-]$_n$ [33] the rate constants appear widely distributed, with the majority of transitions being characterized by rate constants of the order of 10^{-1}–10^2 μs^{-1}, an average jump length of 5.3 Å and activation energies in the range of 2–11 kcal/mol that fall within range of experimental values of activation energy for diffusion of gases in glassy polymers. The dimensionality of the conducted calculations drastically affects the jump energy barrier, which is significantly decreased by increasing the polymer degrees of freedom in the vicinity of the penetrant included, until an asymptotic behavior in the energy is observed and the energy barrier is no longer influenced by the incorporation of additional polymer degrees of freedom [9,32,33]. Detailed analysis of the CO_2 diffusion mechanism reveals the existence of a concerted motion of the polymer segments along the transition path that allows the penetrant to pass from one sorption state to an adjacent one. Along the pathway, changes in the orientation of the CO_2 penetrant take place that are important for the realization of the elementary jumps [33,92].

Increasing the penetrant concentration has a diverse effect on penetrant diffusivity, depending on the specific penetrant-polymer system [232]. The diffusion coefficient remains unaffected if low sorbing penetrants are incorporated in the polymer matrix. If penetrant clustering phenomena are present in the system, then a decrease in diffusivity may be observed. For highly condensable penetrants, high concentrations cause volume swelling effects and increase the free volume and segmental mobility [248], a fact that leads to an acceleration in the mobility of the penetrants in the matrix, resulting in higher diffusivities. Even in this case, though, the jump-type hopping mechanism still characterizes the penetrant diffusional motion [38,227,231] with shorter residence times at these conditions.

Interdiffusion phenomena in the presence of more than one penetrant species have a direct effect on the permeability properties of the polymeric system [249]. In some cases, the presence of the more mobile penetrant has the tendency to accelerate the kinetics of the slower one [235,250]. As a result the diffusional behavior of the faster penetrant is subsequently hindered by an increase in the concentration of the slower one. This behavior was also detected in the computational study of Tanis et al. [237], for fluorinated polyimide homopolymers, according to which actual permeabilities cannot be reliably estimated by the ideal ones. The inverse conclusion was drawn from the simulation of air sorption in a 6FDA-6FpDA polyimide glassy polymer film from which it was determined that mixed gas conditions for O_2/N_2 separation can be extracted based on single gas data [223], a fact that is expected for the case of light gases such as O_2 and N_2 and in the absence of plasticization and competitive sorption effects [232].

5. New Materials, Challenges and Future Outlook

The need to develop new high-performance task specific functional polymer-based membranes necessitates the investigation and use of hybrid and composite materials [251–271]. Examples of advanced polymer-based separation media include polymer/inorganic membranes incorporating, for example, zeolites, inorganic particles, or nanoparticles in polymer matrices; carbon nanotube (CNT)/polymer composites; glassy polymerized ionic liquids; polymer/ionic liquid composites; and many other multicomponent combinations of materials in polymeric membranes. Several molecular simulation studies in recent years have focused on the investigation of polymer-based composite and mixed matrix membranes [272–275]. Ionic liquid/ionic polyimide composites [276,277] have been studied using molecular simulations as gas separation media and the effect of the anion structure on gas permeability and selectivity has been studied and is to be confirmed by experimental measurements. Moreover, polymer electrolyte membranes have been studied computationally for use as clean water and desalination membranes [278] and the transport of ions in polymer electrolytes has been simulated [258] also in the presence of nanoparticles [170,171,279,280]. Gas barrier properties in several mixed matrix organic-inorganic membranes have been studied computationally such as in MOF/polymer composites [281–283], in zeolite/polymer mixed matrix materials [284], in polyhedral oligomeric silsesquioxane/polymer systems [230,285] and in polymer/nanotube composites [286].

In terms of molecular simulation, for the reliable prediction of properties and the fundamental understanding of the underlying mechanisms in complex and multicomponent materials, it is crucial that the existing hierarchical methods, described in the previous sections, be extended and generalized so that they can be implemented in a straightforward manner. Additionally, unravelling the interfacial interactions [287,288] and characteristics [289–291] in polymer-based composite materials is of great importance for an accurate description of their behavior. Development and implementation of efficient, hybrid and adaptive resolution multiscale molecular simulation methods [125,292–298] may in many cases be necessary for the systematic study of microscopic behavior at the interfaces in composite materials and mixed matrix membranes and for further advancement of the materials-by-design target. In this direction it would be worthwhile to study many technologically and commercially important polymer-based nanocomposite membranes [299–305] such as Matrimid-based ones [306–308], incorporating a number of potential inorganic fillers. Multiscale molecular simulation strategies in the area of next generation multicomponent nanostructured materials are important in identifying how the fillers can be tailored towards an optimum selectivity and permeability behavior, investigating a number of crucial factors such as the effects of the size and the dispersity of the fillers and nanoparticles in the terminal properties as well as the influence of the morphology and surface treatment on the behavior and stability of the multicomponent membrane. Systematic strategies need to be applied for the study of nanocomposite materials [275,309] and for the modeling of complex membranes as for example dense glassy 3D networks, developing in this case efficient cross-linking algorithms [310]. The challenge of modeling materials with increasing complexity and the high demands in terms of system sizes and in the accuracy of predictions is greatly supported by the constant increase in the computational power in the one hand and on the development of novel efficient new algorithms and methods that need to be implemented to enable the in-depth understanding of the composite materials.

In parallel, artificial intelligence and machine learning approaches [311–317] appear as very promising in the direction of development of improved atomistic interaction potentials utilizing directly the quantum mechanical scale, or of more accurate CG force fields extracted on the basis of the atomistic detail. In the field of force field development in general, the open challenge of optimized transferable interaction potentials is still to be met.

Apart from design of new materials with controlled properties, molecular modeling can also significantly contribute to the design and optimization of novel separation processes. For example, membrane technologies that are based in pervaporation [318–320] are used for the dehydration of water-organic mixtures, the separation of organic compounds from water or for separations in organic-organic mixtures. Molecular simulation using NEMD techniques can be utilized to simulate such a process and aid in the optimization of the applied fluxes.

Membrane technology is of great economic, environmental and industrial importance and is directly connected to many societal challenges that need to be faced world-wide in the years to come. Future sustainability relies on the indisputable need for green technologies and security of the water-food-energy nexus around the globe. Efficient membrane materials design is a prerequisite for the development of advanced green cutting-edge separation and barrier processes and the complementary contribution of novel experimental and computational techniques is pivotal in order for the scientific and engineering community to address the critical environmental and sustainability issues.

Author Contributions: Both authors have contributed in all aspects of this review.

Funding: This review received no external funding.

Acknowledgments: We would like to thank the Guest-Editors, Giulio C. Sarti and Maria Grazia De Angelis, for inviting us to contribute to the "Gas Transport in Glassy Polymers" Special Issue and for their patience with our delay in finishing the manuscript.

Conflicts of Interest: The authors declare no conflict of interest.

References

1. Maginn, E.J. From discovery to data: What must happen for molecular simulation to become a mainstream chemical engineering tool. *AIChE J.* **2009**, *55*, 1304–1310. [CrossRef]
2. Weiß, H.; Deglmann, P.; in 't Veld, P.J.; Cetinkaya, M.; Schreiner, E. Multiscale Materials Modeling in an Industrial Environment. *Annu. Rev. Chem. Biomol. Eng.* **2016**, *7*, 65–86. [CrossRef] [PubMed]
3. Goldbeck, G. The Economic Impact of Molecular Modelling of Chemicals and Materials. Goldbeck Consulting: 2012. Available online: https://gerhardgoldbeck.files.wordpress.com/2014/01/the-economic-impact-of-modelling.pdf (accessed on 15 April 2019).

4. Materials Genome Initiative USA. Available online: https://www.mgi.gov (accessed on 15 April 2019).
5. OCTA Project Japan. Available online: http://octa.jp/about-octa/project (accessed on 15 April 2019).
6. European Materials Modelling Council European Union. Available online: https://emmc.info (accessed on 15 April 2019).
7. Gusev, A.A.; Müller-Plathe, F.; van Gunsteren, W.F.; Suter, U.W. Dynamics of small molecules in bulk polymers. In *Atomistic Modeling of Physical Properties*; Monnerie, L., Suter, U.W., Eds.; Springer: Berlin/Heidelberg, Germany, 1994; pp. 207–247. [CrossRef]
8. Theodorou, D.N. Molecular Simulations of Sorption and Diffusion in Amorphous Polymers. In *Diffusion in Polymers*; Neogi, P., Ed.; Marcel Dekker Inc.: New York, NY, USA, 1996; p. 67.
9. Theodorou, D.N. Principles of Molecular Simulation of Gas Transport in Polymers. In *Materials Science of Membranes for Gas and Vapor Separation*; John Wiley & Sons, Ltd.: Hoboken, NJ, USA, 2006; pp. 49–94.
10. Colón, Y.J.; Fairen-Jimenez, D.; Wilmer, C.E.; Snurr, R.Q. High-Throughput Screening of Porous Crystalline Materials for Hydrogen Storage Capacity near Room Temperature. *J. Phys. Chem. C* **2014**, *118*, 5383–5389. [CrossRef]
11. Crank, J.; Park, G.S. *Diffusion in Polymers*; Academic Press: New York, NY, USA, 1968.
12. Crank, J. *The Mathematics of Diffusion*; Oxford University Press: Oxford, UK, 1985.
13. Robeson, L.M. The upper bound revisited. *J. Membr. Sci.* **2008**, *320*, 390–400. [CrossRef]
14. Robeson, L.M. Correlation of separation factor versus permeability for polymeric membranes. *J. Membr. Sci.* **1991**, *62*, 165–185. [CrossRef]
15. Robeson, L.M.; Liu, Q.; Freeman, B.D.; Paul, D.R. Comparison of transport properties of rubbery and glassy polymers and the relevance to the upper bound relationship. *J. Membr. Sci.* **2015**, *476*, 421–431. [CrossRef]
16. Robeson, L.M.; Smith, Z.P.; Freeman, B.D.; Paul, D.R. Contributions of diffusion and solubility selectivity to the upper bound analysis for glassy gas separation membranes. *J. Membr. Sci.* **2014**, *453*, 71–83. [CrossRef]
17. Haile, J.M. *Molecular Dynamics Simulation: Elementary Methods*; John Wiley & Sons, Inc.: New York, NY, USA, 1992; p. 489.
18. Frenkel, D.; Smit, B. *Understanding Molecular Simulation: From Algorithms to Applications*; Academic Press: San Diego, CA, USA, 2002.
19. Rapaport, D.C. *The Art of Molecular Dynamics Simulation*; Cambridge University Press: New York, NY, USA, 2002.
20. Metropolis, N.; Rosenbluth, A.W.; Rosenbluth, M.N.; Teller, A.H.; Teller, E. Equation of State Calculations by Fast Computing Machines. *J. Chem. Phys.* **1953**, *21*, 1087–1092. [CrossRef]
21. Allen, M.P.; Tildesley, D.J. *Computer Simulation of Liquids*; Oxford University Press: New York, NY, USA, 1987.
22. Pant, P.V.K.; Theodorou, D.N. Variable Connectivity Method for the Atomistic Monte Carlo Simulation of Polydisperse Polymer Melts. *Macromolecules* **1995**, *28*, 7224–7234. [CrossRef]
23. Mavrantzas, V.G.; Boone, T.D.; Zervopoulou, E.; Theodorou, D.N. End-Bridging Monte Carlo: A Fast Algorithm for Atomistic Simulation of Condensed Phases of Long Polymer Chains. *Macromolecules* **1999**, *32*, 5072–5096. [CrossRef]
24. Uhlherr, A.; Mavrantzas, V.G.; Doxastakis, M.; Theodorou, D.N. Directed Bridging Methods for Fast Atomistic Monte Carlo Simulations of Bulk Polymers. *Macromolecules* **2001**, *34*, 8554–8568. [CrossRef]
25. Karayiannis, N.C.; Mavrantzas, V.G.; Theodorou, D.N. A novel monte carlo scheme for the rapid equilibration of atomistic model polymer systems of precisely defined molecular architecture. *Phys. Rev. Lett.* **2002**, *88*, 105503. [CrossRef] [PubMed]
26. Karayiannis, N.C.; Giannousaki, A.E.; Mavrantzas, V.G.; Theodorou, D.N. Atomistic Monte Carlo simulation of strictly monodisperse long polyethylene melts through a generalized chain bridging algorithm. *J. Chem. Phys.* **2002**, *117*, 5465–5479. [CrossRef]
27. Peristeras, L.D.; Economou, I.G.; Theodorou, D.N. Structure and Volumetric Properties of Linear and Triarm Star Polyethylenes from Atomistic Monte Carlo Simulation Using New Internal Rearrangement Moves. *Macromolecules* **2005**, *38*, 386–397. [CrossRef]
28. Ramos, F.J.; Peristeras, L.; Theodorou, D.N. Monte Carlo Simulation of Short Chain Branched Polyolefins in the Molten State. *Macromolecules* **2007**, *40*, 9640–9650. [CrossRef]
29. Gusev, A.A.; Suter, U.W. Dynamics of Small Molecules in Dense Polymers Subject to Thermal Motion. *J. Chem. Phys.* **1993**, *99*, 2228–2234. [CrossRef]

30. Gusev, A.A.; Arizzi, S.; Suter, U.W.; Moll, D.J. Dynamics of light gases in rigid matrices of dense polymers. *J. Chem. Phys.* **1993**, *99*, 2221–2227. [CrossRef]
31. Greenfield, M.L.; Theodorou, D.N. Coupling of Penetrant and Polymer Motions during Small-Molecule Diffusion in a Glassy Polymer. *Mol. Simul.* **1997**, *19*, 329–361. [CrossRef]
32. Greenfield, M.L.; Theodorou, D.N. Molecular modeling of methane diffusion in glassy atactic polypropylene via multidimensional transition state theory. *Macromolecules* **1998**, *31*, 7068–7090. [CrossRef]
33. Vergadou, N. Prediction of Gas Permeability of Inflexible Amorphous Polymers via Molecular Simulation. Ph.D. Thesis, University of Athens, Athens, Greece, 2006.
34. Theodorou, D.N. Hierarchical modelling of polymeric materials. *Chem. Eng. Sci.* **2007**, *62*, 5697–5714. [CrossRef]
35. Xi, L.; Shah, M.; Trout, B.L. Hopping of Water in a Glassy Polymer Studied via Transition Path Sampling and Likelihood Maximization. *J. Phys. Chem. B* **2013**, *117*, 3634–3647. [CrossRef] [PubMed]
36. Greenfield, M.L.; Theodorou, D.N. Coarse-Grained Molecular Simulation of Penetrant Diffusion in a Glassy Polymer Using Reverse and Kinetic Monte Carlo. *Macromolecules* **2001**, *34*, 8541–8553. [CrossRef]
37. Neyertz, S.; Brown, D. A Trajectory-Extending Kinetic Monte Carlo (TEKMC) Method for Estimating Penetrant Diffusion Coefficients in Molecular Dynamics Simulations of Glassy Polymers. *Macromolecules* **2010**, *43*, 9210–9214. [CrossRef]
38. Neyertz, S.; Brown, D.; Pandiyan, S.; van der Vegt, N.F. Carbon dioxide diffusion and plasticization in fluorinated polyimides. *Macromolecules* **2010**, *43*, 7813–7827. [CrossRef]
39. Harmandaris, V.A.; Adhikari, N.P.; van der Vegt, N.F.A.; Kremer, K.; Mann, B.A.; Voelkel, R.; Weiss, H.; Liew, C. Ethylbenzene Diffusion in Polystyrene: United Atom Atomistic/Coarse Grained Simulations and Experiments. *Macromolecules* **2007**, *40*, 7026–7035. [CrossRef]
40. Fritz, D.; Herbers, C.R.; Kremer, K.; van der Vegt, N.F.A. Hierarchical Modeling of Polymer Permeation. *Soft Matter* **2009**, *5*, 4556–4563. [CrossRef]
41. Lin, E.; You, X.; Kriegel, R.M.; Moffitt, R.D.; Batra, R.C. Atomistic to coarse grained simulations of diffusion of small molecules into polymeric matrix. *Comput. Mater. Sci.* **2017**, *138*, 448–461. [CrossRef]
42. Zhang, K.; Meng, D.; Müller-Plathe, F.; Kumar, S.K. Coarse-grained molecular dynamics simulation of activated penetrant transport in glassy polymers. *Soft Matter* **2018**, *14*, 440–447. [CrossRef]
43. Müller-Plathe, F.; Rogers, S.C.; van Gunsteren, W.F. Computational Evidence for Anomalous Diffusion of Small Molecules in Amorphous Polymers. *Chem. Phys. Lett.* **1992**, *199*, 237–243. [CrossRef]
44. Raptis, V.E.; Economou, I.G.; Theodorou, D.N.; Petrou, J.; Petropoulos, J.H. Molecular dynamics simulation of structure and thermodynamic properties of poly(dimethylsilamethylene) and hydrocarbon solubility therein: Toward the development of novel membrane materials for hydrocarbon separation. *Macromolecules* **2004**, *37*, 1102–1112. [CrossRef]
45. Yang, Q.; Whiting, W.I. Molecular-level insight of the differences in the diffusion and solubility of penetrants in polypropylene, poly(propylmethylsiloxane) and poly(4-methyl-2-pentyne). *J. Membr. Sci.* **2018**, *549*, 173–183. [CrossRef]
46. Van der Vegt, N.F.A.; Briels, W.J.; Wessling, M.; Strathmann, H. The sorption induced glass transition in amorphous glassy polymers. *J. Chem. Phys.* **1999**, *110*, 11061–11069. [CrossRef]
47. Spyriouni, T.; Boulougouris, G.C.; Theodorou, D.N. Prediction of Sorption of CO_2 in Glassy Atactic Polystyrene at Elevated Pressures Through a New Computational Scheme. *Macromolecules* **2009**, *42*, 1759–1769. [CrossRef]
48. Knopp, B.; Suter, U.W. Atomistically Modeling the Chemical Potential of Small Molecules in Dense Polymer Microstructures. 2. Water Sorption by Polyamides. *Macromolecules* **1997**, *30*, 6114–6119. [CrossRef]
49. Knopp, B.; Suter, U.W.; Gusev, A.A. Atomistically Modeling the Chemical Potential of Small Molecules in Dense Polymer Microstructures. 1. Method. *Macromolecules* **1997**, *30*, 6107–6113. [CrossRef]
50. Martyna, G.J.; Tuckerman, M.E.; Tobias, D.J.; Klein, M.L. Explicit reversible integrators for extended systems dynamics. *Mol. Phys.* **1996**, *87*, 1117–1157. [CrossRef]
51. Tuckerman, M.; Berne, B.J.; Martyna, G.J. Reversible multiple time scale molecular dynamics. *J. Chem. Phys.* **1992**, *97*, 1990–2001. [CrossRef]
52. McQuarrie, D.A. *Statistical Mechanics*; Harper & Row: New York, NY, USA, 1975.
53. Hansen, J.-P.; McDonald, I.R. *Theory of Simple Liquids*, 3rd ed.; Academic Press: London, UK, 2006.

54. Müller-Plathe, F.; Rogers, S.C.; van Gunsteren, W.F. Gas sorption and transport in polyisobutylene: Equilibrium and nonequilibrium molecular dynamics simulations. *J. Chem. Phys.* **1993**, *98*, 9895–9904. [CrossRef]
55. Van der Vegt, N.F.A.; Briels, W.J.; Wessling, M.; Strathmann, H. A Nonequilibrium Simulation Method for Calculating Tracer Diffusion Coefficients of Small Solutes in N-Alkane Liquids and Polymers. *J. Chem. Phys.* **1998**, *108*, 9558–9565. [CrossRef]
56. Frentrup, H.; Hart, E.K.; Colina, M.C.; Müller, A.E. In Silico Determination of Gas Permeabilities by Non-Equilibrium Molecular Dynamics: CO_2 and He through PIM-1. *Membranes* **2015**, *5*, 99–119. [CrossRef]
57. Anderson, L.R.; Yang, Q.; Ediger, A.M. Comparing gas transport in three polymers via molecular dynamics simulation. *Phys. Chem. Chem. Phys.* **2018**, *20*, 22123–22133. [CrossRef] [PubMed]
58. Theodorou, D.N. Variable-Connectivity Monte Carlo Algorithms for the Atomistic Simulation of Long-Chain Polymer Systems. In *Bridging Time Scales: Molecular Simulations for the Next Decade*; Nielaba, P., Mareschal, M., Ciccotti, G., Eds.; Springer: Berlin/Heidelberg, Germany, 2002; pp. 67–127.
59. Rosenbluth, M.N.; Rosenbluth, A.W. Monte Carlo Calculation of the Average Extension of Molecular Chains. *J. Chem. Phys.* **1955**, *23*, 356–359. [CrossRef]
60. Siepmann, J.I.; Frenkel, D. Configurational bias Monte Carlo: A new sampling scheme for flexible chains. *Mol. Phys.* **1992**, *75*, 59–70. [CrossRef]
61. De Pablo, J.J.; Laso, M.; Suter, U.W. Estimation of the chemical potential of chain molecules by simulation. *J. Chem. Phys.* **1992**, *96*, 6157–6162. [CrossRef]
62. Dodd, L.R.; Boone, T.D.; Theodorou, D.N. A concerted rotation algorithm for atomistic Monte Carlo simulation of polymer melts and glasses. *Mol. Phys.* **1993**, *78*, 961–996. [CrossRef]
63. Panagiotopoulos, A.Z. Direct determination of phase coexistence properties of fluids by Monte Carlo simulation in a new ensemble. *Mol. Phys.* **1987**, *61*, 813–826. [CrossRef]
64. Briano, J.G.; Glandt, E.D. Statistical thermodynamics of polydisperse fluids. *J. Chem. Phys.* **1984**, *80*, 3336–3343. [CrossRef]
65. Kofke, D.A.; Glandt, E.D. Monte carlo simulation of multicomponent equilibria in a semigrand canonical ensemble. *Mol. Phys.* **1988**, *64*, 1105–1131. [CrossRef]
66. Widom, B. Some Topics in the Theory of Fluids. *J. Chem. Phys.* **1963**, *39*, 2808–2812. [CrossRef]
67. Zervopoulou, E.; Mavrantzas, V.G.; Theodorou, D.N. A new Monte Carlo simulation approach for the prediction of sorption equilibria of oligomers in polymer melts: Solubility of long alkanes in linear polyethylene. *J. Chem. Phys.* **2001**, *115*, 2860–2875. [CrossRef]
68. Lyubartsev, A.P.; Martsinovski, A.A.; Shevkunov, S.V.; Vorontsov-Velyaminov, P.N. New approach to Monte Carlo calculation of the free energy: Method of expanded ensembles. *J. Chem. Phys.* **1992**, *96*, 1776–1783. [CrossRef]
69. De Pablo, J.J.; Yan, Q.; Escobedo, F.A. Simulation of Phase Transitions in Fluids. *Annu. Rev. Phys. Chem.* **1999**, *50*, 377–411. [CrossRef] [PubMed]
70. Cuthbert, T.R.; Wagner, N.J.; Paulaitis, M.E. Molecular Simulation of Glassy Polystyrene: Size Effects on Gas Solubilities. *Macromolecules* **1997**, *30*, 3058–3065. [CrossRef]
71. Deitrick, G.L.; Scriven, L.E.; Davis, H.T. Efficient molecular simulation of chemical potentials. *J. Chem. Phys.* **1989**, *90*, 2370–2385. [CrossRef]
72. Tamai, Y.; Tanaka, H.; Nakanishi, K. Molecular Simulation of Permeation of Small Penetrants through Membranes. 2. Solubilities. *Macromolecules* **1995**, *28*, 2544–2554. [CrossRef]
73. Fukuda, M. Solubilities of small molecules in polyethylene evaluated by a test-particle-insertion method. *J. Chem. Phys.* **1999**, *112*, 478–486. [CrossRef]
74. Dömötör, G.; Hentschke, R. Atomistically Modeling the Chemical Potential of Small Molecules in Dense Systems. *J. Phys. Chem. B* **2004**, *108*, 2413–2417. [CrossRef]
75. Dömötör, G.; Hentschke, R. Equilibrium Swelling of an Epoxy-Resin in Contact with Water—A Molecular Dynamics Simulation Study. *Macromol. Theory Simul.* **2004**, *13*, 506–511. [CrossRef]
76. Boulougouris, G.C.; Economou, I.G.; Theodorou, D.N. On the calculation of the chemical potential using the particle deletion scheme. *Mol. Phys.* **1999**, *96*, 905–913. [CrossRef]
77. Boulougouris, G.C.; Economou, I.G.; Theodorou, D.N. Calculation of the chemical potential of chain molecules using the staged particle deletion scheme. *J. Chem. Phys.* **2001**, *115*, 8231–8237. [CrossRef]

78. Van der Vegt, N.F.A.; Briels, W.J. Efficient sampling of solvent free energies in polymers. *J. Chem. Phys.* **1998**, *109*, 7578–7582. [CrossRef]
79. Vrabec, J.; Hasse, H. Grand Equilibrium: Vapour-liquid equilibria by a new molecular simulation method. *Mol. Phys.* **2002**, *100*, 3375–3383. [CrossRef]
80. Hess, B.; Peter, C.; Ozal, T.; van der Vegt, N.F.A. Fast-Growth Thermodynamic Integration: Calculating Excess Chemical Potentials of Additive Molecules in Polymer Microstructures. *Macromolecules* **2008**, *41*, 2283–2289. [CrossRef]
81. Glasstone, S.; Laidler, K.J.; Eyring, H. *The Theory of Rate Processes: The Kinetics of Chemical Reactions, Viscosity, Diffusion and Electrochemical Phenomena*; McGraw-Hill: New York, NY, USA, 1941.
82. Voter, A.F.; Doll, J.D. Dynamical Corrections to Transition-State Theory for Multistate Systems—Surface Self-Diffusion in the Rare-Event Regime. *J. Chem. Phys.* **1985**, *82*, 80–92. [CrossRef]
83. Eyring, H. The Activated Complex in Chemical Reactions. *J. Chem. Phys.* **1935**, *3*, 107–115. [CrossRef]
84. van Kampen, N.G. *Stochastic Processes in Physics and Chemistry*, 3rd ed.; North Holland Publishing Company: Amsterdam, The Netherlands, 1983.
85. Fukui, K. The Path of Chemical Reactions—The IRC Approach. *Acc. Chem. Res.* **1981**, *14*, 363–368. [CrossRef]
86. Kotelyanskii, M.; Theodorou, D.N. *Simulation Methods for Polymers*; Marcel Dekker Inc: New York, NY, USA, 2004.
87. Chandler, D. Statistical mechanics of isomerization dynamics in liquids and the transition state approximation. *J. Chem. Phys.* **1978**, *68*, 2959–2970. [CrossRef]
88. Hofmann, D.; Fritz, L.; Ulbrich, J.; Paul, D. Molecular simulation of small molecule diffusion and solution in dense amorphous polysiloxanes and polyimides. *Comp. Theor. Polymer Sci.* **2000**, *10*, 419–436. [CrossRef]
89. Hofmann, D.; Fritz, L.; Ulbrich, J.; Schepers, C.; Bohning, M. Detailed-Atomistic Molecular Modeling of Small Molecule Diffusion and Solution Processes in Polymeric Membrane Materials. *Macromol. Theory Simul.* **2000**, *9*, 293–327. [CrossRef]
90. Karayiannis, N.C.; Mavrantzas, V.G.; Theodorou, D.N. Detailed Atomistic Simulation of the Segmental Dynamics and Barrier Properties of Amorphous Poly(ethylene Terephthalate) and Poly(ethylene Isophthalate). *Macromolecules* **2004**, *37*, 2978–2995. [CrossRef]
91. Greenfield, M.L.; Theodorou, D.N. Geometric analysis of diffusion pathways in glassy and melt atactic polypropylene. *Macromolecules* **1993**, *26*, 5461–5472. [CrossRef]
92. Vergadou, N.; Theodorou, D.N. Molecular simulation of CO_2 Diffusion in Glassy Polymers using Multidimensional Transition State Theory. **2019**. In preparation.
93. Bolhuis, P.G. Transition path sampling on diffusive barriers. *J. Phys. Condens. Matter* **2002**, *15*, S113–S120. [CrossRef]
94. Bolhuis, P.G.; Dellago, C.; Chandler, D. Sampling Ensembles of Deterministic Transition Pathways. *Faraday Discuss.* **1998**, *110*, 421–436. [CrossRef]
95. Peters, B.; Trout, B.L. Obtaining Reaction Coordinates by Likelihood Maximization. *J. Chem. Phys.* **2006**, *125*, 054108. [CrossRef] [PubMed]
96. Peters, B.; Beckham, G.T.; Trout, B.L. Extensions to the Likelihood Maximization Approach for Finding Reaction Coordinates. *J. Chem. Phys.* **2007**, *127*, 034109. [CrossRef] [PubMed]
97. Grünwald, M.; Dellago, C.; Geissler, P.L. Precision shooting: Sampling long transition pathways. *J. Chem. Phys.* **2008**, *129*, 194101. [CrossRef] [PubMed]
98. Mullen, R.G.; Shea, J.-E.; Peters, B. Easy Transition Path Sampling Methods: Flexible-Length Aimless Shooting and Permutation Shooting. *J. Chem. Theory Comput.* **2015**, *11*, 2421–2428. [CrossRef]
99. Pan, A.C.; Chandler, D. Dynamics of nucleation in the ising model. *J. Phys. Chem. B* **2004**, *108*, 19681. [CrossRef]
100. Moroni, D.; Ten Wolde, P.R.; Bolhuis, P.G. Interplay between structure and size in a critical crystal nucleus. *Phys. Rev. Lett.* **2005**, *94*, 235703. [CrossRef]
101. Lechner, W.; Dellago, C.; Bolhuis, P.G. Role of the Prestructured Surface Cloud in Crystal Nucleation. *Phys. Rev. Lett.* **2011**, *106*, 085701. [CrossRef] [PubMed]
102. Beckham, G.T.; Peters, B.J. Optimizing Nucleus Size Metrics for Liquid-Solid Nucleation from Transition Paths of Near-Nanosecond Duration. *Phys. Chem. Lett.* **2011**, *2*, 1133–1138. [CrossRef] [PubMed]
103. Chong, L.T.; Saglam, A.S.; Zuckerman, D.M. Path-sampling strategies for simulating rare events in biomolecular systems. *Curr. Opin. Struct. Biol.* **2017**, *43*, 88–94. [CrossRef] [PubMed]

104. Juraszek, J.; Bolhuis, P.G. Rate Constant and Reaction Coordinate of Trp-Cage Folding in Explicit Water. *Biophys. J.* **2008**, *95*, 4246–4257. [CrossRef] [PubMed]
105. Zhang, B.W.; Jasnow, D.; Zuckerman, D.M. Efficient and verified simulation of a path ensemble for conformational change in a united-residue model of calmodulin. *Proc. Natl. Acad. Sci. USA* **2007**, *104*, 18043. [CrossRef]
106. Roy, A.; Taraphder, S. Transition Path Sampling Study of the Conformational Fluctuation of His-64 in Human Carbonic Anhydrase II. *J. Phys. Chem. B* **2009**, *113*, 12555–12564. [CrossRef]
107. Keys, A.S.; Hedges, L.O.; Garrahan, J.P.; Glotzer, S.C.; Chandler, D. Excitations Are Localized and Relaxation Is Hierarchical in Glass-Forming Liquids. *Phys. Rev. X* **2011**, *1*, 021013.
108. Pan, B.; Ricci, M.S.; Trout, B.L. A molecular mechanism of hydrolysis of peptide bonds at neutral pH using a model compound. *J. Phys. Chem. B* **2011**, *115*, 5958–5970. [CrossRef]
109. Quaytman, S.L.; Schwartz, S.D. Reaction coordinate of an enzymatic reaction revealed by transition path sampling. *Proc. Natl. Acad. Sci. USA* **2007**, *104*, 12253–12258. [CrossRef]
110. Basner, J.E.; Schwartz, S.D. How Enzyme Dynamics Helps Catalyze a Reaction in Atomic Detail: A Transition Path Sampling Study. *J. Am. Chem. Soc.* **2005**, *127*, 13822–13831. [CrossRef]
111. Vlugt, T.J.H.; Dellago, C.; Smit, B. Diffusion of Isobutane in Silicalite Studied by Transition Path Sampling. *J. Chem. Phys.* **2000**, *113*, 8791–8799. [CrossRef]
112. Peters, B.; Zimmermann, N.E.R.; Beckham, G.T.; Tester, J.W.; Trout, B.L. Path Sampling Calculation of Methane Diffusivity in Natural Gas Hydrates from a Water-Vacancy Assisted Mechanism. *J. Am. Chem. Soc.* **2008**, *130*, 17342–17350. [CrossRef] [PubMed]
113. Boulfelfel, S.E.; Ravikovitch, P.I.; Sholl, D.S. Modeling Diffusion of Linear Hydrocarbons in Silica Zeolite LTA Using Transition Path Sampling. *J. Phys. Chem. C* **2015**, *119*, 15643–15653. [CrossRef]
114. Auhl, R.; Everaers, R.; Grest, G.S.; Kremer, K.; Plimpton, S.J. Equilibration of long chain polymer melts in computer simulations. *J. Chem. Phys.* **2003**, *119*, 12718–12728. [CrossRef]
115. Theodorou, D.N. Equilibration and Coarse-Graining Methods for Polymers. In *Computer Simulations in Condensed Matter Systems: From Materials to Chemical Biology Volume 2*; Ferrario, M., Ciccotti, G., Binder, K., Eds.; Springer: Berlin/Heidelberg, Germany, 2006; pp. 419–448.
116. Brown, D.; Clarke, J.H.R.; Okuda, M.; Yamazaki, T. The preparation of polymer melt samples for computer simulation studies. *J. Chem. Phys.* **1994**, *100*, 6011–6018. [CrossRef]
117. Gao, J. An efficient method of generating dense polymer model melts by computer simulation. *J. Chem. Phys.* **1995**, *102*, 1074–1077. [CrossRef]
118. Mondello, M.; Grest, G.S.; Webb, E.B.; Peczak, P. Dynamics of n-alkanes: Comparison to Rouse model. *J. Chem. Phys.* **1998**, *109*, 798–805. [CrossRef]
119. Kröger, M. Efficient hybrid algorithm for the dynamic creation of wormlike chains in solutions, brushes, melts and glasses. *Comput. Phys. Commun.* **1999**, *118*, 278–298. [CrossRef]
120. Sliozberg, Y.R.; Kröger, M.; Chantawansri, T.L. Fast equilibration protocol for million atom systems of highly entangled linear polyethylene chains. *J. Chem. Phys.* **2016**, *144*, 154901. [CrossRef]
121. Martínez, L.; Andrade, R.; Birgin, E.G.; Martínez, J.M. PACKMOL: A package for building initial configurations for molecular dynamics simulations. *J. Comput. Chem.* **2009**, *30*, 2157–2164. [CrossRef]
122. Flory, P. *Statistical Mechanics of Chain Molecules*; Interscience Publishers: New York, NY, USA, 1969.
123. Theodorou, D.N.; Suter, U.W. Detailed Molecular-Structure of a Vinyl Polymer Glass. *Macromolecules* **1985**, *18*, 1467–1478. [CrossRef]
124. Rigby, D.; Roe, R.J. Molecular dynamics simulation of polymer liquid and glass. I. Glass transition. *J. Chem. Phys.* **1987**, *87*, 7285–7292. [CrossRef]
125. Gooneie, A.; Schuschnigg, S.; Holzer, C. A Review of Multiscale Computational Methods in Polymeric Materials. *Polymers* **2017**, *9*, 16. [CrossRef] [PubMed]
126. Müller-Plathe, F. Coarse-Graining in Polymer Simulation: From the Atomistic to the Mesoscopic Scale and Back. *ChemPhysChem* **2002**, *3*, 754–769. [CrossRef]
127. Tschöp, W.; Kremer, K.; Batoulis, J.; Bürger, T.; Hahn, O. Simulation of polymer melts. I. Coarse-graining procedure for polycarbonates. *Acta Polym.* **1998**, *49*, 61–74. [CrossRef]
128. Karimi-Varzaneh, H.A.; van der Vegt, N.F.A.; Müller-Plathe, F.; Carbone, P. How Good Are Coarse-Grained Polymer Models? A Comparison for Atactic Polystyrene. *ChemPhysChem* **2012**, *13*, 3428–3439. [CrossRef]

129. Zacharopoulos, N.; Vergadou, N.; Theodorou, D.N. Coarse graining using pretabulated potentials: Liquid benzene. *J. Chem. Phys.* **2005**, *122*, 244111. [CrossRef]
130. Deichmann, G.; Dallavalle, M.; Rosenberger, D.; van der Vegt, N.F.A. Phase Equilibria Modeling with Systematically Coarse-Grained Models—A Comparative Study on State Point Transferability. *J. Phys. Chem. B* **2019**, *123*, 504–515. [CrossRef]
131. Sun, Q.; Faller, R. Systematic Coarse-Graining of a Polymer Blend: Polyisoprene and Polystyrene. *J. Chem. Theory Comput.* **2006**, *2*, 607–615. [CrossRef]
132. Mognetti, B.M.; Virnau, P.; Yelash, L.; Paul, W.; Binder, K.; Müller, M.; MacDowell, L.G. Coarse-grained models for fluids and their mixtures: Comparison of Monte Carlo studies of their phase behavior with perturbation theory and experiment. *J. Chem. Phys.* **2009**, *130*, 044101. [CrossRef]
133. Marrink, S.J.; Risselada, H.J.; Yefimov, S.; Tieleman, D.P.; de Vries, A.H. The MARTINI Force Field: Coarse Grained Model for Biomolecular Simulations. *J. Phys. Chem. B* **2007**, *111*, 7812–7824. [CrossRef] [PubMed]
134. Maerzke, K.A.; Siepmann, J.I. Transferable Potentials for Phase Equilibria−Coarse-Grain Description for Linear Alkanes. *J. Phys. Chem. B* **2011**, *115*, 3452–3465. [CrossRef] [PubMed]
135. Allison, J.R.; Riniker, S.; van Gunsteren, W.F. Coarse-grained models for the solvents dimethyl sulfoxide, chloroform, and methanol. *J. Chem. Phys.* **2012**, *136*, 054505. [CrossRef] [PubMed]
136. Brini, E.; Algaer, E.A.; Ganguly, P.; Li, C.; Rodríguez-Ropero, F.; van der Vegt, N.F.A. Systematic coarse-graining methods for soft matter simulations—A review. *Soft Matter* **2013**, *9*, 2108–2119. [CrossRef]
137. Huang, H.; Wu, L.; Xiong, H.; Sun, H. A Transferrable Coarse-Grained Force Field for Simulations of Polyethers and Polyether Blends. *Macromolecules* **2019**, *52*, 249–261. [CrossRef]
138. Kuo, A.-T.; Okazaki, S.; Shinoda, W. Transferable coarse-grained model for perfluorosulfonic acid polymer membranes. *J. Chem. Phys.* **2017**, *147*, 094904. [CrossRef] [PubMed]
139. Reith, D.; Pütz, M.; Müller-Plathe, F. Deriving effective mesoscale potentials from atomistic simulations. *J. Comput. Chem.* **2003**, *24*, 1624–1636. [CrossRef]
140. Ercolessi, F.; Adams, J.B. Interatomic Potentials from First-Principles Calculations: The Force-Matching Method. *Europhys. Lett. (EPL)* **1994**, *26*, 583–588. [CrossRef]
141. Izvekov, S.; Voth, G.A. Modeling real dynamics in the coarse-grained representation of condensed phase systems. *J. Chem. Phys.* **2006**, *125*, 151101. [CrossRef]
142. Lyubartsev, A.P.; Laaksonen, A. Calculation of effective interaction potentials from radial distribution functions: A reverse Monte Carlo approach. *Phys. Rev. E* **1995**, *52*, 3730–3737. [CrossRef]
143. Chaimovich, A.; Shell, M.S. Coarse-graining errors and numerical optimization using a relative entropy framework. *J. Chem. Phys.* **2011**, *134*, 094112. [CrossRef] [PubMed]
144. Shell, M.S. The relative entropy is fundamental to multiscale and inverse thermodynamic problems. *J. Chem. Phys.* **2008**, *129*, 144108. [CrossRef] [PubMed]
145. Villa, A.; Peter, C.; van der Vegt, N.F.A. Self-assembling dipeptides: Conformational sampling in solvent-free coarse-grained simulation. *Phys. Chem. Chem. Phys.* **2009**, *11*, 2077–2086. [CrossRef] [PubMed]
146. Li, C.; Shen, J.; Peter, C.; van der Vegt, N.F.A. A Chemically Accurate Implicit-Solvent Coarse-Grained Model for Polystyrenesulfonate Solutions. *Macromolecules* **2012**, *45*, 2551–2561. [CrossRef]
147. De Nicola, A.; Kawakatsu, T.; Milano, G. Generation of Well-Relaxed All-Atom Models of Large Molecular Weight Polymer Melts: A Hybrid Particle-Continuum Approach Based on Particle-Field Molecular Dynamics Simulations. *J. Chem. Theory Comput.* **2014**, *10*, 5651–5667. [CrossRef] [PubMed]
148. Rosenberger, D.; van der Vegt, N.F.A. Addressing the temperature transferability of structure based coarse graining models. *Phys. Chem. Chem. Phys.* **2018**, *20*, 6617–6628. [CrossRef] [PubMed]
149. Potter, T.D.; Tasche, J.; Wilson, M.R. Assessing the transferability of common top-down and bottom-up coarse-grained molecular models for molecular mixtures. *Phys. Chem. Chem. Phys.* **2019**, *21*, 1912–1927. [CrossRef] [PubMed]
150. Dunn, N.J.H.; Foley, T.T.; Noid, W.G. Van der Waals Perspective on Coarse-Graining: Progress toward Solving Representability and Transferability Problems. *Acc. Chem. Res.* **2016**, *49*, 2832–2840. [CrossRef]
151. Guenza, M.G.; Dinpajooh, M.; McCarty, J.; Lyubimov, I.Y. Accuracy, Transferability, and Efficiency of Coarse-Grained Models of Molecular Liquids. *J. Phys. Chem. B* **2018**, *122*, 10257–10278. [CrossRef]
152. Tschöp, W.; Kremer, K.; Hahn, O.; Batoulis, J.; Bürger, T. Simulation of polymer melts. II. From coarse-grained models back to atomistic description. *Acta Polym.* **1998**, *49*, 75–79. [CrossRef]

153. Hess, B.; León, S.; van der Vegt, N.; Kremer, K. Long time atomistic polymer trajectories from coarse grained simulations: Bisphenol-A polycarbonate. *Soft Matter* **2006**, *2*, 409–414. [CrossRef]
154. Wu, C. Multiscale simulations of the structure and dynamics of stereoregular poly(methyl methacrylate)s. *J. Mol. Modeling* **2014**, *20*. [CrossRef] [PubMed]
155. Ghanbari, A.; Böhm, M.C.; Müller-Plathe, F. A Simple Reverse Mapping Procedure for Coarse-Grained Polymer Models with Rigid Side Groups. *Macromolecules* **2011**, *44*, 5520–5526. [CrossRef]
156. Krajniak, J.; Pandiyan, S.; Nies, E.; Samaey, G. Generic Adaptive Resolution Method for Reverse Mapping of Polymers from Coarse-Grained to Atomistic Descriptions. *J. Chem. Theory Comput.* **2016**, *12*, 5549–5562. [CrossRef] [PubMed]
157. Krajniak, J.; Zhang, Z.; Pandiyan, S.; Nies, E.; Samaey, G. Reverse mapping method for complex polymer systems. *J. Comput. Chem.* **2018**, *39*, 648–664. [CrossRef]
158. Rzepiela, A.J.; Schäfer, L.V.; Goga, N.; Risselada, H.J.; De Vries, A.H.; Marrink, S.J. Reconstruction of atomistic details from coarse-grained structures. *J. Comput. Chem.* **2010**, *31*, 1333–1343. [CrossRef]
159. Wassenaar, T.A.; Pluhackova, K.; Böckmann, R.A.; Marrink, S.J.; Tieleman, D.P. Going Backward: A Flexible Geometric Approach to Reverse Transformation from Coarse Grained to Atomistic Models. *J. Chem. Theory Comput.* **2014**, *10*, 676–690. [CrossRef]
160. Ohkuma, T.; Kremer, K.; Daoulas, K. Equilibrating high-molecular-weight symmetric and miscible polymer blends with hierarchical back-mapping. *J. Phys. Condens. Matter* **2018**, *30*, 174001. [CrossRef]
161. Zhang, G.; Chazirakis, A.; Harmandaris, V.A.; Stuehn, T.; Daoulas, K.C.; Kremer, K. Hierarchical modelling of polystyrene melts: From soft blobs to atomistic resolution. *Soft Matter* **2019**, *15*, 289–302. [CrossRef]
162. Santangelo, G.; Di Matteo, A.; Müller-Plathe, F.; Milano, G. From Mesoscale Back to Atomistic Models: A Fast Reverse-Mapping Procedure for Vinyl Polymer Chains. *J. Phys. Chem. B* **2007**, *111*, 2765–2773. [CrossRef]
163. Bortz, A.; Kalos, M.; Lebowitz, J. A new algorithm for Monte Carlo simulation of Ising spin systems. *J. Comput. Phys.* **1975**, *17*, 10–18. [CrossRef]
164. Young, W.M.; Elcock, E.W. Monte Carlo studies of vacancy migration in binary ordered alloys: I. *Proc. Phys. Soc. Lond.* **1966**, *89*, 735. [CrossRef]
165. Fichthorn, K.A.; Weinberg, W.H. Theoretical foundations of dynamical Monte Carlo simulations. *J. Chem. Phys.* **1991**, *95*, 1090–1096. [CrossRef]
166. Kolokathis, P.D.; Theodorou, D.N. On solving the master equation in spatially periodic systems. *J. Chem. Phys.* **2012**, *137*, 034112. [CrossRef] [PubMed]
167. Tien, W.-J.; Chiu, C.-C. Generic parameters of trajectory-extending kinetic Monte Carlo for calculating diffusion coefficients. *AIP Adv.* **2018**, *8*, 065311. [CrossRef]
168. Hanson, B.; Pryamitsyn, V.; Ganesan, V. Computer Simulations of Gas Diffusion in Polystyrene–C60 Fullerene Nanocomposites Using Trajectory Extending Kinetic Monte Carlo Method. *J. Phys. Chem. B* **2012**, *116*, 95–103. [CrossRef] [PubMed]
169. Mogurampelly, S.; Borodin, O.; Ganesan, V. Computer Simulations of Ion Transport in Polymer Electrolyte Membranes. *Annu. Rev. Chem. Biomol. Eng.* **2016**, *7*, 349–371. [CrossRef] [PubMed]
170. Hanson, B.; Pryamitsyn, V.; Ganesan, V. Mechanisms Underlying Ionic Mobilities in Nanocomposite Polymer Electrolytes. *ACS Macro Lett.* **2013**, *2*, 1001–1005. [CrossRef]
171. Mogurampelly, S.; Ganesan, V. Effect of Nanoparticles on Ion Transport in Polymer Electrolytes. *Macromolecules* **2015**, *48*, 2773–2786. [CrossRef]
172. Mogurampelly, S.; Ganesan, V. Structure and mechanisms underlying ion transport in ternary polymer electrolytes containing ionic liquids. *J. Chem. Phys.* **2017**, *146*, 074902. [CrossRef]
173. Marrink, S.J.; de Vries, A.H.; Mark, A.E. Coarse Grained Model for Semiquantitative Lipid Simulations. *J. Phys. Chem. B* **2004**, *108*, 750–760. [CrossRef]
174. Lee, H.; de Vries, A.H.; Marrink, S.-J.; Pastor, R.W. A Coarse-Grained Model for Polyethylene Oxide and Polyethylene Glycol: Conformation and Hydrodynamics. *J. Phys. Chem. B* **2009**, *113*, 13186–13194. [CrossRef] [PubMed]
175. Bos, A.; Pünt, I.G.M.; Wessling, M.; Strathmann, H. CO_2-induced plasticization phenomena in glassy polymers. *J. Membr. Sci.* **1999**, *155*, 67–78. [CrossRef]
176. Wessling, M.; Huisman, I.; Boomgaard, T.v.d.; Smolders, C.A. Dilation kinetics of glassy, aromatic polyimides induced by carbon dioxide sorption. *J. Polym. Sci. Part B: Polym. Phys.* **1995**, *33*, 1371–1384. [CrossRef]

177. Coleman, M.R.; Koros, W.J. Conditioning of Fluorine Containing Polyimides. 1. Effect of Exposure to High Pressure Carbon Dioxide on Permeability. *Macromolecules* **1997**, *30*, 6899–6905. [CrossRef]
178. Coleman, M.R.; Koros, W.J. Conditioning of Fluorine-Containing Polyimides. 2. Effect of Conditioning Protocol at 8 Volume Dilation on Gas-Transport Properties. *Macromolecules* **1999**, *32*, 3106–3113. [CrossRef]
179. Ismail, A.F.; Lorna, W. Penetrant-induced plasticization phenomenon in glassy polymers for gas separation membrane. *Sep. Purif. Technol.* **2002**, *27*, 173–194. [CrossRef]
180. Visser, T.; Wessling, M. When Do Sorption-Induced Relaxations in Glassy Polymers Set In? *Macromolecules* **2007**, *40*, 4992–5000. [CrossRef]
181. Reijerkerk, S.R.; Nijmeijer, K.; Ribeiro, C.P.; Freeman, B.D.; Wessling, M. On the effects of plasticization in CO_2/light gas separation using polymeric solubility selective membranes. *J. Membr. Sci.* **2011**, *367*, 33–44. [CrossRef]
182. Duthie, X.; Kentish, S.; Powell, C.; Nagai, K.; Qiao, G.; Stevens, G. Operating temperature effects on the plasticization of polyimide gas separation membranes. *J. Membr. Sci.* **2007**, *294*, 40–49. [CrossRef]
183. Zhang, L.; Xiao, Y.; Chung, T.S.; Jiang, J. Mechanistic understanding of CO_2-induced plasticization of a polyimide membrane: A combination of experiment and simulation study. *Polymer* **2010**, *51*, 4439–4447. [CrossRef]
184. Hölck, O.; Böhning, M.; Heuchel, M.; Siegert, M.R.; Hofmann, D. Gas sorption isotherms in swelling glassy polymers—Detailed atomistic simulations. *J. Membr. Sci.* **2013**, *428*, 523–532. [CrossRef]
185. Heuchel, M.; Boehning, M.; Hoelck, O.; Siegert, M.R.; Hofmann, D. Atomistic Packing Models for Experimentally Investigated Swelling States Induced by CO_2 in Glassy Polysulfone and Poly(ether sulfone). *J. Polym. Sci. Part B: Polym. Phys.* **2006**, *44*, 1874–1897. [CrossRef]
186. Van der Vegt, N.F.A. Molecular dynamics simulations of sorption and diffusion in rubbery and glassy polymers. Ph.D. Thesis, University of Twente, Enschede, The Netherlands, 1998.
187. Lock, S.S.M.; Lau, K.K.; Shariff, A.M.; Yeong, Y.F.; Bustam, M.A.; Jusoh, N.; Ahmad, F. An atomistic simulation towards elucidation of operating temperature effect in CO_2 swelling of polysulfone polymeric membranes. *J. Nat. Gas Sci. Eng.* **2018**, *57*, 135–154. [CrossRef]
188. Balçık, M.; Ahunbay, M.G. Prediction of CO_2-induced plasticization pressure in polyimides via atomistic simulations. *J. Membr. Sci.* **2018**, *547*, 146–155. [CrossRef]
189. Kupgan, G.; Demidov, A.G.; Colina, C.M. Plasticization behavior in polymers of intrinsic microporosity (PIM-1): A simulation study from combined Monte Carlo and molecular dynamics. *J. Membr. Sci.* **2018**, *565*, 95–103. [CrossRef]
190. Pandiyan, S.; Brown, D.; Neyertz, S.; van der Vegt, N.F.A. Carbon Dioxide Solubility in Three Fluorinated Polyimides Studied by Molecular Dynamics Simulations. *Macromolecules* **2010**, *43*, 2605–2621. [CrossRef]
191. Spyriouni, T.; Tzoumanekas, C.; Theodorou, D.; Müller-Plathe, F.; Milano, G. Coarse-Grained and Reverse-Mapped United-Atom Simulations of Long-Chain Atactic Polystyrene Melts: Structure, Thermodynamic Properties, Chain Conformation, and Entanglements. *Macromolecules* **2007**, *40*, 3876–3885. [CrossRef]
192. Velioğlu, S.; Ahunbay, M.G.; Tantekin-Ersolmaz, S.B. Investigation of CO_2-induced plasticization in fluorinated polyimide membranes via molecular simulation. *J. Membr. Sci.* **2012**, *417-418*, 217–227. [CrossRef]
193. Velioğlu, S.; Ahunbay, M.G.; Tantekin-Ersolmaz, S.B. Propylene/propane plasticization in polyimide membranes. *J. Membr. Sci.* **2016**, *501*, 179–190. [CrossRef]
194. Tullos, G.; Mathias, L. Unexpected thermal conversion of hydroxy-containing polyimides to polybenzoxazoles. *Polymer* **1999**, *40*, 3463–3468. [CrossRef]
195. Liu, Q.; Borjigin, H.; Paul, D.R.; Riffle, J.S.; McGrath, J.E.; Freeman, B.D. Gas permeation properties of thermally rearranged (TR) isomers and their aromatic polyimide precursors. *J. Membr. Sci.* **2016**, *518*, 88–99. [CrossRef]
196. Park, H.B.; Jung, C.H.; Lee, Y.M.; Hill, A.J.; Pas, S.J.; Mudie, S.T.; Van Wagner, E.; Freeman, B.D.; Cookson, D.J. Polymers with Cavities Tuned for Fast Selective Transport of Small Molecules and Ions. *Science* **2007**, *318*, 254–258. [CrossRef] [PubMed]

197. Sanders, D.F.; Smith, Z.P.; Ribeiro, C.P.; Guo, R.; McGrath, J.E.; Paul, D.R.; Freeman, B.D. Gas permeability, diffusivity, and free volume of thermally rearranged polymers based on 3,3′-dihydroxy-4,4′-diamino-biphenyl (HAB) and 2,2′-bis-(3,4-dicarboxyphenyl) hexafluoropropane dianhydride (6FDA). *J. Membr. Sci.* **2012**, *409–410*, 232–241. [CrossRef]
198. Scholes, C.A.; Freeman, B.D. Thermal rearranged poly(imide-co-ethylene glycol) membranes for gas separation. *J. Membr. Sci.* **2018**, *563*, 676–683. [CrossRef]
199. Scholes, C.A.; Freeman, B.D.; Kentish, S.E. Water vapor permeability and competitive sorption in thermally rearranged (TR) membranes. *J. Membr. Sci.* **2014**, *470*, 132–137. [CrossRef]
200. Smith, Z.P.; Hernández, G.; Gleason, K.L.; Anand, A.; Doherty, C.M.; Konstas, K.; Alvarez, C.; Hill, A.J.; Lozano, A.E.; Paul, D.R.; et al. Effect of polymer structure on gas transport properties of selected aromatic polyimides, polyamides and TR polymers. *J. Membr. Sci.* **2015**, *493*, 766–781. [CrossRef]
201. Stevens, K.A.; Smith, Z.P.; Gleason, K.L.; Galizia, M.; Paul, D.R.; Freeman, B.D. Influence of temperature on gas solubility in thermally rearranged (TR) polymers. *J. Membr. Sci.* **2017**, *533*, 75–83. [CrossRef]
202. Calle, M.; Doherty, C.M.; Hill, A.J.; Lee, Y.M. Cross-Linked Thermally Rearranged Poly(benzoxazole-co-imide) Membranes for Gas Separation. *Macromolecules* **2013**, *46*, 8179–8189. [CrossRef]
203. Park, H.B.; Han, S.H.; Jung, C.H.; Lee, Y.M.; Hill, A.J. Thermally rearranged (TR) polymer membranes for CO_2 separation. *J. Membr. Sci.* **2010**, *359*, 11–24. [CrossRef]
204. Kim, S.; Jo, H.J.; Lee, Y.M. Sorption and transport of small gas molecules in thermally rearranged (TR) polybenzoxazole membranes based on 2,2-bis(3-amino-4-hydroxyphenyl)-hexafluoropropane (bisAPAF) and 4,4′-hexafluoroisopropylidene diphthalic anhydride (6FDA). *J. Membr. Sci.* **2013**, *441*, 1–8. [CrossRef]
205. Jiang, Y.; Willmore, F.T.; Sanders, D.; Smith, Z.P.; Ribeiro, C.P.; Doherty, C.M.; Thornton, A.; Hill, A.J.; Freeman, B.D.; Sanchez, I.C. Cavity size, sorption and transport characteristics of thermally rearranged (TR) polymers. *Polymer* **2011**, *52*, 2244–2254. [CrossRef]
206. Park, C.H.; Tocci, E.; Lee, Y.M.; Drioli, E. Thermal Treatment Effect on the Structure and Property Change between Hydroxy-Containing Polyimides (HPIs) and Thermally Rearranged Polybenzoxazole (TR-PBO). *J. Phys. Chem. B* **2012**, *116*, 12864–12877. [CrossRef] [PubMed]
207. Chang, K.-S.; Wu, Z.-C.; Kim, S.; Tung, K.-L.; Lee, Y.M.; Lin, Y.-F.; Lai, J.-Y. Molecular modeling of poly(benzoxazole-co-imide) membranes: A structure characterization and performance investigation. *J. Membr. Sci.* **2014**, *454*, 1–11. [CrossRef]
208. Park, C.H.; Tocci, E.; Kim, S.; Kumar, A.; Lee, Y.M.; Drioli, E. A Simulation Study on OH-Containing Polyimide (HPI) and Thermally Rearranged Polybenzoxazoles (TR-PBO): Relationship between Gas Transport Properties and Free Volume Morphology. *J. Phys. Chem. B* **2014**, *118*, 2746–2757. [CrossRef] [PubMed]
209. Rizzuto, C.; Caravella, A.; Brunetti, A.; Park, C.H.; Lee, Y.M.; Drioli, E.; Barbieri, G.; Tocci, E. Sorption and Diffusion of CO_2/N_2 in gas mixture in thermally-rearranged polymeric membranes: A molecular investigation. *J. Membr. Sci.* **2017**, *528*, 135–146. [CrossRef]
210. Velioğlu, S.; Ahunbay, M.G.; Tantekin-Ersolmaz, S.B. An atomistic insight on CO_2 plasticization resistance of thermally rearranged 6FDA-bisAPAF. *J. Membr. Sci.* **2018**, *556*, 23–33. [CrossRef]
211. Brunetti, A.; Tocci, E.; Cersosimo, M.; Kim, J.S.; Lee, W.H.; Seong, J.G.; Lee, Y.M.; Drioli, E.; Barbieri, G. Mutual influence of mixed-gas permeation in thermally rearranged poly(benzoxazole-co-imide) polymer membranes. *J. Membr. Sci.* **2019**, *580*, 202–213. [CrossRef]
212. Hart, K.E.; Colina, C.M. Estimating gas permeability and permselectivity of microporous polymers. *J. Membr. Sci.* **2014**, *468*, 259–268. [CrossRef]
213. Hart, K.E.; Springmeier, J.M.; McKeown, N.B.; Colina, C.M. Simulated swelling during low-temperature N_2 adsorption in polymers of intrinsic microporosity. *Phys. Chem. Chem. Phys.* **2013**, *15*, 20161–20169. [CrossRef]
214. Merkel, T.C.; Bondar, V.I.; Nagai, K.; Freeman, B.D.; Pinnau, I. Gas sorption, diffusion, and permeation in poly(dimethylsiloxane). *J. Polym. Sci. Part B: Polym. Phys.* **2000**, *38*, 415–434. [CrossRef]
215. Zhou, C.; Chung, T.-S.; Wang, R.; Liu, Y.; Goh, S.H. The accelerated CO_2 plasticization of ultra-thin polyimide films and the effect of surface chemical cross-linking on plasticization and physical aging. *J. Membr. Sci.* **2003**, *225*, 125–134. [CrossRef]
216. Horn, N.R.; Paul, D.R. Carbon dioxide plasticization and conditioning effects in thick vs. thin glassy polymer films. *Polymer* **2011**, *52*, 1619–1627. [CrossRef]

217. Rowe, B.W.; Pas, S.J.; Hill, A.J.; Suzuki, R.; Freeman, B.D.; Paul, D.R. A variable energy positron annihilation lifetime spectroscopy study of physical aging in thin glassy polymer films. *Polymer* **2009**, *50*, 6149–6156. [CrossRef]
218. Kim, J.H.; Koros, W.J.; Paul, D.R. Physical aging of thin 6FDA-based polyimide membranes containing carboxyl acid groups. Part I. Transport properties. *Polymer* **2006**, *47*, 3094–3103. [CrossRef]
219. Kim, J.H.; Koros, W.J.; Paul, D.R. Physical aging of thin 6FDA-based polyimide membranes containing carboxyl acid groups. Part II. Optical properties. *Polymer* **2006**, *47*, 3104–3111. [CrossRef]
220. Kim, J.H.; Koros, W.J.; Paul, D.R. Effects of CO_2 exposure and physical aging on the gas permeability of thin 6FDA-based polyimide membranes: Part 1. Without crosslinking. *J. Membr. Sci.* **2006**, *282*, 21–31. [CrossRef]
221. Rowe, B.W.; Freeman, B.D.; Paul, D.R. Physical aging of ultrathin glassy polymer films tracked by gas permeability. *Polymer* **2009**, *50*, 5565–5575. [CrossRef]
222. Murphy, T.M.; Freeman, B.D.; Paul, D.R. Physical aging of polystyrene films tracked by gas permeability. *Polymer* **2013**, *54*, 873–880. [CrossRef]
223. Neyertz, S.; Brown, D. Air Sorption and Separation by Polymer Films at the Molecular Level. *Macromolecules* **2018**, *51*, 7077–7092. [CrossRef]
224. Lock, S.S.M.; Lau, K.K.; Shariff, A.M.; Yeong, Y.F.; Bustam, M.A. Thickness dependent penetrant gas transport properties and separation performance within ultrathin polysulfone membrane: Insights from atomistic molecular simulation. *J. Polym. Sci. Part B Polym. Phys.* **2018**, *56*, 131–158. [CrossRef]
225. Neyertz, S.; Douanne, A.; Brown, D. Effect of Interfacial Structure on Permeation Properties of Glassy Polymers. *Macromolecules* **2005**, *38*, 10286–10298. [CrossRef]
226. Neyertz, S.; Douanne, A.; Brown, D. A molecular dynamics simulation study of surface effects on gas permeation in free-standing polyimide membranes. *J. Membr. Sci.* **2006**, *280*, 517–529. [CrossRef]
227. Neyertz, S.; Brown, D. Molecular Dynamics Simulations of Oxygen Transport through a Fully Atomistic Polyimide Membrane. *Macromolecules* **2008**, *41*, 2711–2721. [CrossRef]
228. Neyertz, S.; Brown, D. The effect of structural isomerism on carbon dioxide sorption and plasticization at the interface of a glassy polymer membrane. *J. Membr. Sci.* **2014**, *460*, 213–228. [CrossRef]
229. Neyertz, S.; Brown, D. Nanosecond-time-scale reversibility of dilation induced by carbon dioxide sorption in glassy polymer membranes. *J. Membr. Sci.* **2016**, *520*, 385–399. [CrossRef]
230. Gopalan, P.; Brachet, P.; Kristiansen, A.; Männle, F.; Brown, D. Oxygen Transport in Amino-Functionalized Polyhedral Oligomeric Silsesquioxanes (POSS) AU—Neyertz, S. *Soft Mater.* **2014**, *12*, 113–123. [CrossRef]
231. Neyertz, S.; Brown, D. Molecular Dynamics Study of Carbon Dioxide Sorption and Plasticization at the Interface of a Glassy Polymer Membrane. *Macromolecules* **2013**, *46*, 2433–2449. [CrossRef]
232. Yampol'skii, Y.P.; Pinnau, I.; Freeman, B.D. *Materials Science of Membranes for Gas and Vapor Separation*; John Wiley: Hoboken, NJ, USA, 2006.
233. Xia, J.; Chung, T.-S.; Paul, D.R. Physical aging and carbon dioxide plasticization of thin polyimide films in mixed gas permeation. *J. Membr. Sci.* **2014**, *450*, 457–468. [CrossRef]
234. O'Brien, K.C.; Koros, W.J.; Barbari, T.A.; Sanders, E.S. A new technique for the measurement of multicomponent gas transport through polymeric films. *J. Membr. Sci.* **1986**, *29*, 229–238. [CrossRef]
235. Genduso, G.; Litwiller, E.; Ma, X.; Zampini, S.; Pinnau, I. Mixed-gas sorption in polymers via a new barometric test system: Sorption and diffusion of CO_2-CH_4 mixtures in polydimethylsiloxane (PDMS). *J. Membr. Sci.* **2019**, *577*, 195–204. [CrossRef]
236. Fraga, S.C.; Monteleone, M.; Lanč, M.; Esposito, E.; Fuoco, A.; Giorno, L.; Pilnáček, K.; Friess, K.; Carta, M.; McKeown, N.B.; et al. A novel time lag method for the analysis of mixed gas diffusion in polymeric membranes by on-line mass spectrometry: Method development and validation. *J. Membr. Sci.* **2018**, *561*, 39–58. [CrossRef]
237. Tanis, I.; Brown, D.; Neyertz, S.; Heck, R.; Mercier, R.; Vaidya, M.; Ballaguet, J.-P. A comparison of pure and mixed-gas permeation of nitrogen and methane in 6FDA-based polyimides as studied by molecular dynamics simulations. *Comput. Mater. Sci.* **2018**, *141*, 243–253. [CrossRef]
238. Coleman, M.R.; Koros, W.J. Isomeric polyimides based on fluorinated dianhydrides and diamines for gas separation applications. *J. Membr. Sci.* **1990**, *50*, 285–297. [CrossRef]
239. Tanaka, K.; Kita, H.; Okano, M.; Okamoto, K.-I. Permeability and permselectivity of gases in fluorinated and non-fluorinated polyimides. *Polymer* **1992**, *33*, 585–592. [CrossRef]

240. Coleman, M.R.; Koros, W.J. The transport properties of polyimide isomers containing hexafluoroisopropylidene in the diamine residue. *J. Polym. Sci. Part B: Polym. Phys.* **1994**, *32*, 1915–1926. [CrossRef]
241. Cornelius, C.J.; Marand, E. Hybrid silica-polyimide composite membranes: Gas transport properties. *J. Membr. Sci.* **2002**, *202*, 97–118. [CrossRef]
242. Takeuchi, H. A jump motion of small molecules in glassy polymers: A molecular dynamics simulation. *J. Chem. Phys.* **1990**, *93*, 2062–2067. [CrossRef]
243. Müller-Plathe, F.; Laaksonen, L.; van Gunsteren, W.F. Cooperative Effects in the Transport of Small Molecules Through an Amorphous Polymer Matrix. *J. Mol. Graph.* **1993**, *11*, 118. [CrossRef]
244. Haus, J.W.; Kehr, K.W. Diffusion in regular and disordered lattices. *Phys. Rep.* **1987**, *150*, 263–406. [CrossRef]
245. Havlin, S.; Ben-Avraham, D. Diffusion in disordered media. *Adv. Phys.* **1987**, *36*, 695–798. [CrossRef]
246. Stauffer, D.; Aharony, A. *Introduction to Percolation Theory*; Taylor & Francis: London, UK, 1992.
247. Rallabandi, P.S.; Thompson, A.P.; Ford, D.M. A Molecular Modeling Study of Entropic and Energetic Selectivities in Air Separation with Glassy Polymers. *Macromolecules* **2000**, *33*, 3142–3152. [CrossRef]
248. Wind, J.D.; Sirard, S.M.; Paul, D.R.; Green, P.F.; Johnston, K.P.; Koros, W.J. Carbon Dioxide-Induced Plasticization of Polyimide Membranes: Pseudo-Equilibrium Relationships of Diffusion, Sorption, and Swelling. *Macromolecules* **2003**, *36*, 6433–6441. [CrossRef]
249. Pinnau, I.; Toy, L.G. Transport of organic vapors through poly(1-trimethylsilyl-1-propyne). *J. Membr. Sci.* **1996**, *116*, 199–209. [CrossRef]
250. Koros, W.J.; Chern, R.T.; Stannett, V.; Hopfenberg, H.B. A model for permeation of mixed gases and vapors in glassy polymers. *J. Polym. Sci. Polym. Phys. Ed.* **1981**, *19*, 1513–1530. [CrossRef]
251. Park, H.B.; Kamcev, J.; Robeson, L.M.; Elimelech, M.; Freeman, B.D. Maximizing the right stuff: The trade-off between membrane permeability and selectivity. *Science* **2017**, *356*, eaab0530. [CrossRef] [PubMed]
252. Du, Y.-C.; Huang, L.-J.; Wang, Y.-X.; Yang, K.; Tang, J.-G.; Wang, Y.; Cheng, M.-M.; Zhang, Y.; Kipper, M.J.; Belfiore, L.A.; et al. Recent developments in graphene-based polymer composite membranes: Preparation, mass transfer mechanism, and applications. *J. Appl. Polym. Sci.* **2019**, *136*, 47761. [CrossRef]
253. Deng, L.; Hägg, M.-B. Carbon nanotube reinforced PVAm/PVA blend FSC nanocomposite membrane for CO_2/CH_4 separation. *Int. J. Greenh. Gas Control* **2014**, *26*, 127–134. [CrossRef]
254. Ihsanullah. Carbon nanotube membranes for water purification: Developments, challenges, and prospects for the future. *Sep. Purif. Technol.* **2019**, *209*, 307–337. [CrossRef]
255. Qu, X.; Alvarez, P.J.J.; Li, Q. Applications of nanotechnology in water and wastewater treatment. *Water Res.* **2013**, *47*, 3931–3946. [CrossRef]
256. Celik, E.; Park, H.; Choi, H.; Choi, H. Carbon nanotube blended polyethersulfone membranes for fouling control in water treatment. *Water Res.* **2011**, *45*, 274–282. [CrossRef] [PubMed]
257. Gong, H.; Chuah, C.Y.; Yang, Y.; Bae, T.-H. High performance composite membranes comprising Zn(pyrz)2(SiF6) nanocrystals for CO_2/CH_4 separation. *J. Ind. Eng. Chem.* **2018**, *60*, 279–285. [CrossRef]
258. Dettori, R.; Yang, Q.; Whiting, W.I. Exploration of anion transport in a composite membrane via experimental and theoretical methods. *J. Membr. Sci.* **2018**, *563*, 270–276. [CrossRef]
259. Vinoba, M.; Bhagiyalakshmi, M.; Alqaheem, Y.; Alomair, A.A.; Pérez, A.; Rana, M.S. Recent progress of fillers in mixed matrix membranes for CO_2 separation: A review. *Sep. Purif. Technol.* **2017**, *188*, 431–450. [CrossRef]
260. Favvas, E.P.; Katsaros, F.K.; Papageorgiou, S.K.; Sapalidis, A.A.; Mitropoulos, A.C. A review of the latest development of polyimide based membranes for CO_2 separations. *React. Funct. Polym.* **2017**, *120*, 104–130. [CrossRef]
261. Etxeberria-Benavides, M.; David, O.; Johnson, T.; Łozińska, M.M.; Orsi, A.; Wright, P.A.; Mastel, S.; Hillenbrand, R.; Kapteijn, F.; Gascon, J. High performance mixed matrix membranes (MMMs) composed of ZIF-94 filler and 6FDA-DAM polymer. *J. Membr. Sci.* **2018**, *550*, 198–207. [CrossRef]
262. Li, W.; Samarasinghe, S.A.S.C.; Bae, T.-H. Enhancing CO_2/CH_4 separation performance and mechanical strength of mixed-matrix membrane via combined use of graphene oxide and ZIF-8. *J. Ind. Eng. Chem.* **2018**, *67*, 156–163. [CrossRef]
263. Barooah, M.; Mandal, B. Synthesis, characterization and CO_2 separation performance of novel PVA/PG/ZIF-8 mixed matrix membrane. *J. Membr. Sci.* **2019**, *572*, 198–209. [CrossRef]
264. Li, H.; Han, L.; Hou, J.; Liu, J.; Zhang, Y. Oriented Zeolitic imidazolate framework membranes within polymeric matrices for effective $N2/CO_2$ separation. *J. Membr. Sci.* **2019**, *572*, 82–91. [CrossRef]

265. Ding, X.; Tan, F.; Zhao, H.; Hua, M.; Wang, M.; Xin, Q.; Zhang, Y. Enhancing gas permeation and separation performance of polymeric membrane by incorporating hollow polyamide nanoparticles with dense shell. *J. Membr. Sci.* **2019**, *570–571*, 53–60. [CrossRef]
266. Xie, K.; Fu, Q.; Qiao, G.G.; Webley, P.A. Recent progress on fabrication methods of polymeric thin film gas separation membranes for CO_2 capture. *J. Membr. Sci.* **2019**, *572*, 38–60. [CrossRef]
267. Akther, N.; Phuntsho, S.; Chen, Y.; Ghaffour, N.; Shon, H.K. Recent advances in nanomaterial-modified polyamide thin-film composite membranes for forward osmosis processes. *J. Membr. Sci.* **2019**, *584*, 20–45. [CrossRef]
268. Liu, G.; Labreche, Y.; Chernikova, V.; Shekhah, O.; Zhang, C.; Belmabkhout, Y.; Eddaoudi, M.; Koros, W.J. Zeolite-like MOF nanocrystals incorporated 6FDA-polyimide mixed-matrix membranes for CO_2/CH_4 separation. *J. Membr. Sci.* **2018**, *565*, 186–193. [CrossRef]
269. Liu, Y.; Liu, G.; Zhang, C.; Qiu, W.; Yi, S.; Chernikova, V.; Chen, Z.; Belmabkhout, Y.; Shekhah, O.; Eddaoudi, M.; et al. Enhanced CO_2/CH_4 Separation Performance of a Mixed Matrix Membrane Based on Tailored MOF-Polymer Formulations. *Adv. Sci.* **2018**, *5*, 1800982. [CrossRef] [PubMed]
270. Teodoro, R.M.; Tomé, L.C.; Mantione, D.; Mecerreyes, D.; Marrucho, I.M. Mixing poly(ionic liquid)s and ionic liquids with different cyano anions: Membrane forming ability and CO_2/N2 separation properties. *J. Membr. Sci.* **2018**, *552*, 341–348. [CrossRef]
271. Kanehashi, S.; Kishida, M.; Kidesaki, T.; Shindo, R.; Sato, S.; Miyakoshi, T.; Nagai, K. CO_2 separation properties of a glassy aromatic polyimide composite membranes containing high-content 1-butyl-3-methylimidazolium bis(trifluoromethylsulfonyl)imide ionic liquid. *J. Membr. Sci.* **2013**, *430*, 211–222. [CrossRef]
272. Li, J.; Liu, B.; Zhang, X.; Cao, D.; Chen, G. Hydrogen Bond Networks of Glycol Molecules on ZIF-8 Surfaces as Semipermeable Films for Efficient Carbon Capture. *J. Phys. Chem. C* **2017**, *121*, 25347–25352. [CrossRef]
273. Gonciaruk, A.; Althumayri, K.; Harrison, W.J.; Budd, P.M.; Siperstein, F.R. PIM-1/graphene composite: A combined experimental and molecular simulation study. *Microporous Mesoporous Mater.* **2015**, *209*, 126–134. [CrossRef]
274. Kupgan, G.; Abbott, L.J.; Hart, K.E.; Colina, C.M. Modeling Amorphous Microporous Polymers for CO_2 Capture and Separations. *Chem. Rev.* **2018**, *118*, 5488–5538. [CrossRef]
275. Vogiatzis, G.G.; Theodorou, D.N. Multiscale Molecular Simulations of Polymer-Matrix Nanocomposites. *Arch. Comput. Methods Eng.* **2018**, *25*, 591–645. [CrossRef]
276. Abedini, A.; Crabtree, E.; Bara, J.E.; Turner, C.H. Molecular Simulation of Ionic Polyimides and Composites with Ionic Liquids as Gas-Separation Membranes. *Langmuir* **2017**, *33*, 11377–11389. [CrossRef] [PubMed]
277. Abedini, A.; Crabtree, E.; Bara, J.E.; Turner, C.H. Molecular analysis of selective gas adsorption within composites of ionic polyimides and ionic liquids as gas separation membranes. *Chem. Phys.* **2019**, *516*, 71–83. [CrossRef]
278. Aryal, D.; Ganesan, V. Reversal of Salt Concentration Dependencies of Salt and Water Diffusivities in Polymer Electrolyte Membranes. *ACS Macro Lett.* **2018**, *7*, 739–744. [CrossRef]
279. Mogurampelly, S.; Sethuraman, V.; Pryamitsyn, V.; Ganesan, V. Influence of nanoparticle-ion and nanoparticle-polymer interactions on ion transport and viscoelastic properties of polymer electrolytes. *J. Chem. Phys.* **2016**, *144*, 154905. [CrossRef] [PubMed]
280. Mogurampelly, S.; Ganesan, V. Influence of nanoparticle surface chemistry on ion transport in polymer nanocomposite electrolytes. *Solid State Ion.* **2016**, *286*, 57. [CrossRef]
281. Zhang, L.; Hu, Z.; Jiang, J. Metal–Organic Framework/Polymer Mixed-Matrix Membranes for H_2/CO_2 Separation: A Fully Atomistic Simulation Study. *J. Phys. Chem. C* **2012**, *116*, 19268–19277. [CrossRef]
282. Azar, A.N.V.; Velioglu, S.; Keskin, S. Large-Scale Computational Screening of Metal Organic Framework (MOF) Membranes and MOF-Based Polymer Membranes for H_2/N_2 Separations. *ACS Sustain. Chem. Eng.* **2019**, *7*, 9525–9536. [CrossRef]
283. Yilmaz, G.; Keskin, S. Predicting the Performance of Zeolite Imidazolate Framework/Polymer Mixed Matrix Membranes for CO_2, CH_4, and H_2 Separations Using Molecular Simulations. *Ind. Eng. Chem. Res.* **2012**, *51*, 14218–14228. [CrossRef]
284. Dutta, R.C.; Bhatia, S.K. Structure and Gas Transport at the Polymer–Zeolite Interface: Insights from Molecular Dynamics Simulations. *ACS Appl. Mater. Interfaces* **2018**, *10*, 5992–6005. [CrossRef]

285. Zhang, Q.G.; Liu, Q.L.; Wu, J.Y.; Chen, Y.; Zhu, A.M. Structure-related diffusion in poly(methyl methacrylate)/polyhedral oligomeric silsesquioxanes composites: A molecular dynamics simulation study. *J. Membr. Sci.* **2009**, *342*, 105–112. [CrossRef]
286. Wang, C.; Jagirdar, P.; Naserifar, S.; Sahimi, M. Molecular Simulation Study of Gas Solubility and Diffusion in a Polymer-Boron Nitride Nanotube Composite. *J. Phys. Chem. B* **2016**, *120*, 1273–1284. [CrossRef]
287. Bonakala, S.; Lalitha, A.; Shin, J.E.; Moghadam, F.; Semino, R.; Park, H.B.; Maurin, G. Understanding of the Graphene Oxide/Metal-Organic Framework Interface at the Atomistic Scale. *ACS Appl. Mater. Interfaces* **2018**, *10*, 33619–33629. [CrossRef] [PubMed]
288. Hwang, S.; Semino, R.; Seoane, B.; Zahan, M.; Chmelik, C.; Valiullin, R.; Bertmer, M.; Haase, J.; Kapteijn, F.; Gascon, J.; et al. Revealing the Transient Concentration of CO_2 in a Mixed-Matrix Membrane by IR Microimaging and Molecular Modeling. *Angew. Chem. Int. Ed.* **2018**, *57*, 5156–5160. [CrossRef]
289. Semino, R.; Dürholt, J.P.; Schmid, R.; Maurin, G. Multiscale Modeling of the HKUST-1/Poly(vinyl alcohol) Interface: From an Atomistic to a Coarse Graining Approach. *J. Phys. Chem. C* **2017**, *121*, 21491–21496. [CrossRef]
290. Lin, R.; Villacorta Hernandez, B.; Ge, L.; Zhu, Z. Metal organic framework based mixed matrix membranes: An overview on filler/polymer interfaces. *J. Mater. Chem. A* **2018**, *6*, 293–312. [CrossRef]
291. Zhao, J.; He, G.; Liu, G.; Pan, F.; Wu, H.; Jin, W.; Jiang, Z. Manipulation of interactions at membrane interfaces for energy and environmental applications. *Prog. Polym. Sci.* **2018**, *80*, 125–152. [CrossRef]
292. Zavadlav, J.; Bevc, S.; Praprotnik, M. Adaptive resolution simulations of biomolecular systems. *Eur. Biophys. J.* **2017**, *46*, 821–835. [CrossRef] [PubMed]
293. Kreis, K.; Tuckerman, M.E.; Donadio, D.; Kremer, K.; Potestio, R. From Classical to Quantum and Back: A Hamiltonian Scheme for Adaptive Multiresolution Classical/Path-Integral Simulations. *J. Chem. Theory Comput.* **2016**, *12*, 3030–3039. [CrossRef]
294. Kreis, K.; Potestio, R.; Kremer, K.; Fogarty, A.C. Adaptive Resolution Simulations with Self-Adjusting High-Resolution Regions. *J. Chem. Theory Comput.* **2016**, *12*, 4067–4081. [CrossRef] [PubMed]
295. Kreis, K.; Kremer, K.; Potestio, R.; Tuckerman, M.E. From classical to quantum and back: Hamiltonian adaptive resolution path integral, ring polymer, and centroid molecular dynamics. *J. Chem. Phys.* **2017**, *147*, 244104. [CrossRef]
296. Praprotnik, M.; Delle Site, L.; Kremer, K. Adaptive resolution molecular-dynamics simulation: Changing the degrees of freedom on the fly. *J. Chem. Phys.* **2005**, *123*, 224106. [CrossRef] [PubMed]
297. Delle Site, L.; Abrams, C.F.; Alavi, A.; Kremer, K. Polymers near Metal Surfaces: Selective Adsorption and Global Conformations. *Phys. Rev. Lett.* **2002**, *89*, 156103. [CrossRef] [PubMed]
298. Dong, D.; Zhang, W.; Barnett, A.; Lu, J.; Van Duin, C.A.; Molinero, V.; Bedrov, D. Multiscale Modeling of Structure, Transport and Reactivity in Alkaline Fuel Cell Membranes: Combined Coarse-Grained, Atomistic and Reactive Molecular Dynamics Simulations. *Polymers* **2018**, *10*, 1289. [CrossRef] [PubMed]
299. Maghami, S.; Mehrabani-Zeinabad, A.; Sadeghi, M.; Sánchez-Laínez, J.; Zornoza, B.; Téllez, C.; Coronas, J. Mathematical modeling of temperature and pressure effects on permeability, diffusivity and solubility in polymeric and mixed matrix membranes. *Chem. Eng. Sci.* **2019**, *205*, 58–73. [CrossRef]
300. Castro-Muñoz, R.; Iglesia, Ó.d.; Fíla, V.; Téllez, C.; Coronas, J. Pervaporation-Assisted Esterification Reactions by Means of Mixed Matrix Membranes. *Ind. Eng. Chem. Res.* **2018**, *57*, 15998–16011. [CrossRef]
301. Ursino, C.; Castro-Muñoz, R.; Drioli, E.; Gzara, L.; Albeirutty, H.M.; Figoli, A. Progress of Nanocomposite Membranes for Water Treatment. *Membranes* **2018**, *8*, 18. [CrossRef] [PubMed]
302. Sarango, L.; Paseta, L.; Navarro, M.; Zornoza, B.; Coronas, J. Controlled deposition of MOFs by dip-coating in thin film nanocomposite membranes for organic solvent nanofiltration. *J. Ind. Eng. Chem.* **2018**, *59*, 8–16. [CrossRef]
303. Navarro, M.; Benito, J.; Paseta, L.; Gascón, I.; Coronas, J.; Téllez, C. Thin-Film Nanocomposite Membrane with the Minimum Amount of MOF by the Langmuir–Schaefer Technique for Nanofiltration. *ACS Appl. Mater. Interfaces* **2018**, *10*, 1278–1287. [CrossRef] [PubMed]
304. Echaide-Górriz, C.; Navarro, M.; Téllez, C.; Coronas, J. Simultaneous use of MOFs MIL-101(Cr) and ZIF-11 in thin film nanocomposite membranes for organic solvent nanofiltration. *Dalton Trans.* **2017**, *46*, 6244–6252. [CrossRef]

305. Sánchez-Laínez, J.; Zornoza, B.; Orsi, A.F.; Łozińska, M.M.; Dawson, D.M.; Ashbrook, S.E.; Francis, S.M.; Wright, P.A.; Benoit, V.; Llewellyn, P.L.; et al. Synthesis of ZIF-93/11 Hybrid Nanoparticles via Post-Synthetic Modification of ZIF-93 and Their Use for H2/CO_2 Separation. *Chem. A Eur. J.* **2018**, *24*, 11211–11219. [CrossRef]
306. Castro-Muñoz, R.; Martin-Gil, V.; Ahmad, M.Z.; Fíla, V. Matrimid® 5218 in preparation of membranes for gas separation: Current state-of-the-art. *Chem. Eng. Commun.* **2018**, *205*, 161–196. [CrossRef]
307. Castro-Muñoz, R.; Fíla, V.; Ahmad, M.Z. Enhancing the CO_2 Separation Performance of Matrimid 5218 Membranes for CO_2/CH_4 Binary Mixtures. *Chem. Eng. Technol.* **2019**, *42*, 645–654. [CrossRef]
308. Castro-Muñoz, R.; Fíla, V.; Martin-Gil, V.; Muller, C. Enhanced CO_2 permeability in Matrimid® 5218 mixed matrix membranes for separating binary CO_2/CH_4 mixtures. *Sep. Purif. Technol.* **2019**, *210*, 553–562. [CrossRef]
309. Mathioudakis, I.G.; Vogiatzis, G.G.; Tzoumanekas, C.; Theodorou, D.N. Molecular modeling and simulation of polymer nanocomposites at multiple length scales. *IEEE Trans. Nanotechnol.* **2016**, *15*, 416–422. [CrossRef]
310. Li, C.; Strachan, A. Molecular simulations of crosslinking process of thermosetting polymers. *Polymer* **2010**, *51*, 6058–6070. [CrossRef]
311. Chmiela, S.; Sauceda, H.E.; Poltavsky, I.; Müller, K.-R.; Tkatchenko, A. sGDML: Constructing accurate and data efficient molecular force fields using machine learning. *Comput. Phys. Commun.* **2019**. [CrossRef]
312. Bejagam, K.K.; Singh, S.; An, Y.; Deshmukh, S.A. Machine-Learned Coarse-Grained Models. *J. Phys. Chem. Lett.* **2018**, *9*, 4667–4672. [CrossRef]
313. Chan, H.; Narayanan, B.; Cherukara, M.J.; Sen, F.G.; Sasikumar, K.; Gray, S.K.; Chan, M.K.Y.; Sankaranarayanan, S.K.R.S. Machine Learning Classical Interatomic Potentials for Molecular Dynamics from First-Principles Training Data. *J. Phys. Chem. C* **2019**, *123*, 6941–6957. [CrossRef]
314. Wang, J.; Olsson, S.; Wehmeyer, C.; Pérez, A.; Charron, N.E.; de Fabritiis, G.; Noé, F.; Clementi, C. Machine Learning of Coarse-Grained Molecular Dynamics Force Fields. *ACS Cent. Sci.* **2019**, *5*, 755–767. [CrossRef]
315. Zhang, L.; Han, J.; Wang, H.; Car, R.; Weinan, E. Deep Potential Molecular Dynamics: A Scalable Model with the Accuracy of Quantum Mechanics. *Phys. Rev. Lett.* **2018**, *120*, 143001. [CrossRef]
316. Zhang, L.; Han, J.; Wang, H.; Car, R.; Weinan, E. DeePCG: Constructing coarse-grained models via deep neural networks. *J. Chem. Phys.* **2018**, *149*, 034101. [CrossRef] [PubMed]
317. Wang, H.; Zhang, L.; Han, J.; Weinan, E. DeePMD-kit: A deep learning package for many-body potential energy representation and molecular dynamics. *Comput. Phys. Commun.* **2018**, *228*, 178–184. [CrossRef]
318. Castro-Muñoz, R.; Galiano, F.; de la Iglesia, Ó.; Fíla, V.; Téllez, C.; Coronas, J.; Figoli, A. Graphene oxide—Filled polyimide membranes in pervaporative separation of azeotropic methanol–MTBE mixtures. *Sep. Purif. Technol.* **2019**, *224*, 265–272. [CrossRef]
319. Baker, R.W.; Wijmans, J.G.; Huang, Y. Permeability, permeance and selectivity: A preferred way of reporting pervaporation performance data. *J. Membr. Sci.* **2010**, *348*, 346–352. [CrossRef]
320. Castro-Muñoz, R.; Buera-González, J.; de la Iglesia, Ó.; Galiano, F.; Fíla, V.; Malankowska, M.; Rubio, C.; Figoli, A.; Téllez, C.; Coronas, J. Towards the dehydration of ethanol using pervaporation cross-linked poly(vinyl alcohol)/graphene oxide membranes. *J. Membr. Sci.* **2019**, *582*, 423–434. [CrossRef]

MDPI
St. Alban-Anlage 66
4052 Basel
Switzerland
Tel. +41 61 683 77 34
Fax +41 61 302 89 18
www.mdpi.com

Membranes Editorial Office
E-mail: membranes@mdpi.com
www.mdpi.com/journal/membranes

www.ingramcontent.com/pod-product-compliance
Lightning Source LLC
LaVergne TN
LVHW070123170726
843515LV00007B/1739

* 9 7 8 3 0 3 6 5 0 2 1 2 0 *